Power Systems

Electrical power has been the technological foundation of industrial societies for many years. Although the systems designed to provide and apply electrical energy have reached a high degree of maturity, unforeseen problems are constantly encountered, necessitating the design of more efficient and reliable systems based on novel technologies. The book series Power Systems is aimed at providing detailed, accurate and sound technical information about these new developments in electrical power engineering. It includes topics on power generation, storage and transmission as well as electrical machines. The monographs and advanced textbooks in this series address researchers, lecturers, industrial engineers and senior students in electrical engineering.

Power Systems is indexed in Scopus

More information about this series at http://www.springer.com/series/4622

Francisco M. Gonzalez-Longatt ·
José Luis Rueda Torres
Editors

Modelling and Simulation of Power Electronic Converter Dominated Power Systems in PowerFactory

Springer

Editors
Francisco M. Gonzalez-Longatt
University of South-Eastern Norway
Porsgrunn, Norway

José Luis Rueda Torres
Department of Electrical Sustainable
Delft University of Technology
Delft, Zuid-Holland, The Netherlands

ISSN 1612-1287 ISSN 1860-4676 (electronic)
Power Systems
ISBN 978-3-030-54123-1 ISBN 978-3-030-54124-8 (eBook)
https://doi.org/10.1007/978-3-030-54124-8

This Springer imprint is published by the registered company Springer Nature Switzerland AG
The registered company address is: Gewerbestrasse 11, 6330 Cham, Switzerland

Foreword

To ensure the system stability of both large interconnected grids and smaller isolated grids, it is imperative that appropriate characteristics of generating plants be defined. Wind turbines that came into operation twenty years ago are now gradually being decommissioned or repowered. At the same time, ever more new renewable-based generation is coming online. As a result, the ratio of synchronous generation to inverter-based generation is changing dramatically.

Developments in grid codes in recent years have not adequately taken this changed situation into account. In essence, it is recognised that even the latest requirements regarding LVRT-behaviour and especially participation in reactive power and active power control are still lacking comprehensive definitions. The grid code requirements applicable in many countries currently are therefore still not designed such that one can speak of comprehensive system stability.

There are essentially no stipulations regarding grid-forming properties with regard to frequency control as well as voltage control. In principle, it can be observed that the grid codes are poorly thought out with regard to grid stability and are overall insufficient in this respect. This can lead to serious problems for the reliability and robustness of electrical networks and will result either in limitation of the renewable generation or in compensation investments to ensure grid stability.

In this context, the initiative of the publishers to address the topic of "Power Electronic Converter dominated Power Systems" in another PowerFactory application publication is to be very much welcomed. The collected contributions show the wide range of issues, whereby the search for the right approach to grid-forming characteristics and the associated questions play a central role.

The developers of the PowerFactory software have had to handle many complex problems in recent years. It is quite obvious that increasing numbers of power electronic converters leads to a clear overlap of the EMT- and RMS-based analysis functions, whereby the complexity of models in conjunction with the model scope leads to calculation problems that are sometimes difficult to solve. Here, the experience of the engineer is of paramount importance when it comes to defining the system under consideration.

We thank all the authors who have contributed their own work and experience on these topics for this book and who have shared their ideas and solutions. Our thanks also go to the editors Francisco M. Gonzalez-Longatt and José Luis Rueda Torres for their contribution to creating appropriate conditions for grid stability, in power systems increasing dominated by renewable energy.

Tübingen, Germany
April 2020

Dr.-Ing. Martin Schmieg
Chairman of Advisory Board, DIgSILENT GmbH

Contents

Chapter 1
Power Converters Dominated Power Systems

Francisco M. Gonzalez-Longatt, Martha Nohemi Acosta, Harold R. Chamorro, and José Luis Rueda Torres

Abstract The transition to a low-carbon society is the driving force pushing the traditional power system to increase the volume of non-synchronous technologies which mainly use power electronic converters (PEC) as an interface to the power network. Today, PEC is found in many applications ranging from generation, transmission, storage, and even active loads and protections. The variety of types, applications is almost unlimited. However, the behaviour of PEC is radically different to the typical devices used in traditional power systems, and its specificity in terms of control, overload characteristics, etc. entails an entirely different approach for modelling and simulation. This chapter presents a general review of the main concepts related to power electronic converters and their implementation in DIgSILNET PowerFactory. The chapter finalises with a discussion about the modern tendencies in power electronic applications in power systems: grid following and grid-forming converters.

F. M. Gonzalez-Longatt (✉) · M. N. Acosta
Institutt for elektro, IT og kybernetikk, Universitetet I Sørøst-Norge, Porsgrunn, Norway
e-mail: fglongatt@fglongatt.org

M. N. Acosta
e-mail: martha.acostamnt@uanl.edu.mx

M. N. Acosta
School of Mechanical and Electrical Engineering, Universidad Autónoma de Nuevo León, Nuevo León, México

H. R. Chamorro
KU Leuven, Katholieke Universiteit Leuven, Leuven, Belgium
e-mail: hr.chamo@ieee.org

J. L. Rueda Torres
Department of Electrical Sustainable Energy, Delft University of Technology (TU Delft), Delft, Netherlands
e-mail: J.L.RuedaTorres@tudelft.nl

F. M. Gonzalez-Longatt and J. L. Rueda Torres (eds.), *Modelling and Simulation of Power Electronic Converter Dominated Power Systems in PowerFactory*, Power Systems,
https://doi.org/10.1007/978-3-030-54124-8_1

Keywords Back-to-back connection · Current-source converters · DIgSILENT PowerFactory software · HVDC · Line commutated converters · Multi-modular converter · Power electronic converters · Power semiconductor switches · Voltage-source converters

1.1 Introduction

Power systems are evolving, and now, the changes are faster and more profound. The modern power system is changing the way that electricity is produced, transported, and consumed. From the generation side, there is a clear tendency to increase the use of environmentally friendly resources like solar power and wind power; all those sources have some sort of intermittency or high variability and also, the modern, efficient ones use *power electronic converters* (PECs) as interfaces to the power network. The power system has changed from the classical model, where the demand dictates the pattern of generation and fully dispatchable generators were able to follow changes in the demand easily. On the other side of the modern power system, the demand side has changed a lot in recent time; novel technologies like electric vehicles (EVs) and small battery energy storage systems (BESSs) are adding. Nowadays, the demand has a more complex pattern as never before requiring new ways to see the transmission and distribution system; however, the new PEC-based technologies are more controllable enabling flexibility from the demand point of view [1]. The operation and control of the modern and future power systems have increased complexity, generation, and demand have very different patterns, and the transmission system must define appropriate balancing mechanism. The situation at generation and demand sides also makes complex by the pressure created by the market that is increasing stress and narrowly conditions on the transmission system [2].

One common tendency in the revolution of the power system is the massive integration and deployment of PECs at very different levels (transmission and distribution) and applications (generation, transport and use) [3]. The use of PEC has many consequences for the traditional power system; one of them is the massive decoupling between generation/load caused by the use of a DC link. However, there are several other features expected into the future electricity networks will negatively affect the system security: lack of reactive power support, minimum to none overload capacity, limited short-circuit contribution, etc.

The PECs is one of the key elements on the energy transition toward a low-carbon power system by allowing a change in generation technology, from fossil fuel-based thermal generation to converter interfaced renewable generation. However, as the penetration of PEC-based technologies increases and the number of classical synchronous generation tends to decline, the power system is looking many positive features coming from the reduced volume of synchronous machines (SMs) (see Fig. 1.1): reduced rotating inertia, loss of natural control mechanisms, loss of short-circuit contribution, etc.

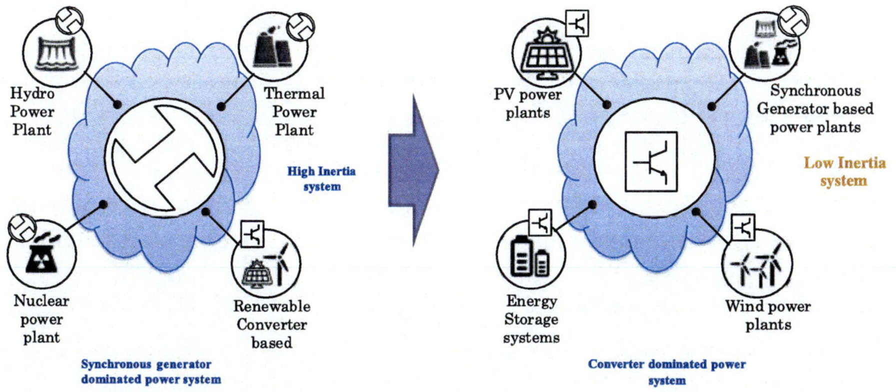

Fig. 1.1 Schematic diagram of a large scale photovoltaic system interconnected to a medium and high voltage network

1.2 Power Converter Fundamentals

The classical power electronics is dedicated to the study of electronic circuits intended to control the flow of electrical energy. A general energy conversion system is shown in Fig. 1.2.

The core of the conversion system in Fig. 1.2 is the power electronic converter PEC); it is the energy interface between the components and controls de energy flow (magnitude and direction).

A power electronic converter, usually abbreviated as PEC, is a power electronic-based device that mainly deals with the conversion of power from one form to another; it can involve modifying several electrical parameters like the change from one voltage level to another, differ in frequency and number of phases. Power electronics (PE) deals with conversion and control of electrical power with the help of *electronic switching devices.*

The PEC is an enabling technology, and it has been a critical element in the integration of new low-carbon technologies providing the needed interface between two or more energy systems; initial applications of PECs were dedicated to interface

Fig. 1.2 Schematic diagram of a generic energy conversion system showing the main signals and components

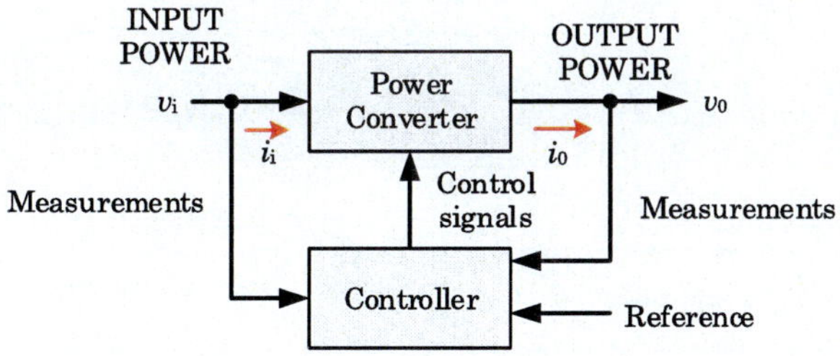

source and load. However, modern application of PEC allows energy flow in both directions as a consequence some of the power converters require bi-directionality.

Low-carbon technologies based on power electronics conversion are used to guarantee grid and load requirements in terms of controllability and efficiency of the electrical energy demanded. *PEC topologies* convert energy from a primary source to another source (or a load) requiring any level of processed energy [4].

It is useful to categorise the PEC in terms of their input and output form or frequency. In most power electronic systems, the input is from the electric utility source.

Depending on the application, the PEC output may have any of the following forms [5]:

- Direct current (DC):
  - Regulated (constant) magnitude
  - Adjustable magnitude
- Alternate current (AC):
  - Constant frequency, adjustable magnitude
  - Adjustable frequency and adjustable magnitude.

The input and the output side of the PEC are independent of each other, maybe a single-phase or three-phase. The power flow direction is typically from the side connected to the source and going to a load. However, the new low-carbon technologies impose exceptions. For example, a classical top-roof photovoltaic (PV) system interfaced with the utility grid via a PEC; the power flow is from the PVs (a DC input source) to the AC utility (as the output load). In some systems, the direction of power flow is reversible, depending on the operating conditions, that is the case of battery energy storage systems (BESSs).

PEC topologies convert energy from a primary source to another source (or a load) requiring any level of processed energy [4]. Classifications of the PEC topologies can be done in terms of the type of variable under control (i.e., AC or DC), as well as the number of stages of power conversions used, as observed in Fig. 1.3.

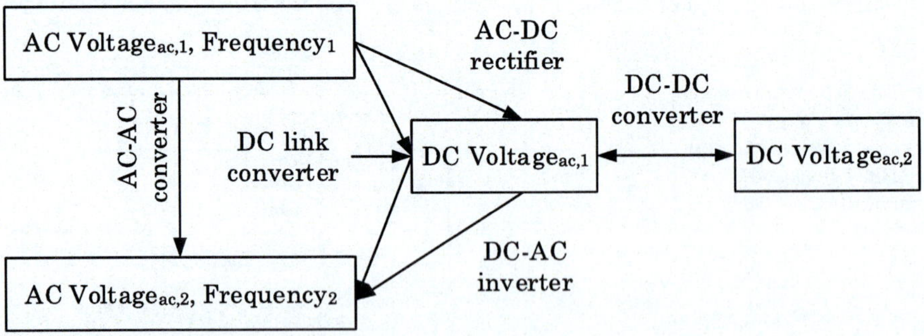

Fig. 1.3 Schematic diagram depicting the relationship between variables at the conversion process

Figure 1.3 presents, in a general way, the possible energy conversions available by using PECs. Typically the word converter is used as a generic term for a PEC that can perform energy conversion based on the form (frequency) on the two sides: AC–AC, AC–DC, DC–AC and DC–DC.

A direct AC–AC converter takes an AC waveform and converter that into another AC waveform, where the output voltage and frequency can be set arbitrarily. In this specific case, the PEC converts an AC voltage ($\text{Voltage}_{ac,1}$) with a specific frequency (Frequency_1) to another AC voltage with a different (or same) voltage ($\text{Voltage}_{ac,2}$) and frequency (Frequency_2); AC–AC PEC with no intermediate DC link came in the form of cyclo converters and matrix converters.

Special attention is given to the PECs dealing with configurations that process AC voltage (at input and/or output converter sides) with one DC stage. In this case, the term *rectifier* refers to a PEC when the average power flow is from the AC ($\text{Voltage}_{ac,1}$, Frequency_1) to the DC side ($\text{Voltahe}_{dc,1}$). *Inverter* refers to the converter when the average power flow is from DC ($\text{Voltahe}_{dc,1}$) to the AC side ($\text{Voltage}_{ac,2}$, Frequency_2) (Fig. 1.4).

Finally, the DC-to-DC converters convert one DC voltage level ($\text{Voltage}_{dc,1}$) to another ($\text{Voltage}_{dc,2}$), which may be higher or lower. The term *buck converter* (step-down converter) is used when the steps down voltage ($\text{Voltage}_{dc,1} > \text{Voltage}_{dc,2}$) and if the PEC steps up voltage (while stepping down current) from its input (supply) to its output (load), it is referred as a *boost converter* (step-up converter, $\text{Voltage}_{dc,1} < \text{Voltage}_{dc,2}$). DC–DC converter has multiple applications in power systems, optimising the performance of low-carbon technologies is a classic example, where the DC voltage of produced by the PV array is controlled to optimise the power production; a similar situation is found at wind turbines. On the other hand, DC–DC converters have the potential to contribute with the efficient operation of high voltage direct current (HVDC) transmission systems, but at the moment this book is written, one of the main challenges is reaching very high conversion ratios ($a = \text{Voltage}_{dc,1}/\text{Voltage}_{dc,2}$) (Fig. 1.5).

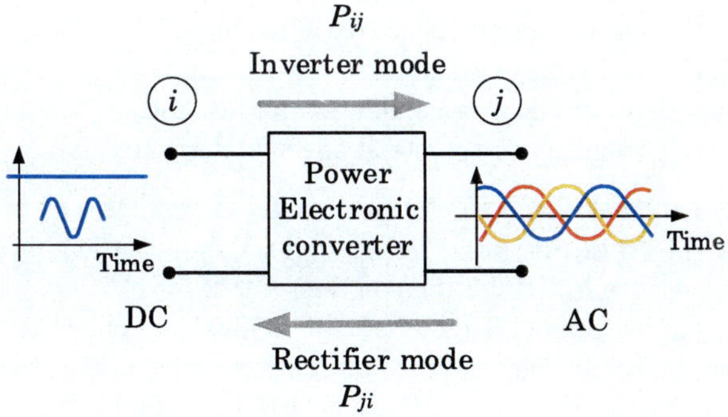

Fig. 1.4 AC–DC converter operation. Operation mode considering the power flow direction between AC and DC

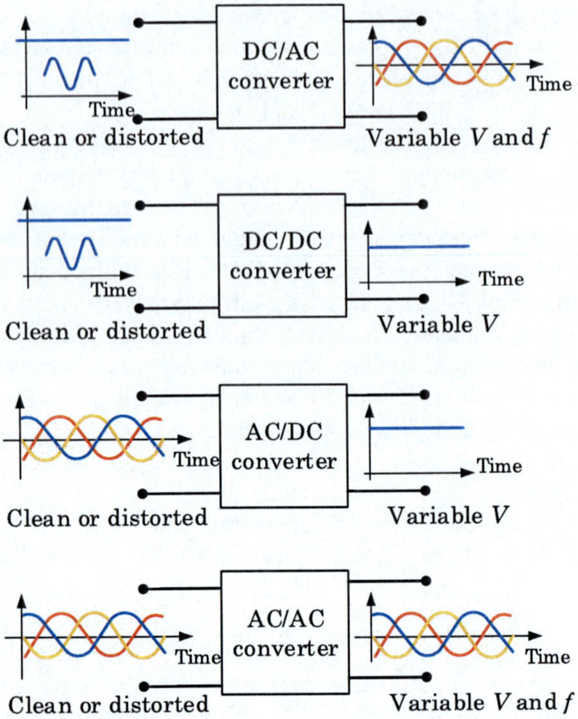

Fig. 1.5 Illustrative diagram showing the power electronic conversion and the relationship with the signals

1.3 Power Electronic Switch

An *ideal switch* is a switch that not consume or dissipate any power from its sources. It as has the following characteristics [4]:

(i) Infinite blocking voltage capability,
(ii) No current while the switch is off,
(iii) Infinite current capability when on,
(iv) Drop voltage equal to zero while on,
(v) No switching or conduction losses, and
(vi) Capability to operate at any switching frequency.

An ideal power electronic switch can be represented as a three terminals device, as shown in Fig. 1.6. The input (1), the output (2), and a control terminal (3) that imposes on/off conditions on the switch.

An important characteristic of an ideal switch is that it exhibits zero-power dissipation, carries bidirectional current, and can support bidirectional voltage. A plot of the switch current (i) with respect to its terminal voltage (v) produces a v-i plane with four quadrants (Fig. 1.6). By definition, an power electronic switch can operate in all four quadrants.

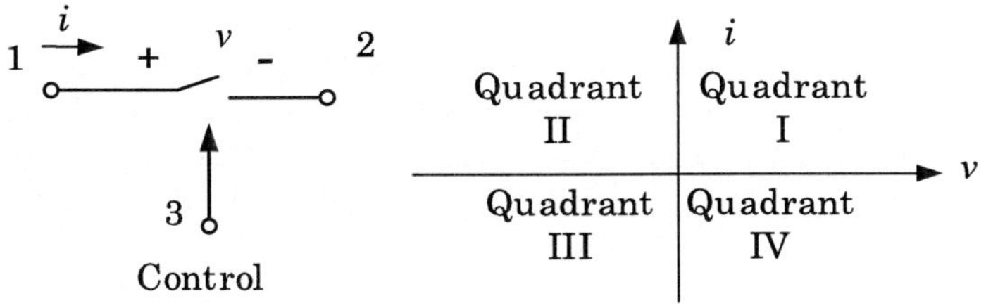

Fig. 1.6 Symbol representing an ideal power electronic switch (left) and four-quadrant switch *v*-*i* characteristics (right)

Considering the power semiconductor devices as ideal switches provide the advantage of making easier the analysis PECs. Using this approach, the details of device operation will not obscure the basic operation of the circuit, making easier the understanding of the underlying concepts. However, the real power semiconductor switching devices have many features to consider, and they might affect the performance of the PEC in some aspects. However, the design and construction of power semiconductor devices focus on how to improve their performance toward the hypothetical concept of an ideal switch. The modern power electronic switches are capable of handling high voltage (HV) and high current operations at high frequency (HF), they are the most important devices needed in the design of energy conversion systems that use PEC.

There are different ways to classify a power semiconductor switch. One simple classification of the power semiconductor switch is based on the number of terminals:

- *Two-terminal devices:* It is the case of diodes; their state is entirely dependent on the external power circuit they are connected to.
- *Three-terminal devices:* These power semiconductor switches are more frequently fount the PEC and their state is not only dependent on their external power circuit but also the signal on their driving terminal (*gate* or *base*).

Another way to classify the power semiconductor switch considers the type of charge carriers they use:

- Some devices are *majority carrier devices* (Schottky diode, MOSFET, JFET)—use only one type of charge carriers (i.e., either electrons or holes).
- Others are *minority carrier devices* (p-n diode, Thyristor, BJT, IGBT)—use both charge carriers (i.e. electrons and holes).
- Finally, a third classification is based on the degree of controllability:
  - *Uncontrollable switches*: The power semiconductor switch has no controlling terminal, and the state of the switch (on/off) is determined by the external voltage or current conditions of the circuit in which the switch is connected. An example of these power semiconductor switches is the diode; it conduces current primarily in one direction (asymmetric conductance); it has low (ideally zero) resistance in one direction, and high (ideally infinite) resistance in the other.

   - *Semi-controllable switches*: In this case, there is limited control over the power semiconductor switch; there is full control to turn on the switch but uncontrolled turn-off. For instance, the power semiconductor switch can be turned on from the control terminal buy once on; it cannot be turned off from the control signal. The power semiconductor switch can be switched off by the operation of the circuit or by an auxiliary circuit that is added to force the power semiconductor switch to turn-off. A *thyristor* or an *SCR* (silicon controlled rectifier) is an example of this switch type.
   - *Fully controllable switches*: This case allows full control of the power semiconductor switch. The power semiconductor switch can be turned on and off via the control terminal. Examples of this power semiconductor switch are the BJT, the MOSFET, the IGBT, the GTO thyristor, *junction gate field-effect transistor* (JFET), and the MOS-controlled thyristor (MCT).

- An additional classification of power semiconductor devices includes:
   - Continuous gate signal requirement: Examples include BJT, MOSFET, IGBT, static induction transistor (SIT).
   - *Pulse gate requirement*: Some examples of this power semiconductor switches include: SCR, GTO, MCT).
   - *Bipolar voltage withstanding capability*: Examples include SCR, GTO.
   - *Unipolar voltage withstanding capability*: e.g. BJT, MOSFET, GTO, GBT, *MOS-controlled thyristor* (MCT).
   - *Bidirectional current capability*: e.g. Triac, Reverse Conducting Thyristor (RCT).
   - *Unidirectional current capability*: e.g. SCR, GTO, BJT, MOSFET, MCT,

IGBT, SITH, SIT and diode (Fig. 1.7).

The history of power electronics predates the development of semiconductor devices employed nowadays. The first PECs were conceived in the early 1900s when the *mercury arc rectifiers* were introduced. Until the 1950s, the power switches used to build PECs were *grid-controlled vacuum tube rectifier*, *ignitron*, *phanotron*, and *thyratron*. Two critical events in the evolution of the PECs were the development:

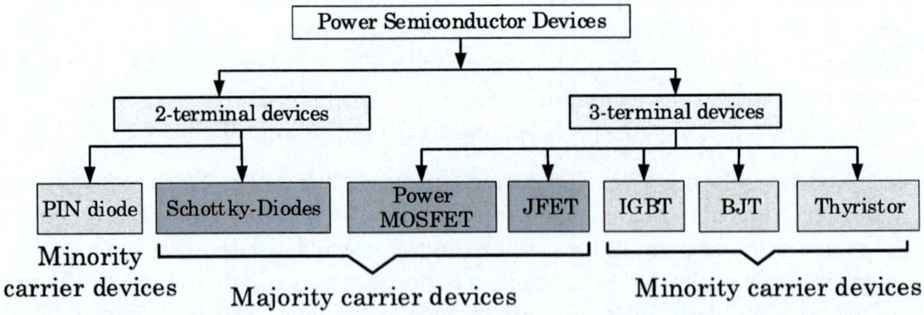

Fig. 1.7 Classification of the power semiconductor switch based on the number of terminals

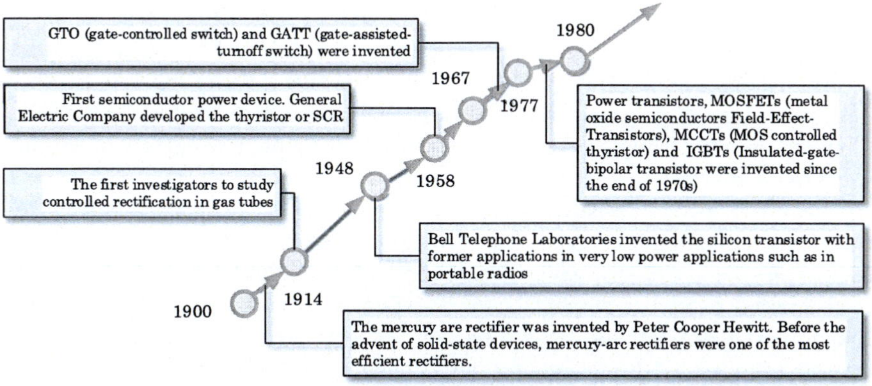

Fig. 1.8 Representative timeline of the most important historical events and developments in the PEC evolution

(i) Bell Telephone Laboratories invented the silicon transistor in 1948; it allows the development of very low power devices such as in portable radios and (ii) General Electric Company developed the thyristors or SCR in 1958. Initially, the SCR was built by using germaniums and later silicon. It was the first semiconductor power device (Fig. 1.8).

One significant step forward in the development of modern PECs was when the thyristor appeared in 1957. The thyristors are power electronic switches with the capacity to withstand very high reverse breakdown voltage at the time that are able to carry very high currents; those characteristics open the possibility of using them at high voltage application as the pioneering HVDC systems. However, one of the major disadvantages of the thyristor is the lack of control to turn off the devices when they are on. It means the switching circuit is that once it is "*latched-on*" in the conducting state (turn on), as a consequence, it cannot be turned off by external control (e.g. gate signal). This characteristic made the thyristor had limited control, useful in some applications but for the more controllable application was a critical challenge.

In the decade of 1960s, a more controllable power semiconductor switches with enhanced power capabilities were introduced; it was the first *bipolar transistors* device. A clear advantage of this power switch was the capability of a fully controllable turn on or off capability using a control signal; this overcomes an important limitation of the thyristor, but power handling was below the thyristors. The sudden development in the late 1970s of the *metal oxide semiconductor* (MOS) technology allows the power semiconductor switch to overcome some of the critical issues related to the bipolar transistors, the power handling! By 1970s, the power MOSFET was available, and it allowed the power semiconductor switch to operate at relatively high power than bipolar transistors at the time to operate at a higher frequency. However, the main issue was the limited application to load voltage. Although the *insulated gate bipolar transistor* (IGBT) was developed in the 1980s, it only becomes widely available by 1990s. The IGBT was able to fully built on top of the power handling capability of the bipolar transistor but adding the advantages of having an isolated gate drive available at the power MOSFET. An illustrative graph is showing a representative timeline of the most important historical events and developments in the PEC evolution and a simplified timeline of the power semiconductor switches and the year of development of each switch are shown in Fig. 1.9.

Figure 1.10 shows a classification of the semiconductor switches based on the semiconductor material used to build the power electronic switch. Classical switches are based on silicon (Si) that material has a breakdown voltage of around 600 V. A new generation of power devices based on *wide band gap semiconductors* such as silicon carbide (SiC) and gallium nitrate (GaN) is used. The modern SiC can withstand voltages 5–10 times higher; also power switches from this material can switch at nearly ten times the rate of silicon, which results in smaller control circuitry.

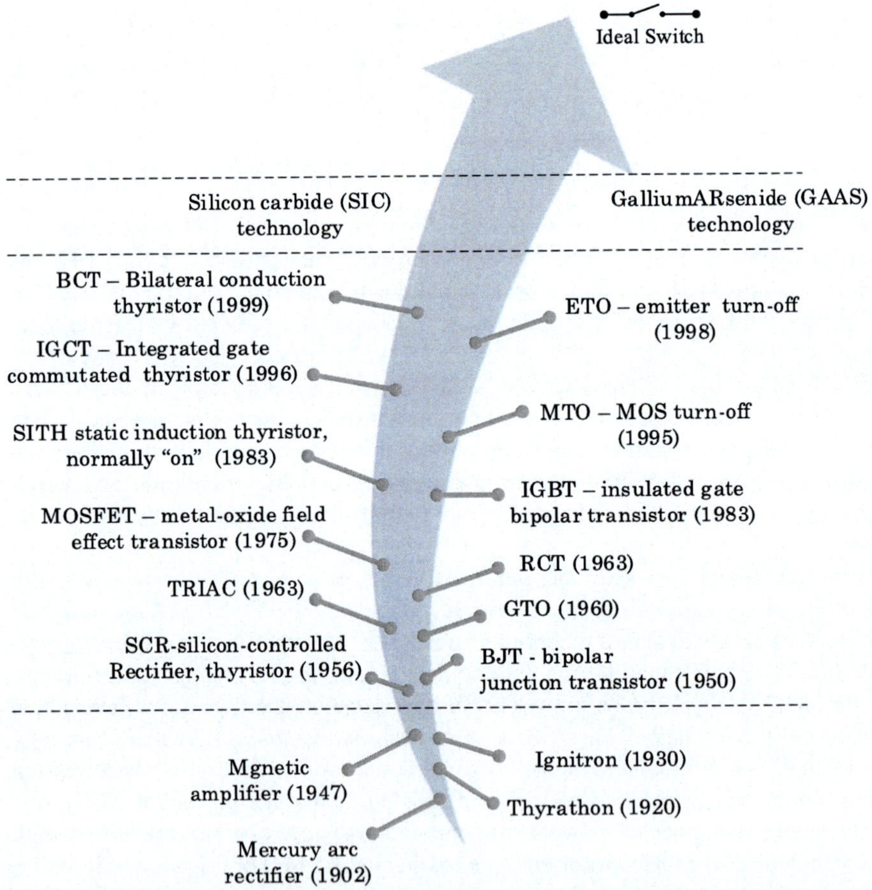

Fig. 1.9 Representative timeline of the most important development of power semiconductor switches toward the development of an ideal switch

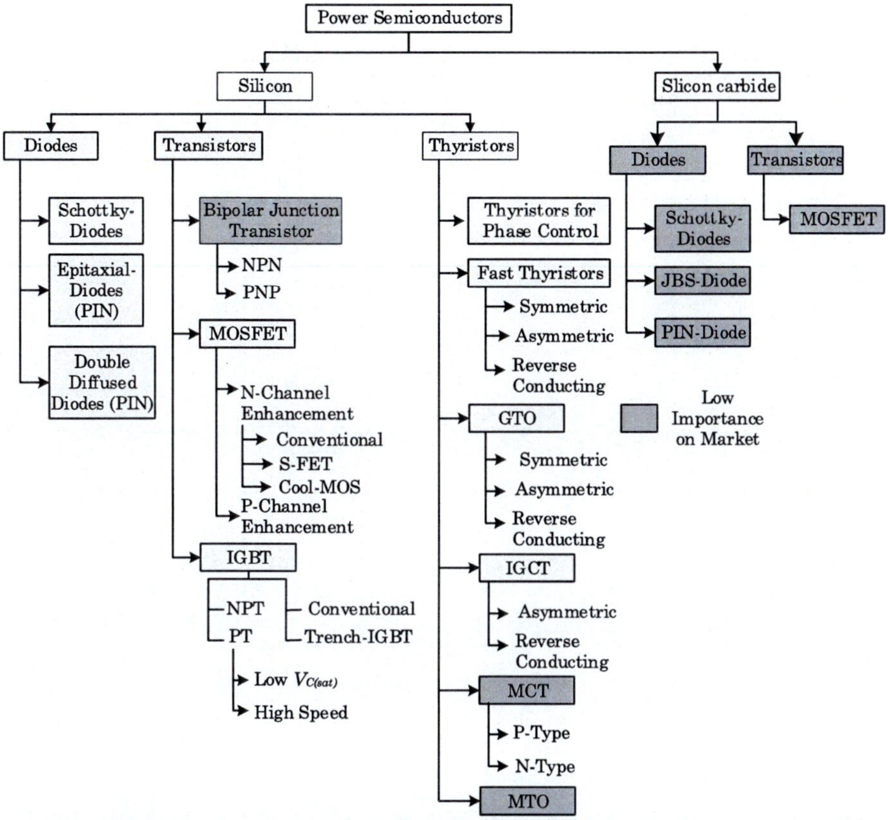

Fig. 1.10 Specific classification power electronic switches based on the semiconductor used: **a** Silicon and **b** Silicon carbide

Additionally, the used of silicon carbide decrease in resistance results in less energy loss during operation, making silicon carbide nearly ten times more efficient at higher voltages than silicon.

Figures 1.11 and 1.12 present an illustrative example of the commonly found power electronic switches indicating the symbols and the representative characteristics.

Finally, Fig. 1.13 shows a comprehensive summary of the primary commonly used power electronic switching devices considering the rated current, rated voltages and the frequency range of use (Fig. 1.14).

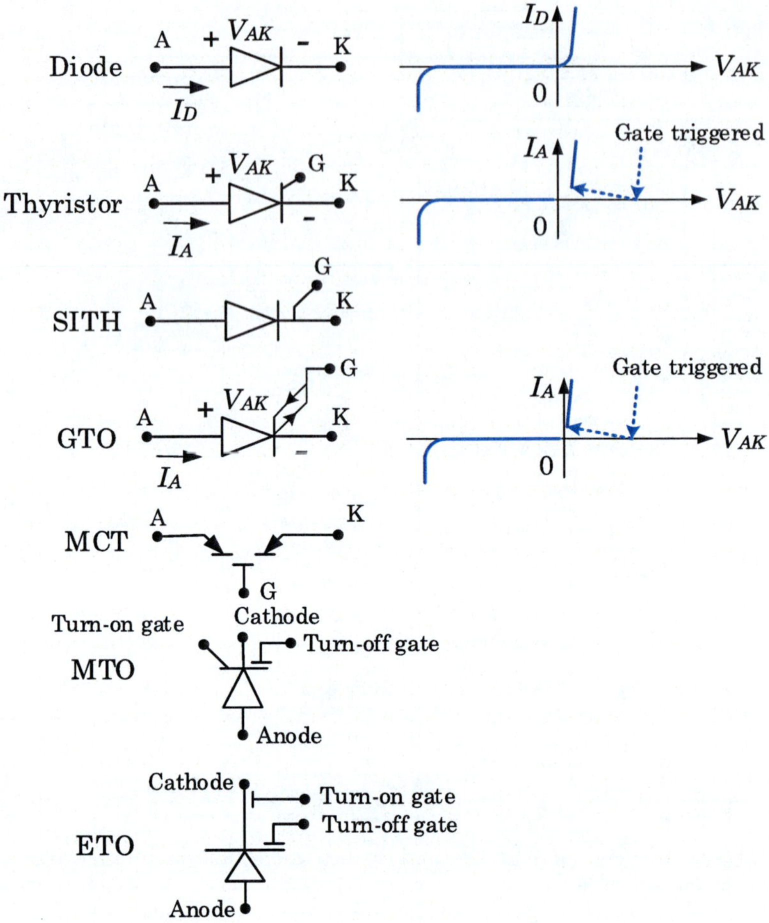

Fig. 1.11 Representative characteristic and symbol of power electronic devices: PART I

1.4 Power Semiconductor Switches in DIgSILENT PowerFactory

DIgSILENT PowerFactory version 2020 includes a model for a power semiconductor switches. It is a network element named ElmValve, and it is used to model and simulated the behaviour of different semiconductor devices. The DC valve element (as called in the *User's Manual of DIgSILENT PowerFactory*) has the limitation that can be connected to DC terminals (*ElmTerm*); as a consequence, it is not possible to use it at AC terminal.

The network element DC valve can be configured to model and simulate three different types of switching: (1) *No controllable device*, as the case of a classical dione at the network element has two terminals, or (2) *semi-controllable* (only turn

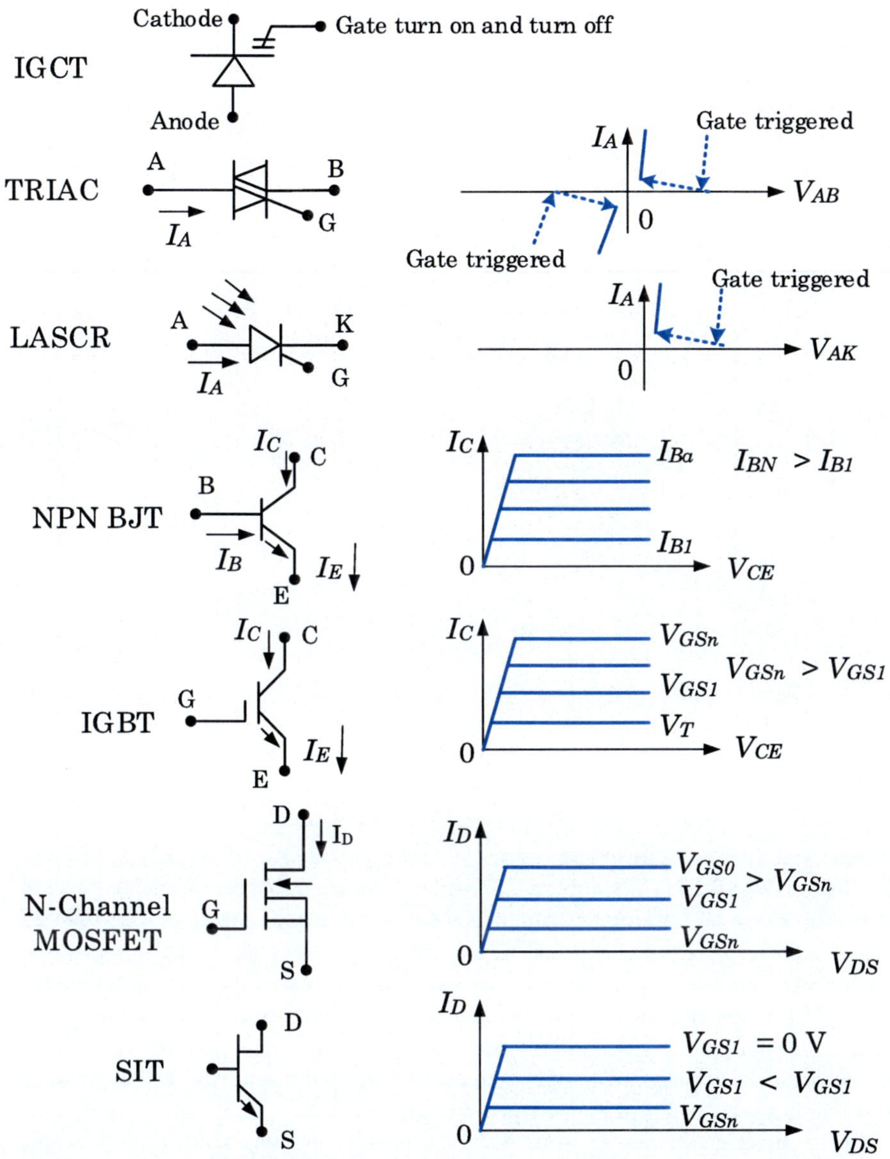

Fig. 1.12 Representative characteristic and symbol of power electronic devices: PART II

on) like the case of classical thyristor, and (3) *fully controllable (turn on/off)* like gate turn-off thyristor (GTO), integrated gate-commutated thyristor (IGCT), isolated gate bipolar transistor (IGBT). The semi-controllable and fully controllable DC valves have a third terminal that is used for control purposes (Fig. 1.15).

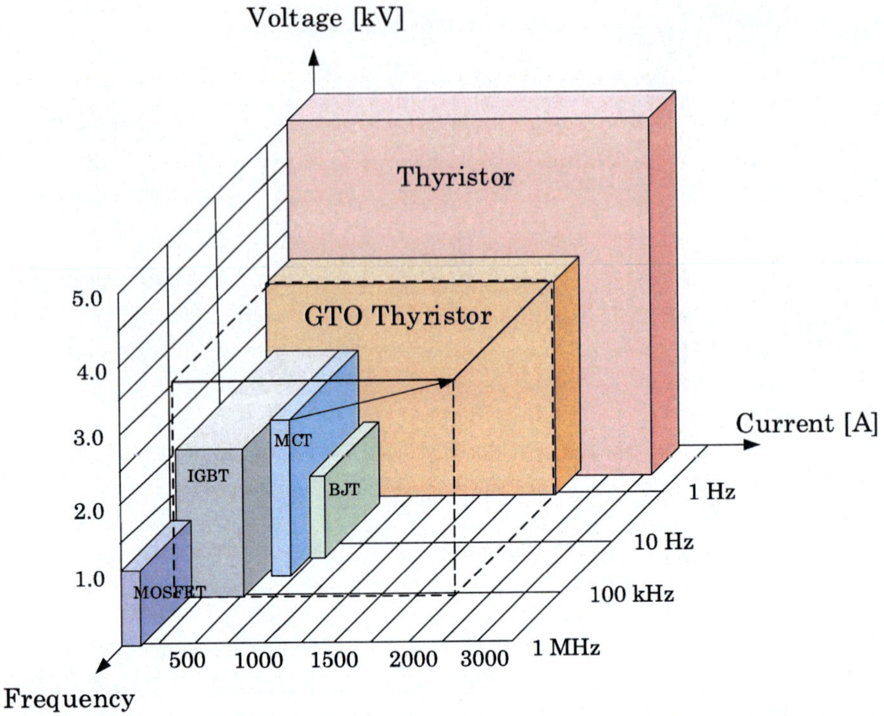

Fig. 1.13 Maximum characteristics (voltage, current, frequency) of some power semiconductor switches

Understanding the state change control of the element DC valve is key in order to use it inside DIgSILENT PowerFactory properly. The model of the DC value element for the time domain simulation includes two main components: (i) the internal conductance (G_{off}) and the internal resistance (R_{on}) dedicated to the switch position and (ii) a parallel snubber circuit consisting of an internal parallel conductance (G_p) and an internal parallel capacitance (C_p). Semi-controllable and controllable devices have an input signal called "gate"; using a DSL , the signal for controlling the device can be appropriately generated depending on the application. The element DC valve has a state named capacitance–voltage in volts.

A very illustrative example of the use of ElmValve can be found in the example "*Offshore wind farm*"; it shows an offshore windfarm connected via a VSC HVDC link, where a controllable chopper is located at the onshore converter station. Details of the controller generating the signals of the ElmValve are presented in the *OnshoreChopperCtrl.BlkDef*.

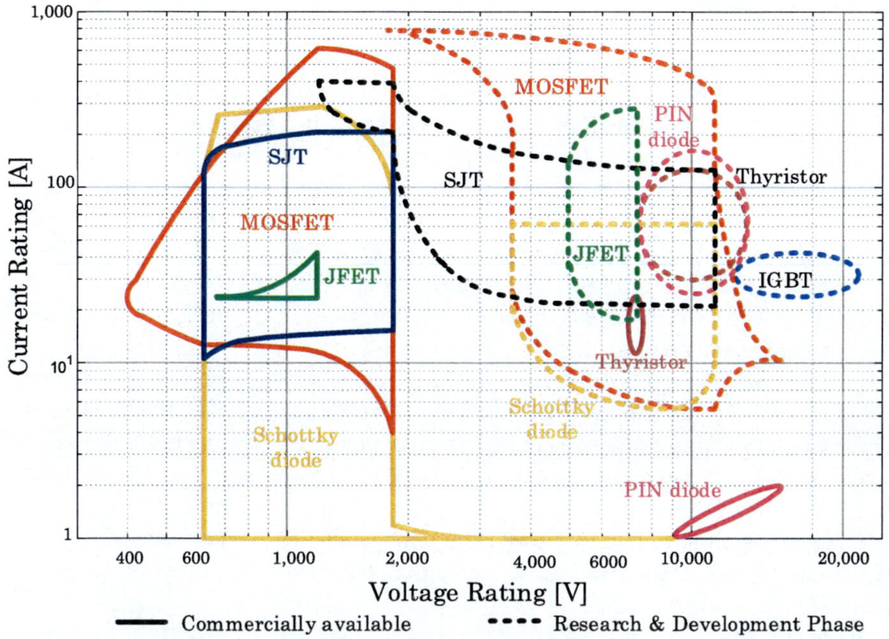

Fig. 1.14 Rating characteristics (voltage and current) of some power semiconductor switches

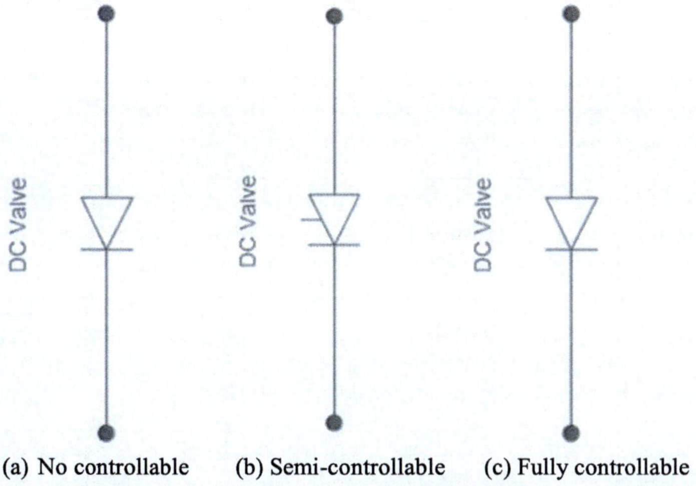

Fig. 1.15 Symbol representation of element DC valve (ElmValve) depending on the types of switching

1.5 Basic PECs

A PEC is typically used to interface a system that can define as an energy provider (source) and energy user (load or source). The energy provider (or source) can be considered either of voltage type or current type. Also, the energy user can be of either voltage type (e.g., a capacitive load) or current type (e.g., an inductive load).

Figure 1.16 shows a general schematic diagram depicting the four possible uses of a PECs as an interface between energy provided and user: (a) *V–I*, (b) *I–V*, (c) *V–V* and (d) *I–I*. The *V–V* (Fig. 1.16c) connection involves the PEC interfacing load and source is expected to have an inductor element inside that block, and the same rationale applies for *I–I* (Fig. 1.16d).

The *commutation cell* is the basic structure in PEC. It is composed of a power switch; nowadays, a high-power semiconductor switch is used instead of a mechanical switch. A PEC makes use of a configuration of power semiconductor devices that function as switches; switches are closed and open in a systematic way allowing the power conversion. Whenever a switch is turned off, the path of current flow changes, that is a commutation of current away from that switch. These are made to turn on and off repetitively in such way as to implement the required conversion function (Fig. 1.17).

The basic commutation cells allow the PECs to convert energy from DC–DC, DC–AC, AC–DC and AC–AC and also to extend them to more complex structures, in a more systematic way.

As a basic illustration in this chapter, a simple PEC topology of a DC–AC converter is shown in Fig. 1.18. This circuit allows the power conversion from DC to AC; it is an inverter. The inverters can be categorised as *voltage-source inverters* (VSIs), *current-source inverters* (CSIs), and *impedance-source inverters* (ZSIs).

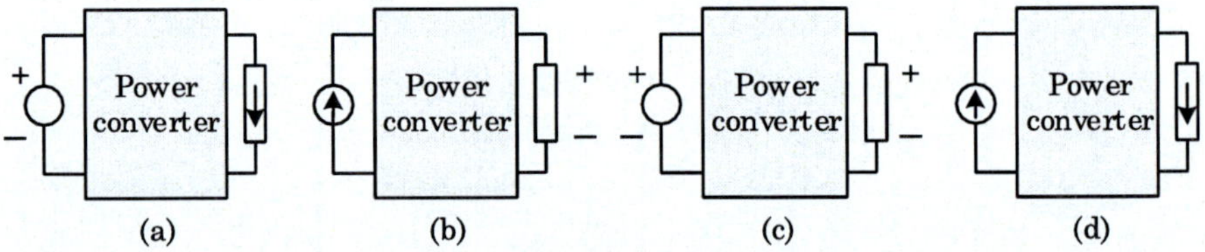

Fig. 1.16 General representation of the four possible uses of PECs

Fig. 1.17 An illustrative example of a PEC showing ideal switches and the possible states (on/off)

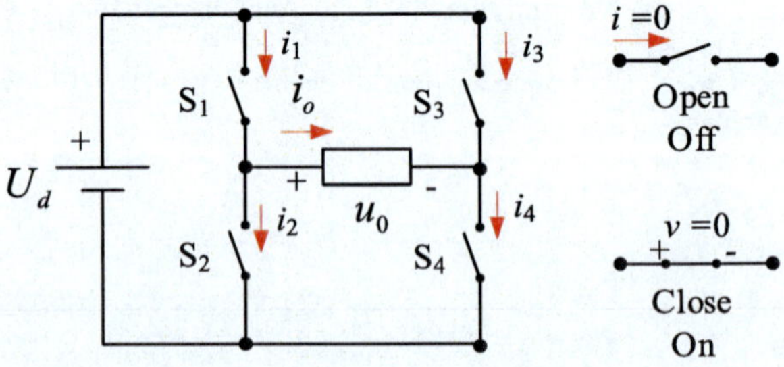

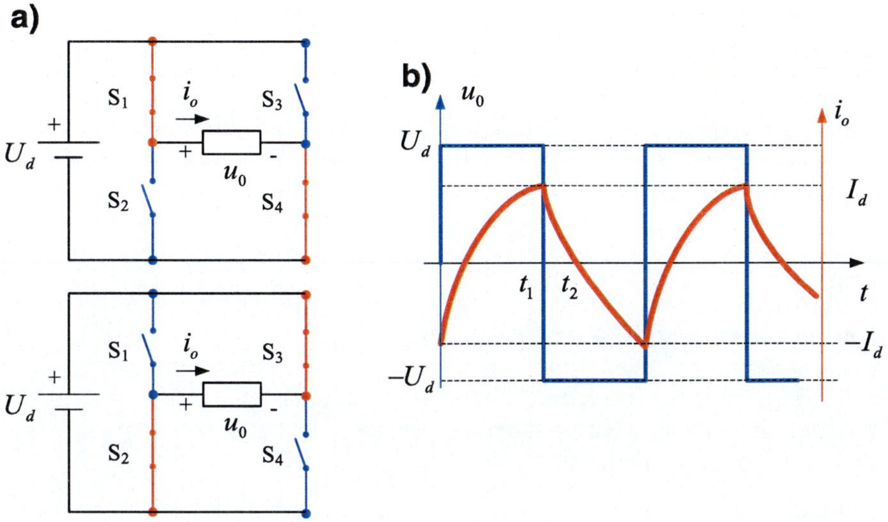

Fig. 1.18 Illustrative topology of a single-phase DC-AC PEC **a** full-bridge inverter topology is shown (left side, top positive semi cycle); **b** the output waveforms of voltage (u_0) and current (i_0) for a *duty cycle* of $d = 0.5$

The term commutation, in the context of PECs, means the transfer of currents from one path to another. The *commutation process* takes places with the appropriate sequencing of closing and opening the power switches. Two main types of communication process can be identified:

- *Circuit-dependent parameters*: it depends on circuit topology and includes components such as transformer leakage (inductor), commutation capacitor, auxiliary switching device, etc.
- *Switch-dependent parameters*: it includes device turn-on and turn-off times, *di*/*dt* and *dv*/*dt* limitations, etc. The most significant switch-dependent parameter is turn-off time since the switches are not perfect; their turn-off times are finite and impact on the commutation process.

Table 1.1 shows summary between the type of devices and the communication type. It is evident that modern power semiconductor switches as the GTO, IGBT and MCT have the possibility of turning on and off by using a gate signal.

Table 1.1 Summary of the main type of power electronic switching devices and its commutation type and characteristic

Type of device	Commutation	Initiated by
Conventional thyristor	Line	AC line voltage
	Circuit (or Capacitor)	Capacitor voltage
GTO, IGBT or MCT	Self	Gate drive

1.6 Power Electronic Converters in Power Systems Applications

An illustrative diagram showing the main application of power electronic converter inside power systems is shown in Fig. 1.19. The diagram shows the uses of power electronic converters in several areas of the power systems: the generation side, transmission, distribution and end-user side. The power electronic converter is a key technology for all those subsystems and has spread in many applications, examples including:

- *Low voltage residential sector*: heat pumps;
- *Commercial sector*: uninterruptible power supply (UPS), battery chargers;
- *Industrial sector*: variable speed of induction machines;
- *Transportation sector*: charger stations of electric vehicles;
- *Transmission sector*: high voltage direct current (HVDC).

Power electronic converter is a key element enabling the transition to a low-carbon power system; as a consequence, it is expected that the penetration of PECs continues increasing in the future. As the technology evolves, the PECs will evolve in three main advances: high power, higher frequency and lower losses will continue to be invented.

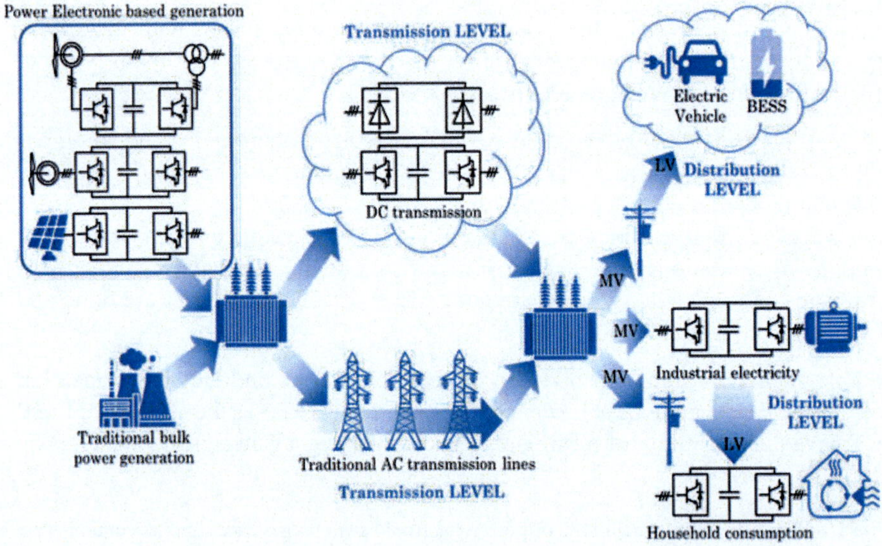

Fig. 1.19 Power electronic converter applications inside modern power systems

1.7 Current-Source Converter (CSC) and Voltage-Source Converter (VSC)

The two major types of power electronic converter can be defined in terms of the main electrical variables: (i) voltage-source inverter (VSI) and current-source inverter (CSI).

The *current-source converter* (CSC) uses thyristors, and a large inductor is connected on the DC side; the large inductor is used to maintain the DC current more or less constant (other than a small ripple) (Fig. 1.20).

The voltage-source converter used fully controllable (turn on/off) power electronic switches allows creating DC voltages voltage is always the same polarity and the current reverses to change the direction of power flow. This type of converters uses a large capacitor on the DC side, thus maintaining the DC voltage more or less constant. The VSC converters use IGBT, GTO and other transistors as power electronic switches (Fig. 1.21).

The CSC has several advantages:

- The CSC can be made up to very high power, and DC voltage ratings and the thyristors are comparatively robust with a significant transient overload capability.
- The CSC uses thyristors, and as they switch off only when the current through them has dropped to zero, switching losses are low.

On the other hand, VSC has several advantages:

- The VSC can operate at any combination of active and reactive power; the power control is very flexible because of the use of fully controllable power electronic switches.
- The VCS has the ability to operate into a weak grid (low short-circuit level) and even black-start an AC network.

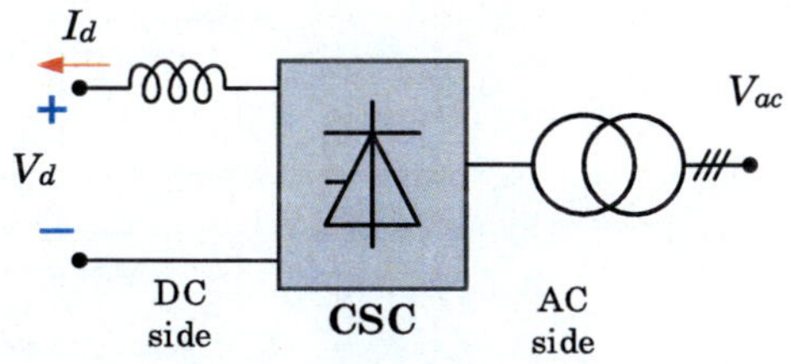

Fig. 1.20 Illustrative example of a current-source converter (CSC)

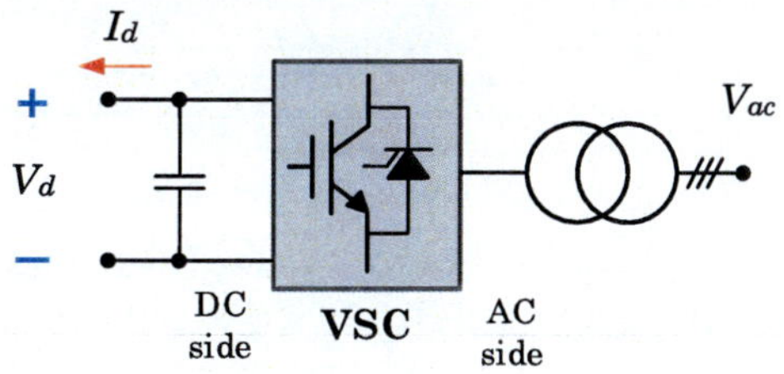

Fig. 1.21 Illustrative example of a voltage-source converter (VSC)

The CSC is a mature technology that arrived into the power systems by 1950s; the active power losses caused by the communication process still lower than the VSC a typical converter station power losses are lower than 0.8% compared with 1.1% for the same size installation of VSC. Modern CSCs are able to deliver 12,000 MW and using ultra-high-voltage DC ±1100 kV (e.g. Changji to Guquan). On the other hand, the state-of-the-art converters reached a level of ±525 kV, 1400 MW (e.g. new Norway-Germany interconnector) (Fig. 1.22).

VSC and CSC have many applications of insider power systems; a very limited summary is shown in Fig. 1.23.

Different HVDC power electronic converter topologies have been proposed, built and utilised all over the world. However, the two dominant types are the *line commutated converter* (LCC) and the *voltage-source converter* (VSC). Figure 1.24 shows a generic point-to-point HVDC link interconnecting two AC power systems. The power converter stations (PEC_1 and PEC_2) can be created using LCC, VSC or combination of them, one of the converter station working as a rectifier and the other as inverter depending on the energy flow between the AC systems.

CSC and VSC have been key enablers on the development of the HVDC systems.

HVDC interconnections can be configured in different forms to suit the different desired performance and operational requirements; a non-exhaustive classification includes (for simplicity, the figures are created using thyristor-based converters):

Back-to-back: The back-to-back connection is one where both, the inverter and the rectifier, are localised at the same location [6]. The back-to-back connection has much application at many applications as PV systems, BESS and full rated

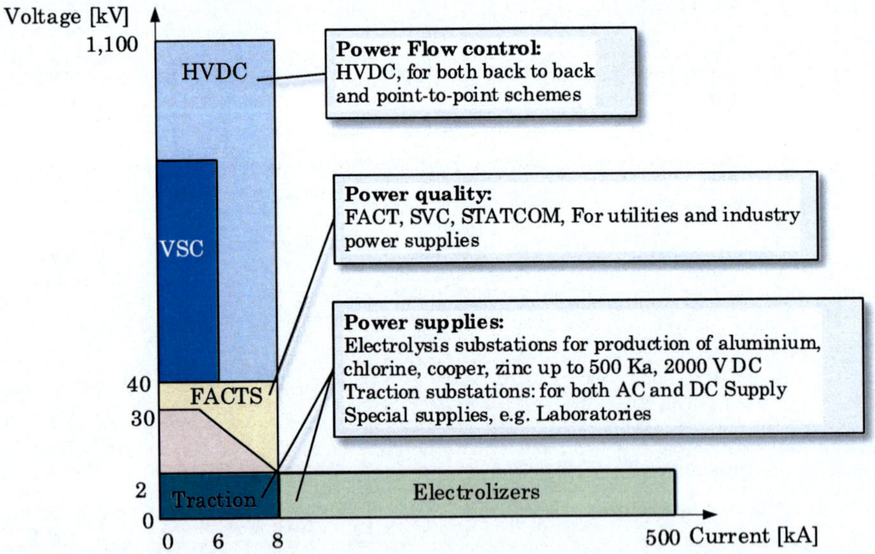

Fig. 1.22 Simplified summary of the power electronic converter application indicating voltage and current indicators

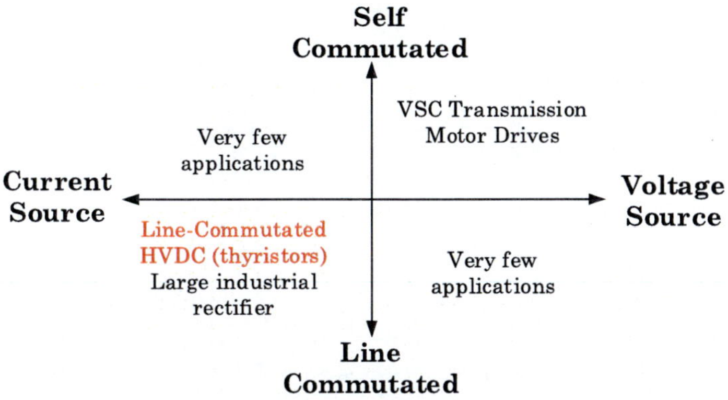

Fig. 1.23 A summary of the main power electronic application depending on the type of power electronic converter (VSC and CSC) and the type of commutation

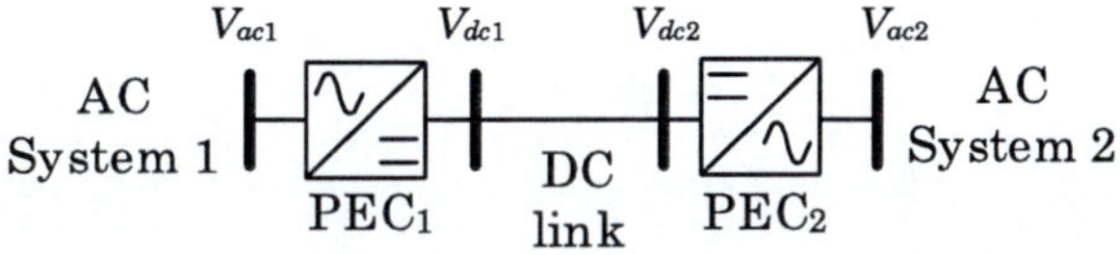

Fig. 1.24 Illustrative point-to-point DC link between two AC power systems

power converters. At HVDC applications, back-to-back configuration the rectifier and inverter at the same location usually interconnected via a short DC line of few metres located inside the same environment. The typical application is the separation of two independent power systems (see Fig. 1.25). These systems are not synchronised or on a different nominal frequency [7]. An example of back-to-back thyristor-based link installed at the River Canada; the back-to-back converter was rated at 320 MW.

Monopolar connection: A monopolar link has one conductor at a high DC voltage. The other conductor can use either the ground (see Fig. 1.26) or sea as the return path or metallic-return wire (see Fig. 1.27). Both converters are separated by a single

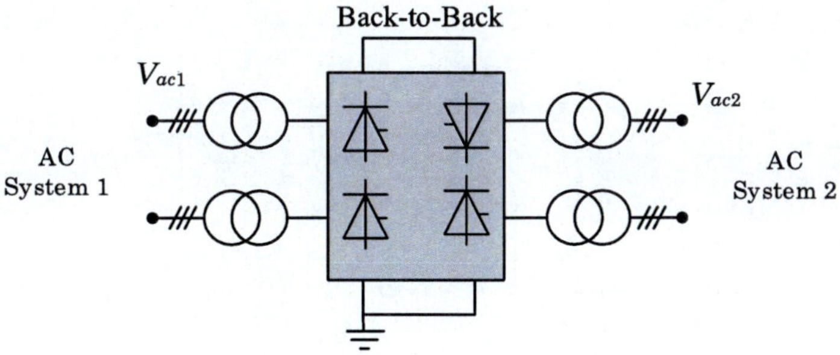

Fig. 1.25 Illustrative diagram of a back-to-back configuration for HVDC

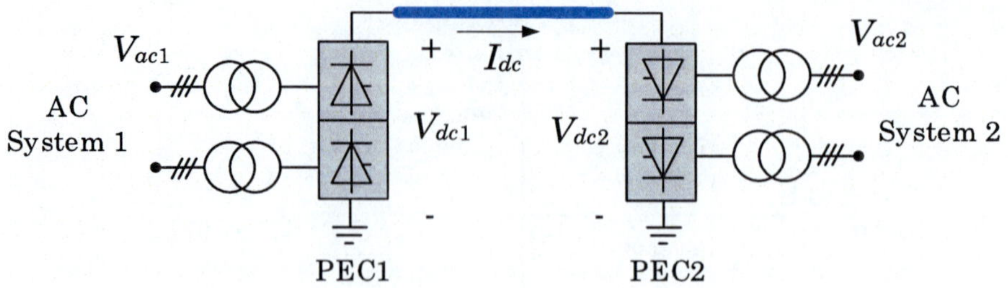

Fig. 1.26 Illustrative diagram of a monopole ground-return configuration for HVDC

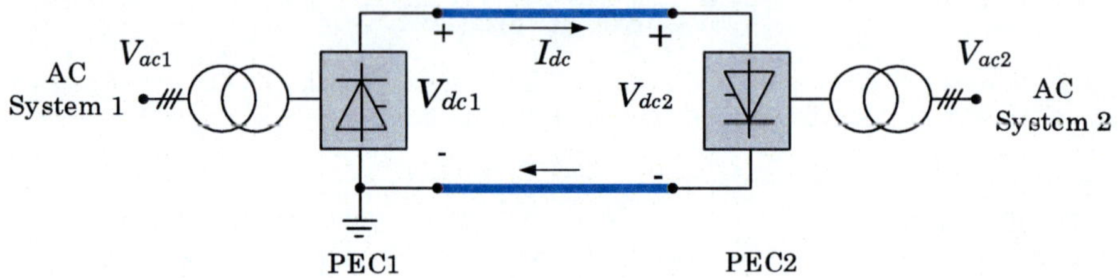

Fig. 1.27 Illustrative diagram of a monopole metallic-return configuration for HVDC

DC pole line, either positive or negative voltage. The ground is used as a current return path. Most submarine cable connections use monopolar systems. Low-voltage electrode lines and sea electrodes to carry the return current in submarine cable crossings.

Homopolar connection: This has two or more DC line of the same polarity connected to the converters (see Fig. 1.28). The negative polarity is typically used for less corona and reactive power loss. The physical ground is used as a return path of the currents. A simple glance, the connection works as a monopole when one pole develops a fault; as a consequence, this connection has a improve reliability. The disadvantage of high cost makes it unpopular and seldom used.

Bipolar connection: The bipolar connection is the most popular method in HVDC interconnection of converters. This connection similar to the homopolar connection,

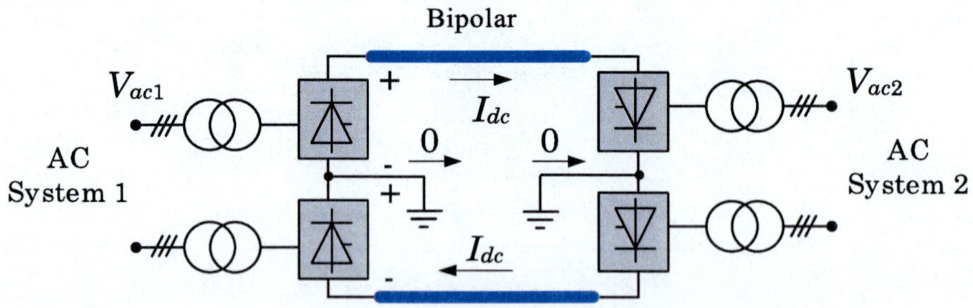

Fig. 1.28 Illustrative diagram of a homopolar configuration for HVDC

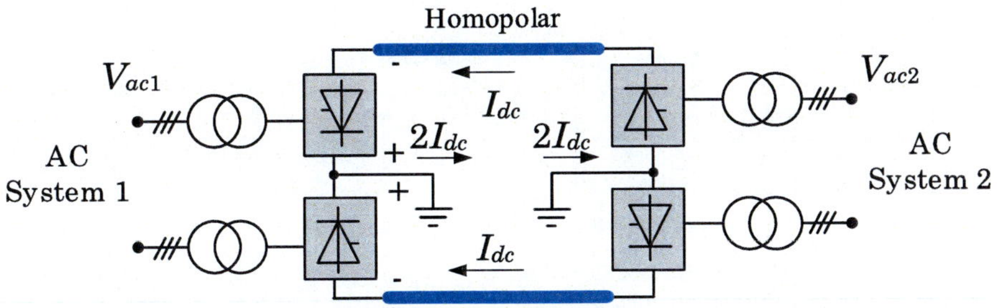

Fig. 1.29 Illustrative diagram of a bipolar configuration for HVDC

but it has different polarities (see Fig. 1.29). In the bipolar connection, each pole is independent, that is, it can operate with a single pole with ground used as a return path (Fig. 1.30).

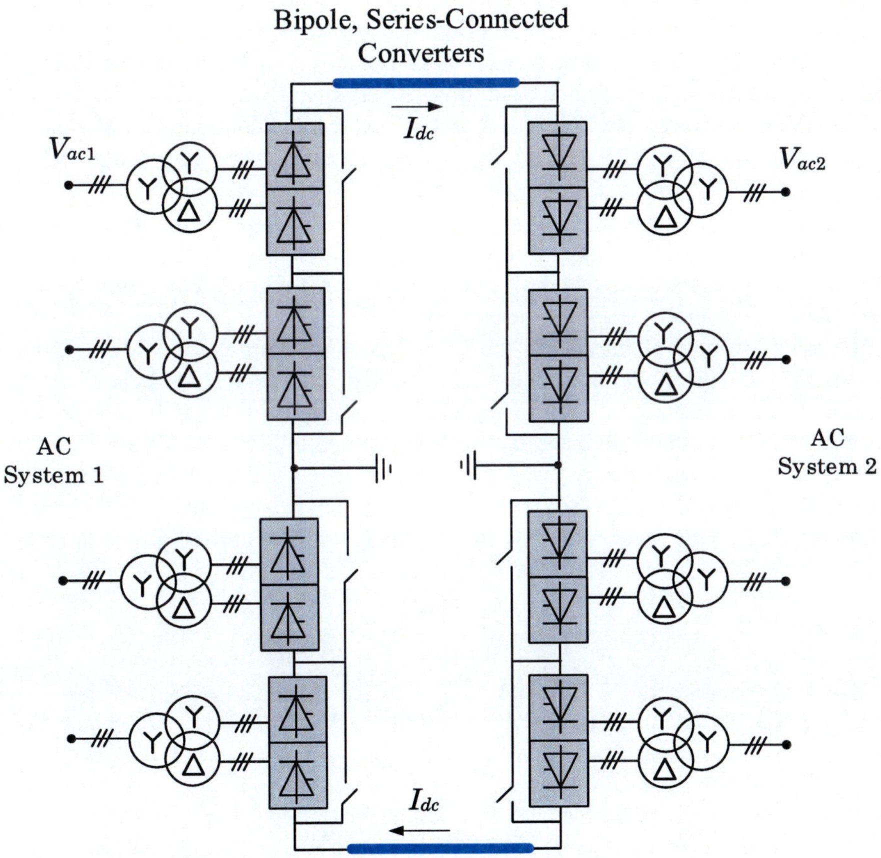

Fig. 1.30 Illustrative diagram of a bipolar, series-connected configuration for HVDC

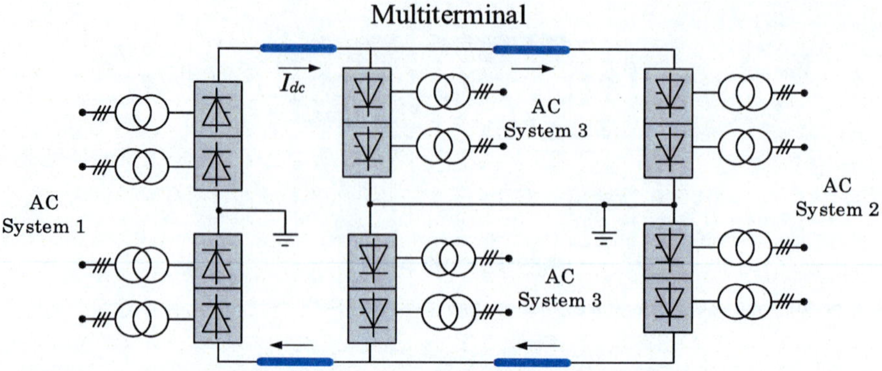

Fig. 1.31 Illustrative diagram of a multi-terminal (three-LCC converter stations) configuration for HVDC

Multi-terminal connection: The most recent development in VSC with enhanced operating flexibility is allowing the development of novel connections. The multi-terminal connection has more than two sets of converters operating independently. Each converter can operate as a rectifier or an inverter (Fig. 1.31).

The HVDC converter stations based on LCC have been dominating by decades in the market. Actual LCC-based HVDC has the highest power rating of all HVDC technologies; LCC current rating reaches up to 6250 A and blocking voltage of 10 kV. Figure 1.32 shows point-to-point DC link showing the details of the LCC power electronic converter stations. The power transfer is from AC System 1 to AC System 2, as a consequence, the PEC_1 is working as a rectifier (AC/DC conversion), and PEC_2 is an inverter (DC-AC conversion).

The main variable of interest in the LCC link is the currents moving through the system; a circuit model can be used to represent the steady-state behaviour. A controlled current source is used to represent the current provided by the power converter; this situation is depicted in the circuit shown in Fig. 1.33.

VSC uses fully controllable power electronic switches suck as IGBT, GTO, etc. The fully controllable turn on/off of the switches allows the current being switched on and off at any time independent of the AC voltage; this feature is quite attractive

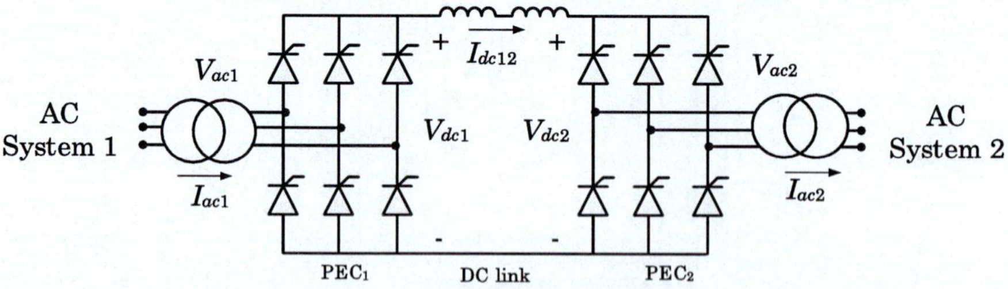

Fig. 1.32 Details of a point-to-point DC link based on LCC power converter stations

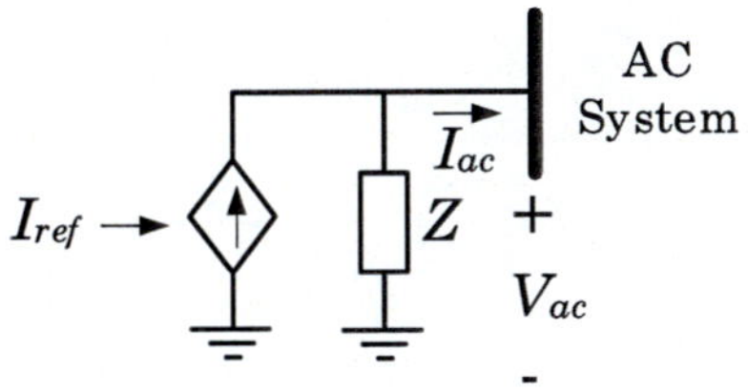

Fig. 1.33 Equivalent circuit representing the AC side of the LCC working as an inverter

because allows the VSC to create its own AC voltages providing black-start capabilities. This technology was developed in the 1990s with the first project commissioned by ABB, 1997. Today, rated capacity of VSC still below the LCC, the largest VSC example is the 1400 MW, ±525 kV Nordlink that interconnect Norway and Germany (distance ~623 km).

Figure 1.34 shows point-to-point DC link showing the details of the VSC power electronic converter stations, in this case, the PEC_1 is working as a rectifier (AC/DC conversion), and PEC_2 is an inverter (DC–AC conversion).

The main variable of interest when modelling a VSC is the voltage; a circuit model can be used to represent steady-state behaviour. The circuit consists of a controlled voltage source is used to represent the voltage provided by the power converter; this situation is depicted in the circuit shown in Fig. 1.35.

The VSC operates at a high frequency using a pulse width modulation (PWM) techniques which allows simultaneous adjustment of the amplitude ($|V_{VSC}|$) and phase angle (δ_{VSC}) of voltage generated by the converter while keeping the voltage constant.

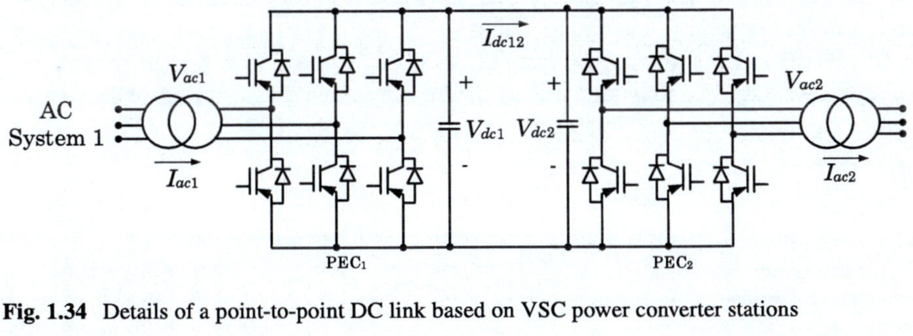

Fig. 1.34 Details of a point-to-point DC link based on VSC power converter stations

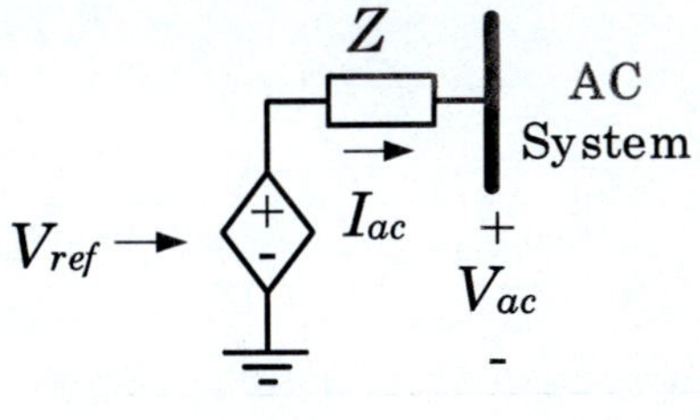

Fig. 1.35 Equivalent circuit representing the AC side of the VSC working as an inverter

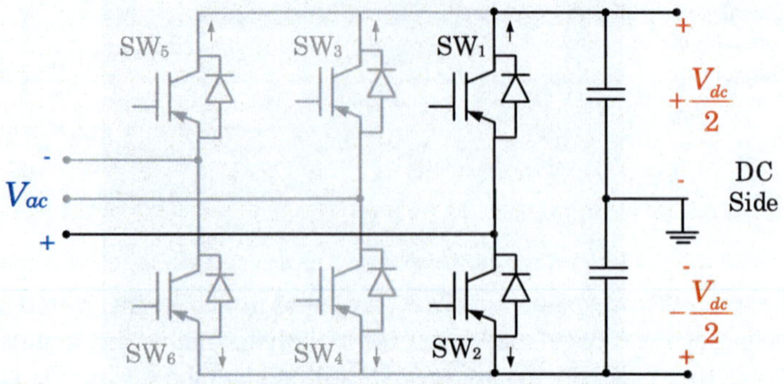

Fig. 1.36 Schematic diagram of a three-phase full-bridge (VSC) inverter

The PWM, or *pulse-duration modulation* (PDM), is a method of reducing the average power delivered by an electrical signal, by effectively chopping it up into discrete parts (Fig. 1.36).

Ideal PWM inverter output voltage (instantaneous component, blue trace) and its averaged counterpart (fundamental component, red trace) are shown in Fig. 1.37.

Many different types of PWM control technique are available: Single pulse width modulation (Single PWM), Multiple pulse width modulation (MPWM), Sinusoidal pulse width modulation (SPWM), Modified Sinusoidal pulse width modulation (MSPWM), etc. Each of those techniques has advantages and applications a full discussion of them is out the scope of this book. However, the reader must be aware that each technique has a specific set of advantages and disadvantages that make them useful at power systems applications.

The PWM is the most used technique at VSC, it operates at a high frequency, and it allows simultaneous adjustment of the amplitude ($|V_{\mathrm{VSC}}|$) and phase angle (δ_{VSC}) of voltage generated by the converter while keeping the voltage constant (see Fig. 1.38).

Fig. 1.37 Ideal voltage and average voltage of three-phase full bridge shown in Fig. 1.36

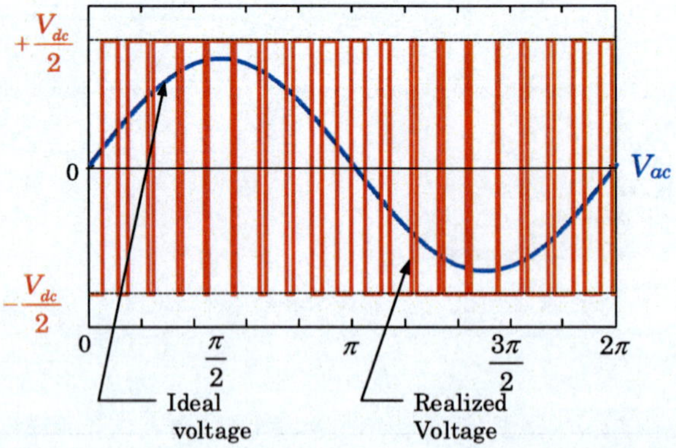

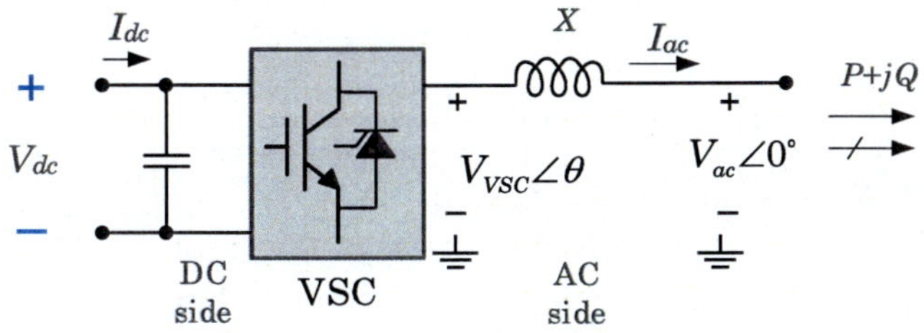

(a) Schematic diagram of the steady-state operation of a VSC inverter (RMS values depicted)

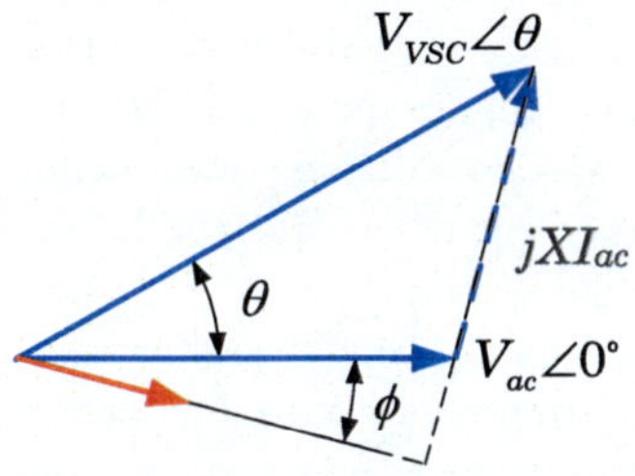

(b) Phasor diagram of the steady-state operation of a VSC inverter

Fig. 1.38 Steady-state operation of a VSC inverter

The controllability of the VSC provides a very high degree of flexibility, making it possible to control the active (P) and reactive power control (Q) [8]:

$$P = \frac{|V_{ac}||V_{VSC}|}{X} \sin\theta \tag{1.1}$$

$$Q = \frac{|V_{ac}|[|V_{VSC}|\cos\theta - |V_{ac}|]}{X} \tag{1.2}$$

1.8 Power Converter in DIgSILENT PowerFactory

DIgSILENT PowerFactory is equipped with a set of network elements with specific applications for power electronic.

On to the power, electronic devices included in PowerFactory is the model of the thyristor-based converter, specifically the *6-pulse power converter* with the possibility of operating as *rectifier* or *inverter*. The element *ElmRec*, *ElmRecMono* are designed as configurable devices that allow the simulation of *three-phase diode rectifier* or *three-phase line-commutated rectifier/inverter*; a general overview of the one HVDC converter station (using *ElmRec*, *ElmRecMono*) [9] is shown in Fig. 1.39.

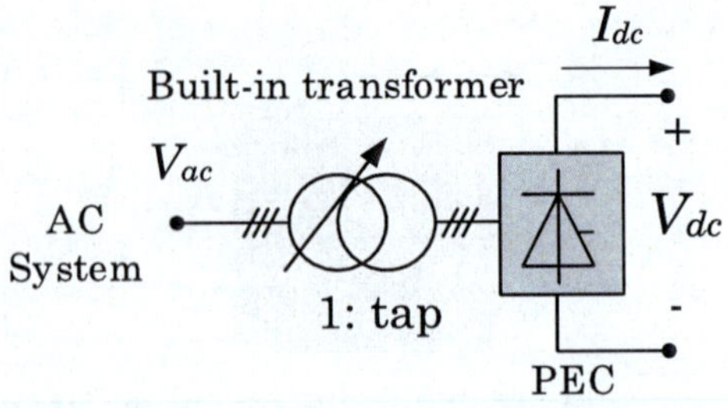

Fig. 1.39 Illustrative schematic of an HVDC converter using *ElmRec*, *ElmRecMono* in DIgSILENT PowerFactory. The model comes with a built-in transformer

Figure 1.39 shows a thyristor-based HVDC converter station that can be specified to operate on rectifier or inverter, but the element allows the use of standard full-bridge diode rectifier. In such a case, the bridge is rectifying the three-phase AC voltage to a 6-pulse DC voltage. Diodes are no-controllable power switch as a consequence they can neither be turned-on or turned-off externally, the DC voltage or DC current of the rectifier cannot be controlled.

Figure 1.40 shows a detailed model of a thyristor-based controlled converter; the model consists of six power thyristors. These power electronic switches can be turned on by an external control signal, but only turned-off when the current flowing through them becomes negative.

Figure 1.41 shows the voltage waveforms on AC and DC side of a 1 MW, 1 kV, 50 Hz (RMS, line-to-line) 6-pulse power converter based on diodes. The peak value of the signal applied at the AC side is:

$$V_p = \frac{\sqrt{2}}{\sqrt{3}} V_{\text{rms}} = \frac{\sqrt{2}}{\sqrt{3}} (1000\,\text{V}) \cong 816.496\,\text{V} \tag{1.3}$$

The DC output voltage (V_o) is obtained as the mean value of the direct-output voltage:

$$V_0 = V_{\text{rms}} \frac{3}{\sqrt{2}\pi} = (1000\,\text{V}) \frac{3}{\sqrt{2}\pi} \cong 675.237\,\text{V} \tag{1.4}$$

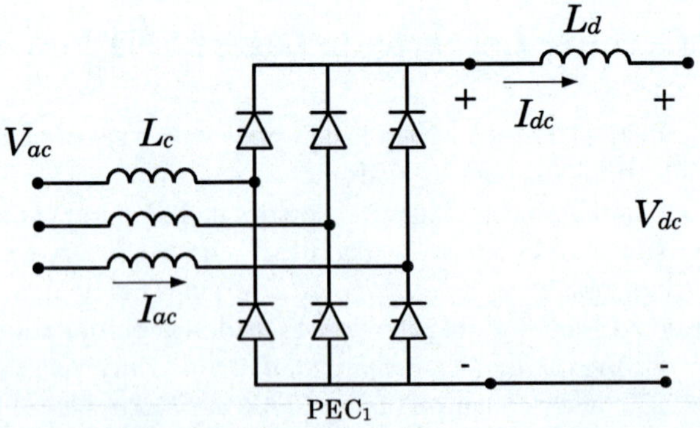

Fig. 1.40 Illustrative schematic the three-phase thyristor bridge used in an HVDC converter using *ElmRec*, *ElmRecMono* in DIgSILENT PowerFactory. The conductance, inductance and the DC reactance are not part of the model

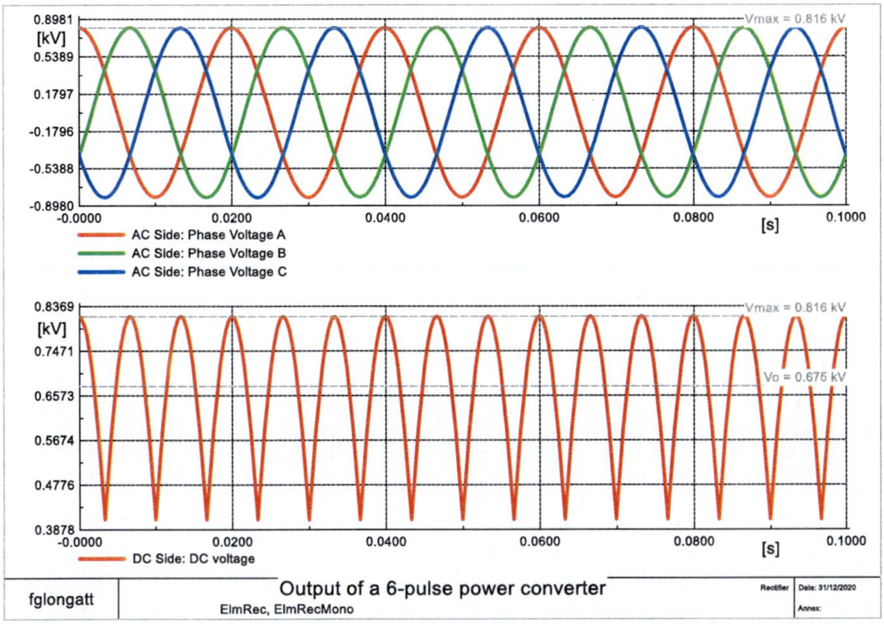

Fig. 1.41 Illustrative example of AC and DC side oltage waveform vof a three-phase thyristor bridge

Thyristor based power converter allows the controllability of the average output voltage by controlling the fining angle. Figure 1.42 shows a comparison of waveform taken from the DC side of a thyristor-based controlled-rectifier considering a firing angle $\alpha = 35°$ and $\alpha = 75°$, as expected the average mean value of the direct-output voltage depends on the cosine of the firing angle α:

$$\hat{V}_0 = V_{\text{rms}} \frac{3}{\sqrt{2}\pi} \cos\alpha = V_0 \cos\alpha \tag{1.5}$$

DIgSILENT PowerFactory is well-equipped with several examples where the user can try several pre-built projects with specific power system studies. One of those examples was implemented in DIgSILENT PowerFactory version 15.0 and is contained in the file "HVDC-Example.pfd". Now, the example is an integral part of the examples in version 2020, and it is a classical thyristor-based HVDC link[1] used for assessing the performance of the system by using power flow and time-domain simulation (RMS and EMT) [9]. The reader is invited to open DIgSILENT PowerFactory, and find the aforementioned example, read the literature provided by DIgSILENT and conduct the proposed simulations.

[1]M. Szechtman, T. Wess, and C. Thio, "A benchmark model for HVDC system studies", in International Conference on AC and DC Power Transmission. IET, 1991, pp. 374–378.

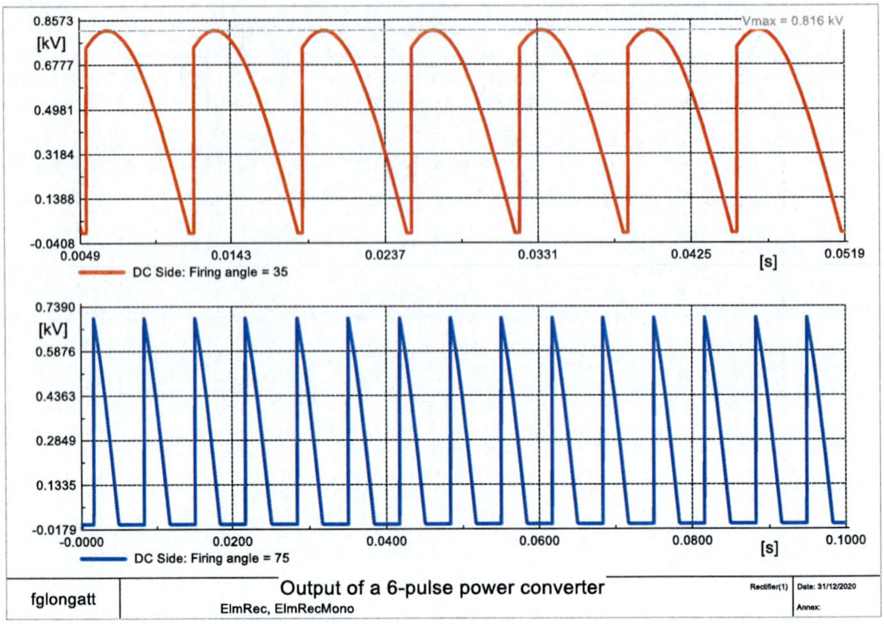

Fig. 1.42 Illustrative example of AC and DC side voltage of a three-phase thyristor bridge

The modern PECs take advantage of the fully controllable power electronic switches, such as IGBT and GTO, and the use of the VSC has escalated for many power system applications up to HVDC systems. DIgSILENT PowerFactory includes the element *ElmVsc* and *ElmVscmono*; both of them are able to simulate VSC.

The element *ElmVsc* and *ElmVscmono* are based on the use of PWM to operate the fully commutable power electronic switches systematically. The model has the option to be configured as a self-commutated, *voltage-source AC/DC two-level converter* and a *modular multilevel converter* (MMC). For those converter types, DIgSILENT PowerFactory offers two types of connections, the so-called *ElmVsc* providing two ports of connection at the DC side, on the other hand, the *ElmVscmono* is designed to offer a single DC port at the return is assumed to be by the ground path (Fig. 1.43).

The internal topology of the PEC is easily configurable, allowing the presentation of the *two-level converter*, and *half-bridge* and *full-bridge MMC*.

The two-level converter is the simplest type of three-phase VSC and can be thought of as a six pulse bridge in which the thyristors have been replaced by power electronic switches with turn-off capability with inverse-parallel diodes, and the DC smoothing reactors have been replaced by DC smoothing capacitors (see Fig. 1.44).

PowerFactory includes the implementation of MMC based PECs, to do so two submodules types of internal topologies are available: half-bridge and full bridge (see Fig. 1.45).

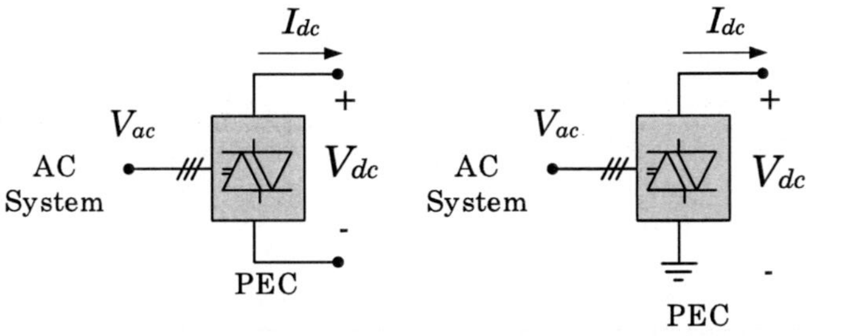

(a) PEC with two DC connection ports: *ElmVsc* (b) PEC with a single DC connection port: *ElmVscmono*

Fig. 1.43 Illustrative examples of the PWM based PEC available in DIgSILENT PowerFactory

Fig. 1.44 Illustrative circuit of *two-level converter* implementation in DIgSILENT PowerFactory (*ElmVsc* and *ElmVscmono*)

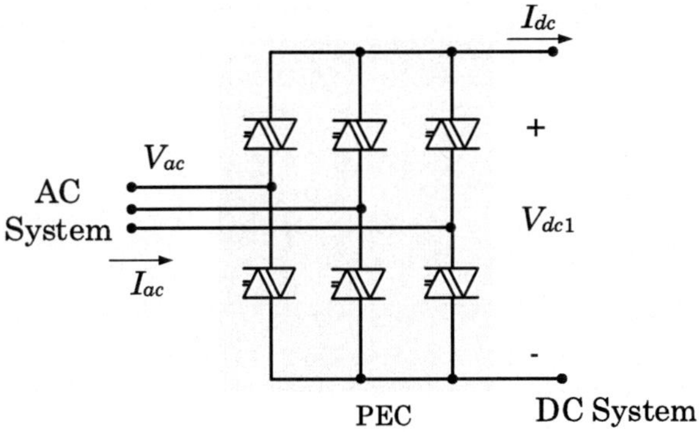

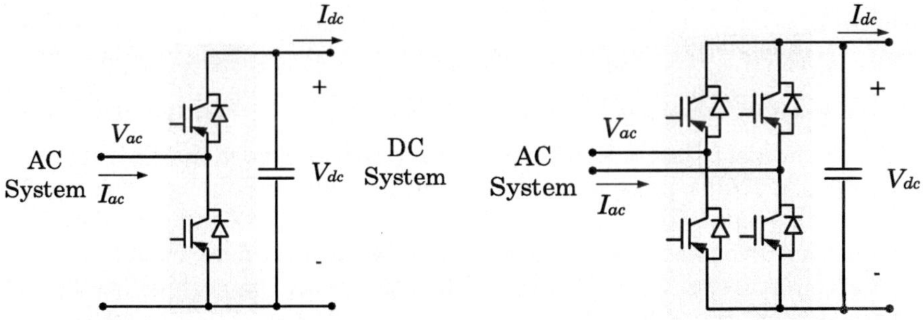

(a) Half-bridge Submodule (b) Full-bridge Submodule

Fig. 1.45 Illustrative circuit of half-bridge and full-bridge MMC implementation in DIgSILENT PowerFactory (*ElmVsc* and *ElmVscmono*)

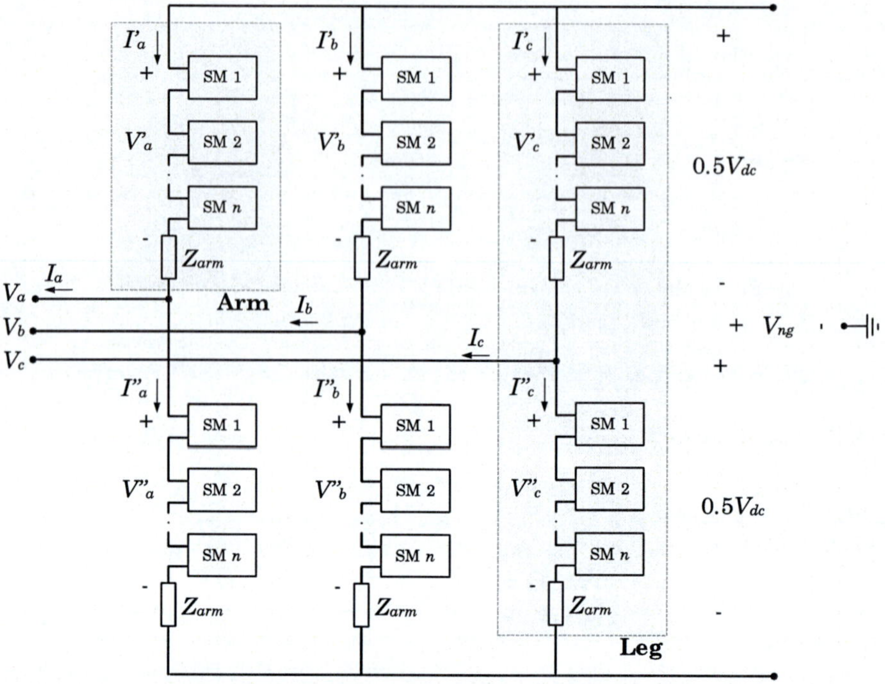

Fig. 1.46 Illustrative circuit of half-bridge and full-bridge MMC implementation in DIgSILENT PowerFactory (*ElmVsc* and *ElmVscmono*) [9]

The MMC configuration enables the uses of half-bridge and full-bridge topologies at each *submodule* (SM), DIgSILENT PowerFactory allows a configuration of SM creating arms and legs, see Fig. 1.46.

DIgSILENT PowerFactory includes several examples one of them is titled "HVDC Connected Offshore Wind Farm". The examples consist of a VSC-HVDC link for a 400 MW offshore wind farm consisting of DFIG wind turbines. The model is very well developed, including several control loops and also a chopper controller to use the ElmValve (the reader is highly encouraged to check those models).

The example is very comprehensive, including short-circuit analysis and RMS and EMT simulations.

1.9 Modern Tendencies in Power Electonic Applications in Power Systems

The modern power system is in a very important transition, moving from a synchronous machine-based system towards an inverter-dominated system, with large-scale penetration of low-carbon technologies, including renewable energy sources (RESs) like wind and PVs.

Traditional *synchronous machines* (SM) are very different from the model PECs. A traditional generation unit using SM directly connected to the grid can be divided into two main areas: the energy conversion process where the electro-mechanical interactions take places, it is basically dictated by the magnetic coupling between the roto at the stator of the SM; and the generation control process where at least two basic controllers are considered: frequency control and AC voltage control (see Fig. 1.47). One very important aspect of the electro-mechanical process is the fact there is a natural energy storage system in the SM, it comes from the kinetic energy stored in the rotating mass: flywheel, rotor and turbine (total H).

The increasing penetration of low-carbon technologies is motivating the growing integration of PECs [10]. The PEC is basically an energy conversion system, where the interconnection to the grid is based on a DC–AC converter, the energy conversion is done through the power electronic switches. It is evident the PEC alone is not able to generate electricity,in fact, the PEC requires a primary energy source and energy storage in order to look similar to a traditional synchronous generator. From the control point of view, the PEC has intentive requirements, the power electronic switches work a very high frequency as a consequence high-frequency signal must be systematically produced to control the PEC appropriately, the typical control is based on the use of the two-axis theory, the d-axis and q-axis components of the pulse-width modulation (PWM) control, which directly actuate the d-axis and q-axis components of the AC current [11].

From the power electronic and machine theory, it is clear one of these components is typically utilised to regulate the grid-side AC voltage. On the other hand, depending on the configuration and the converter topology, the other component can be used, for example, to regulate the DC voltage or current flowing through the converter; this is the typical case of the PV systems and fully related to converter variable speed wind turbines. Lastly, the active power injection and the energy balance across the

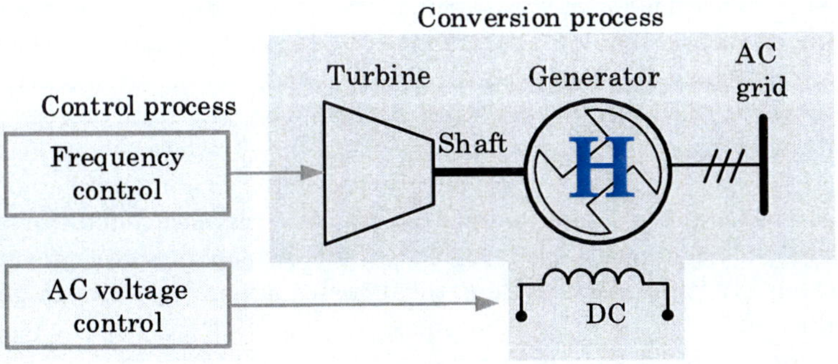

Fig. 1.47 Illustrative scheme of the energy conversion, energy storage and controllers of a synchronous machine

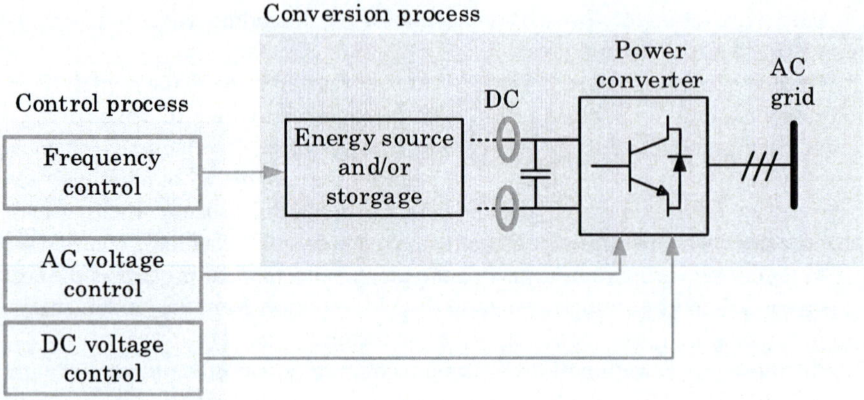

Fig. 1.48 Illustrative scheme of the energy conversion, energy storage and controllers of PEC

converter's DC storage (DC side capacitor) are regulated through the control of the energy source connected to the converter (Fig. 1.48).

The two-axis theory has their fundamentals form the analysis of SM, and it is something very natural to thing and understands; however, when the *dq*-axis theory is applied to the control of PEC, it requires an angle reference for this *dq*-frame. Without any generalisation, two taxonomy of control loops have been formulated for frequency control: *grid following* (reference angle obtained from a PLL) and *grid forming converters.*

1.10 Conclusion

The most recent advancements in the power electronics industry made the integration of low-carbon technologies be a reliable alternative to substitute the technologies base on SM, and consequently, the integration of VSC-based generation units into the transmission system has been increased. The fast dynamic response of power electronic switches provides the solar PEC-based technologies with a quick response that enables these resources to contribute to system stability. However, there are major changes to the way to supply and use energy; building a smarter grid lies at heart to these changes: (i) The significant increases in electricity generation from intermittent/highly variable renewable resource, (ii) the increase of stressing/narrowly conditions on the transmission system, (iii) massive decoupling between generation/load caused by PEC-based system, and several other features expected into the future electricity networks will negatively affect the system security.

The core of the energy transition toward a reduced low-carbon footprint in the electric sector is the change in generation technology, from fossil fuel-based thermal generation to converter interfaced renewable generation. However, a very important issue is rising; the power system is gradually losing synchronous machines (SMs). The installed capacity is increasing because more generation is being installed, but the generation mix is changing; the renewable generation connected via PEC is increasing at the time that many fossil-based and nuclear power plants are decommissioned. The power system is losing synchronous machines with all the excellent properties of them, and they are substitute with very sensible and fragile PEC. The SM provides very useful services and characteristics in natural from, kinetic energy storage, overloading capacity, simple and reliable control mechanisms. On the other hand, power engineers have tried to make the PECs to behave like SM, as a consequence multiple control loops to mimics SM has been developed (e.g. inertial control, droop control etc.), but one of the main disadvantages of the PECs is that they require measurements, signal processing and so many other things that they are not able to naturally do like SM does. As a consequence, the way of PEC being integrated into the power systems must change. More research and development are required for the integration of PEC, and exciting times are coming.

References

1. F. Gonzalez-Longatt, Frequency control and inertial response schemes for the future power networks. in *Large Scale Renewable Power Generation*, eds. by J. Hossain and A. Mahmud (Springer Singapore, Singapore, 2014), pp. 193–231
2. F. Sanchez, J. Cayenne, F. Gonzalez-Longatt, J. L. Rueda, Controller to enable the enhanced frequency response services from a multi-electrical energy storage system. IET Gener. Trans. Distrib. **13**(2), 258–265 (2019)
3. H. Chamorro, F. Gonzalez-Longatt, V.K. Sood, Dynamic measurements of the wind power impact on power system inertia and stability. in *International Conference on Innovative Smart Grid Technologies (ISGT Asia 2018)*, 2018
4. E.C. Dos Santos, E.R. Cabral da Silva, *Advanced Power Electronic Converters: PWM Converter Processing AC Voltages* (John Wiley & Sons Inc., Hoboken, NJ, USA, 2014)
5. V.T. Ranganathan, N. Mohan, T.M. Underland, Power electronics: converters, applications and design **33**(5) (2008)
6. V.K. Sood, *HVDC and FACTS Controllers : Applications of Static Converters in Power Systems* (Kluwer Academic Publishers, Boston; London, 2004)
7. J.L. Dirk Van Hertem, Oriol Gomiz-Bellmunt, *HVDC GRIDS: For Offshore and Supergrid of the Future*, no. c. Hoboken (Wiley, New Jersey, 2016)
8. H.R. Chamorro, A.C. Sanchez, A. Pantoja, I. Zelinka, F. Gonzalez-Longatt, V.K. Sood, A network control system for hydro plants to counteract the non-synchronous generation integration. Int. J. Electr. Power Energy Syst. **105**, 404–419 (2019)
9. DIgSILENT, User's Manual DIgSILENT PowerFactory 2020. (Gomaringen, Germany, 2019)
10. A. Peña Asensio, F. Gonzalez-Longatt, S. Arnaltes, J.L. Rodríguez-Amenedo, Analysis of the converter synchronizing method for the contribution of battery energy storage systems to inertia emulation. *Energies* **13**(6), 1478 (2020).
11. H. Chamorro, F. Gonzalez, K. Rouzbehi, R. Sevilla, H. Chavez, V. Sood, nnovative primary frequency control in low-inertia power systems based on wide-area RoCoF sharing. *IET Energy Syst. Integr.* (2020)

Chapter 2
Integration of Large-Scale Photovoltaic Power Plants into Power Networks to Maintain System Stability

Hamid Khoshkhoo, Reza Ranji Bourachalou, and Aref Pouryekta

Abstract Recently, several large-scale photovoltaic power plants (LS-PVPPs) have been integrated into the high-voltage grids around the globe. On the contrary to the conventional synchronous generators, static generators use voltage source converters (VSCs) as an interface device to connect into the AC network. Since the VSCs consist of power electronic switches to control their output, the response time of these resources is considerably lower compared to the synchronous generators. Consequently, the VSCs are able to react much faster to any disturbance occurred in the network. In this respect, when generation rescheduling is inevitable, LS-PVPPs seem to be more effective as they can perform the required remedial actions within a short time. The study aims to investigate the performance of the LS-PVPP in maintaining system stability as a fast-response power generation unit. In this respect, LS-PVPPs (along with their controllers) and a remedial action scheme (proposed to improve the stability status) have been implemented in DSL and DPL environment of PowerFactory, respectively. The results of the dynamic simulations performed in IEEE 14-bus test system show that the LS-PVPPs effectively decay the deteriorating impacts of disturbances and improve the stability status by performing timely remedial actions.

Keywords DIgSILENT PowerFactory software · DIgSILENT Programming Language (DPL) · DIgSILENT Simulation Language (DSL) · Large-scale

---

Electronic supplementary material The online version of this chapter (https://doi.org/10.1007/978-3-030-54124-8_2) contains supplementary material, which is available to authorized users.

H. Khoshkhoo (✉) · R. R. Bourachalou
Faculty of Electrical Engineering, Sahand University of Technology, Sahand New Town, Tabriz, Iran
e-mail: khoshkhoo@sut.ac.ir

R. R. Bourachalou
e-mail: re_ranji96@sut.ac.ir

A. Pouryekta
Energy Advisory, Clean Technology Center, DNV GL, Singapore, Singapore
e-mail: aref.pouryekta@gmail.com

F. M. Gonzalez-Longatt and J. L. Rueda Torres (eds.), *Modelling and Simulation of Power Electronic Converter Dominated Power Systems in PowerFactory*, Power Systems,
https://doi.org/10.1007/978-3-030-54124-8_2

photovoltaic power plants · Power system dynamic stability · Voltage source converter (VSC)

2.1 Introduction

During the last decades, to stabilise energy prices as well as to achieve lower emission of greenhouse gases and reduce air pollution, renewable energy resources have been extensively used in the power systems all over the world [1]. Among different types of these resources, photovoltaic (PV) ones that directly convert the energy of sunlight into the electricity have been widely integrated into the low-voltage networks to operate in grid-connected or stand-alone modes to provide low-cost electrical energy.

Over the past years, the reduction in the production cost of solar panels resulted in the emergence of large-scale PV power plants (LS-PVPPs) which are connected to the high-voltage network through a series reactance and step-up transformer (Fig. 2.1). Since conventional generation resources connected to the transmission systems usually participate in the voltage and frequency control procedures to maintain the system stability, the operation of LS-PVPP should be controlled in such a way that they participate in maintaining system stability, especially in emergency conditions appear after major disturbances [2–6].

Since from stability point of view, timely execution of remedial actions is of great importance, PV resources that have low response time are proper tools for system operators to perform the required preventive and/or corrective control actions. To achieve this goal, the LS-PVPP may select the remedial action based on the locally measured data or receive required commands/variables from the system operators and/or wide area measurement system (WAMS).

This study aims to investigate the performance of the LS-PVPP when a power rescheduling scheme is carried out to maintain system stability. To this end, a PV system along with its AC–DC converter which is controlled by a decoupled PQ controller has been implemented in DIgSILENT Simulation Language (DSL) to simulate the dynamic behaviour of LS-PVPP in power systems. Then, following

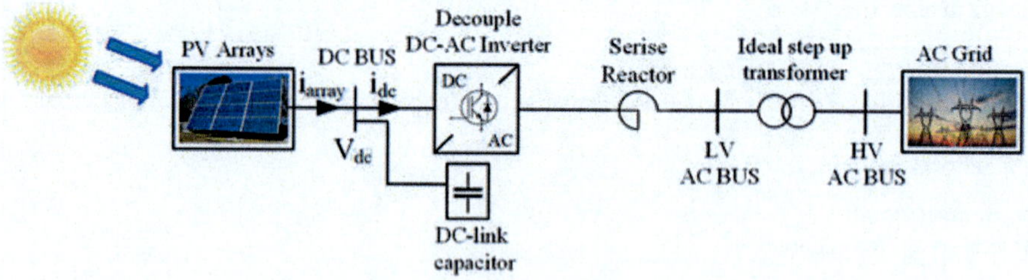

Fig. 2.1 Schematic diagram of a large-scale photovoltaic system interconnected to a medium- and high-voltage network

any critical contingency, a proposed remedial action scheme implemented in DPL environment changes the active and reactive power generation of the PV power station to maintain the voltage/frequency stability of the test system.

2.2 Decoupled PQ Controller

Figure 2.2 shows a LS-PVPP connected to a network through a series reactor and a transformer. The dynamics of the AC side of the VSC is described by [7]:

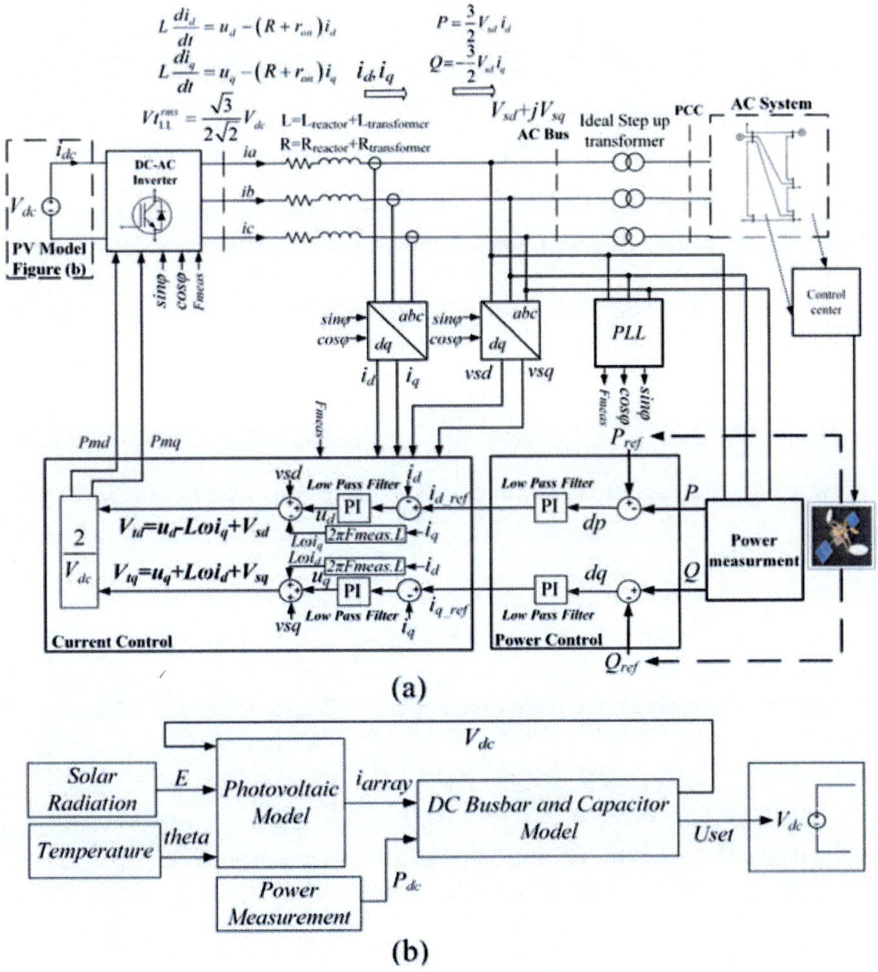

Fig. 2.2 Schematic diagram of a LS-PVPP which will be implemented in DSL environment of PowerFactory software

$$L\frac{\mathrm{d}\vec{i}}{\mathrm{d}t} = -(R + r_{\mathrm{on}})\,\vec{i} + \vec{V_t} - \vec{V_s} \tag{2.1}$$

where r_{on} is the equivalent on-state resistance of the converter, and R and L are the equivalent resistance and inductance of the reactor and transformer through which the converter is connected to the grid. Also, V_s represents the AC system voltage:

$$\vec{V_s}(t) = V_{sm}e^{j\varphi} = \begin{bmatrix} V_{sa} \\ V_{sb} \\ V_{sc} \end{bmatrix} = \begin{bmatrix} V_{sm}\cos(\varphi) \\ V_{sm}\cos(\varphi - 120) \\ V_{sm}\cos(\varphi + 120) \end{bmatrix} \tag{2.2}$$

where $\varphi = \omega t + \varphi_0$. Similarly:

$$\vec{V_t}(t) = \begin{bmatrix} V_{tm}\cos(\varphi + \delta) \\ V_{tm}\cos(\varphi + \delta - 120) \\ V_{tm}\cos(\varphi + \delta + 120) \end{bmatrix};\ \vec{i}(t) = \begin{bmatrix} i_m\cos(\varphi - \theta) \\ i_m\cos(\varphi - \theta - 120) \\ i_m\cos(\varphi - \theta + 120) \end{bmatrix} \tag{2.3}$$

where δ and θ show the phase shift of V_t and i_t with respect to V_s, respectively.

Assuming $\rho(t) = \omega t + \varphi_0$, $\vec{V}_s(t)$ may be transformed into dq-frame as shown in (2.4) and (2.5) [7]:

$$f_d + jf_q = \overrightarrow{f(t)}\,\mathrm{e}^{-j\rho(t)} \tag{2.4}$$

$$V_{sd} + jV_{sq} = V_{sm}\mathrm{e}^{j(\omega t+\varphi_0)}.\,e^{-j\rho(t)} \overset{if\ \rho(t)=\omega t+\varphi_0}{\Rightarrow} V_{sd} = V_{sm};\quad V_{sq} = 0 \tag{2.5}$$

Also, using abc- to dq-frame transformation, (2.1) may be written as follows:

$$\begin{aligned} L\frac{\mathrm{d}i_d}{\mathrm{d}t} &= L\,\omega(t)\,i_q - (R + r_{\mathrm{on}})i_d + V_{td} - V_{sd} \\ L\frac{\mathrm{d}i_q}{\mathrm{d}t} &= -L\,\omega(t)\,i_d - (R + r_{\mathrm{on}})i_q + V_{tq} - V_{sq} \end{aligned} \tag{2.6}$$

where V_{td} and V_{tq} represent the output voltage of VSC in dq-frame [7]. Also:

$$V_{td} = \frac{V_{dc}}{2}P_{md}\quad V_{tq} = \frac{V_{dc}}{2}P_{mq} \tag{2.7}$$

where V_{dc} is the voltage of the DC side and P_{md} and P_{mq} are pulse width modulation indices. Thus, the RMS value of line-to-line voltage is calculated by:

$$V_{t_{L-L}}^{\mathrm{RMS}} = \frac{\sqrt{3}}{2\sqrt{2}}V_{dc}\sqrt{P_{md}^2 + P_{mq}^2} \tag{2.8}$$

Therefore, if P_{md} and P_{mq} are defined by (2.9), (2.6) may be written as (2.10):

$$P_{md} = \frac{2}{V_{DC}}(u_d - L\,\omega(t)\,i_q + V_{sd}) \quad P_{mq} = \frac{2}{V_{DC}}(u_q + L\,\omega(t)i_d + V_{sq}) \tag{2.9}$$

$$L\frac{di_d}{dt} = -(R + r_{on})i_d + u_d \quad L\frac{di_q}{dt} = -(R + r_{on})i_q + u_q \tag{2.10}$$

where u_d and u_q are new control inputs which can control i_d and i_q, respectively. In this condition, assuming $\rho(t) = \omega t + \varphi_0$ which leads to $V_{sq} = 0$ as mentioned in (2.5), the active and reactive power generation of VSC is calculated by [7]:

$$\begin{matrix} P = \frac{3}{2}(V_{sd}i_d + V_{sq}i_q) \\ Q = \frac{3}{2}(-V_{sd}i_q + V_{sq}i_d) \end{matrix} \quad \overset{V_{sq}=0}{\Rightarrow} \quad \begin{matrix} i_{d-\text{ref}} = \frac{2}{3V_{sd}}P_{ref} \\ i_{q-\text{ref}} = -\frac{2}{3V_{sd}}Q_{\text{ref}} \end{matrix} \tag{2.11}$$

which indicates that P and Q can be controlled by i_d and i_q, respectively.

The schematic of the above-mentioned decoupled PQ controller is shown in Fig. 2.2a in which PLL is used to calculate the system frequency (i.e. $F_{\text{meas}} = \omega/2\pi$) and the voltage angle of $\overrightarrow{V}_s(t)$ at PCC, cos(φ) and sin(φ). It should be noted that in this figure, V_{dc} (DC-side voltage) is the output voltage of PV arrays.

2.3 Modelling of Large-Scale Photovoltaic Power Plants

To precisely model, a LS-PVPP in PowerFactory, PV array and the inverter's controller (i.e. decoupled PQ controller) should be implemented in DSL to accurately simulate the dynamic behaviour of LS-PVPP. In the following subsections, based on Fig. 2.3, the implementation of these resources in DSL will be described. It should be noted that in Fig. 2.3, the nominal DC voltage is 0.7 kV and the nominal voltage at the AC side of the inverter is 0.4 kV.

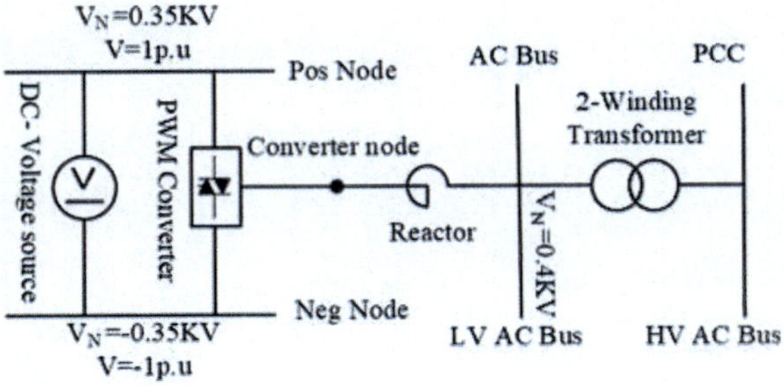

Fig. 2.3 Single line diagram of the LS-PVPP model in PowerFactory

2.3.1 Modelling of Decoupled PQ Controller

As mentioned before, since LS-PVPP is connected to high-voltage network, they should control their active and reactive power generation to participate in remedial action schemes to improve system stability. For this purpose, the decoupled PQ controller, described in the previous section and illustrated in Fig. 2.2, has been implemented in a DSL environment of PowerFactory software. Figure 2.4 shows the

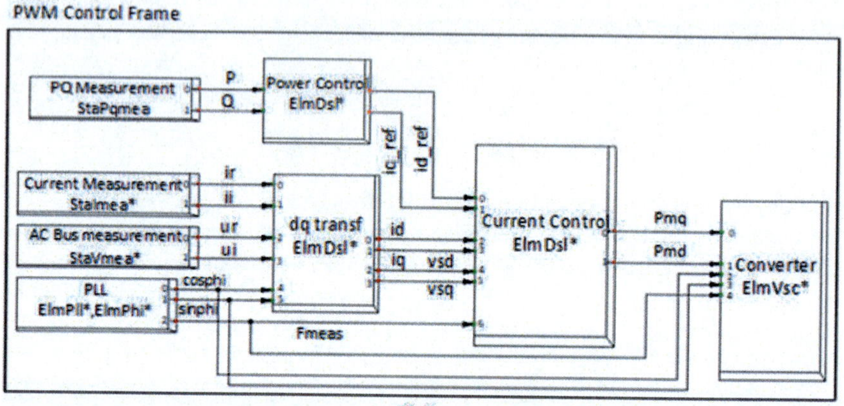

(a)

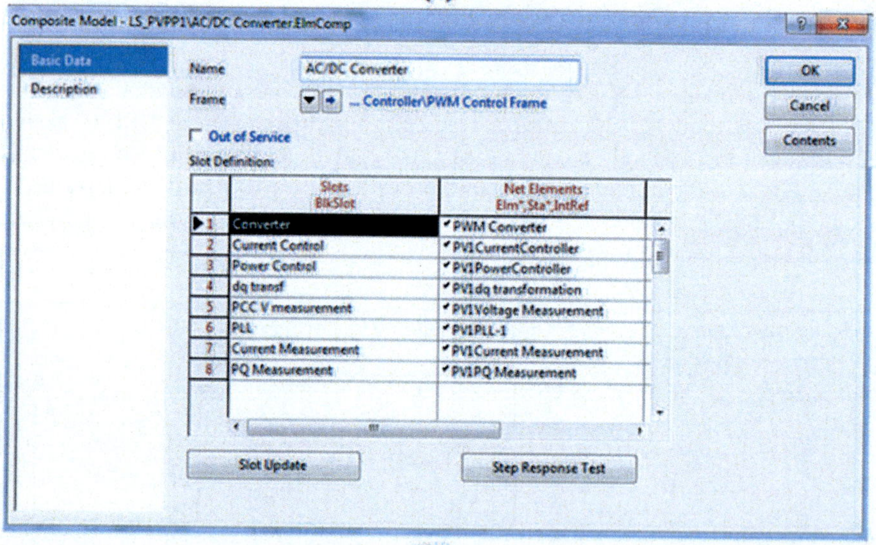

(b)

Fig. 2.4 Composite frame and composite model of the decoupled PQ controller in the DSL environment composite frame and composite model of this controller (*ElmComp*) which includes several components explained in the following sub-subsections.

2.3.1.1 PLL

According to Fig. 2.2, in the decoupled PQ controller, the voltage of *AC bus* and the output current of the inverter should be transformed from *abc*- to *dq*-frame. Therefore, as shown in Fig. 2.4, *PLL (ElmPhi_pll)* is used to measure the voltage at *AC bus* and calculate the system frequency (F_{meas}) and the voltage angle of *AC bus*, $\cos(\varphi)$ and $\sin(\varphi)$.

2.3.1.2 PQ Measurement

In Fig. 2.4, *PQ measurement* is used to measure the active and reactive power (P, Q) passing through *AC bus*. As illustrated in Fig. 2.2, these measured powers are compared with reference values (P_{ref} and Q_{ref}) to control P_{md} and P_{mq}.

2.3.1.3 Current Measurement

According to Fig. 2.4, the *current measurement* model measures the current passing through the AC side of the inverter. The outputs of this model are i_r and i_i, which indicate the real and imaginary parts of the measured current.

2.3.1.4 AC BUS V Measurement

Based on Fig. 2.4, *AC bus V measurement* which is a *voltage measurement* model measures the voltage of *AC bus* (shown in Figs. 2.2 and 2.3). The outputs of this model are u_r and u_i which are the real and imaginary parts of the *AC bus* voltage.

2.3.1.5 *dq* Transf

This component uses the angle of *AC bus* voltage (i.e. $\cos(\varphi)$ and $\sin(\varphi)$) to transform voltage and current from *abc*- to *dq*-frame. The DSL code used for initialisation and implementation of this model is as follows:

```
inc(vsd)=(ur*cosphi+ui*sinphi)*sqrt(2/3)                !1
inc(vsq)=(-ur*sinphi+ui*cosphi)*sqrt(2/3)               !2
inc(id)=-(ir*cosphi+ii*sinphi)*sqrt(2)                  !3
inc(iq)=-(-ir*sinphi+ii*cosphi)*sqrt(2)                 !4
vsd=(ur*cosphi+ui*sinphi)*sqrt(2/3)                     !5
vsq=(-ur*sinphi+ui*cosphi)*sqrt(2/3)                    !6
id=-(ir*cosphi+ii*sinphi)*sqrt(2)                       !7
iq=-(-ir*sinphi+ii*cosphi)*sqrt(2)                      !8
```

It should be noted that since *voltage measurement* measures the RMS value of line-to-line voltage (u_r, u_i), the coefficient *sqrt(2/3)* is used in lines 5–6 to calculate the amplitude of the line-to-ground voltage and then to calculate V_{sd} and V_{sq} (see Eq. (2.1) where V_{sm} is the amplitude of the line-to-ground voltage). Similarly, since *current measurement* measures the RMS value of the current (i_r,i_i), the coefficient *sqrt(2)* is used in lines 7–8 to calculate i_d and i_q.

2.3.1.6 Power Control

As illustrated in Fig. 2.2, in *power control*, the active and reactive powers (*P*, *Q*) passing through *AC bus* are compared with a reference value (P_{ref} and Q_{ref}) and their differences (*dp*, *dq*) are sent as input to PI controllers whose outputs are i_{d_ref} and i_{q_ref}. Figure 2.5 shows the implementation of *power control* in DSL.

Also, the initialisation of this model is as follows:

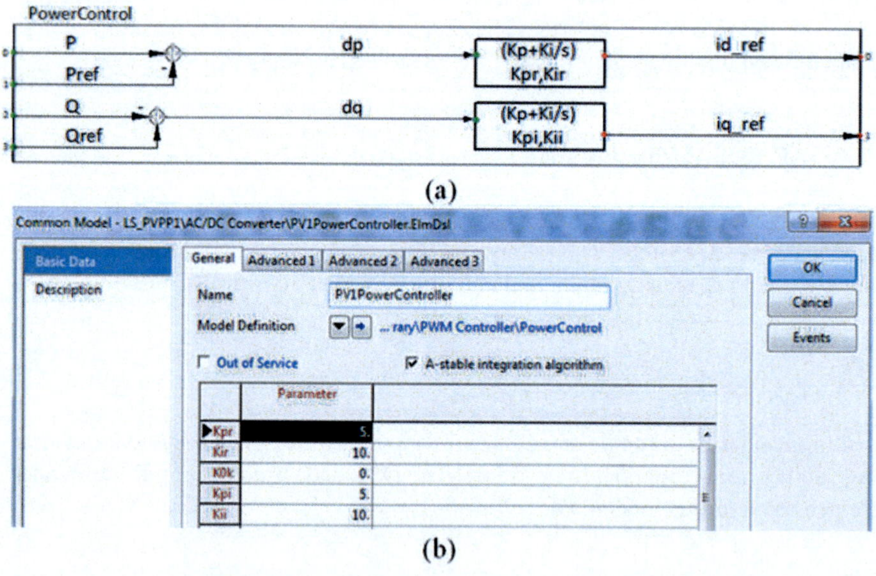

Fig. 2.5 **a** *Power control* in DSL environment and **b** the parameters of *power control* used in an LS-PVPP in the test system.

```
inc(xp)=id_ref
inc(xq)=iq_ref
inc0(Pref)=P
inc0(Qref)=Q
```

2.3.1.7 Current Control

As described in Sect. 2.2 and illustrated in Fig. 2.2, *current control* is used to calculate P_{md} and P_{mq} so that the LS-PVPP generates the pre-specified power (P_{ref} and Q_{ref}). Accordingly, *current control* has been implemented in DSL which is shown in Fig. 2.6. The initialisation of the model is:

```
inc(xr)=Pmd*(0.7/2)-vsd1+Lwiq
inc(xi)=Pmq*(0.7/2)-vsq1-Lwid
inc(id_ref)=id
inc(iq_ref)=iq
inc(xvsd)=vsd
inc(xvsq)=vsq
vardef(Tvsd)='s';'feed-forward filter Time Constant-d axis'
vardef(Tvsq)='s';'feed-forward filter Time Constant-q axis'
vardef(Lr)='mH';'Inductance of the series reactor'
vardef(Ltr)='mH';'Inductance of the transformer'
!2/Vdc=2/0.7=2.857142857142857142857142857142 9
```

2.3.1.8 Converter

In PowerFactory, DC-AC inverter may be controlled by different variables such as P_{md}, P_{mq} or P_{mr}, P_{mi}. As shown in Fig. 2.2, in this work, P_{md} and P_{mq} (the output variables of *current control*) control the AC-side voltage as follows:

$$
\begin{aligned}
V_{t\,LG}^{\max} &= \frac{V_{dc}}{2}\sqrt{P_{md}^2 + P_{mq}^2}; \quad \angle V_{t_{LG}} = \varphi + \delta \\
V_{t\,LL}^{\mathrm{RMS}} &= \frac{\sqrt{3}V_{dc}}{2\sqrt{2}}\sqrt{P_{md}^2 + P_{mq}^2}; \quad \angle V_{t_{LG}} = \varphi + \delta + 30^{\circ}
\end{aligned}
\tag{2.12}
$$

where $\delta = \tan^{-1}(P_{mq}/P_{md})$, φ is the voltage angle of *AC bus* calculated by *PLL* and V_{dc} is the voltage of the DC side. It should be noted that $\vec{V}_t$ leads $\vec{V}_s$ by δ. Figure 2.7 shows the PWM converter window of an LS-PVPP in the test system used for balanced dynamic simulation.

2.3.2 *Modelling of PV Array in DSL*

As shown in Fig. 2.2, to accurately simulate the dynamic behaviour of PV arrays, the effect of the temperature and solar radiation changes as well as the inverter's

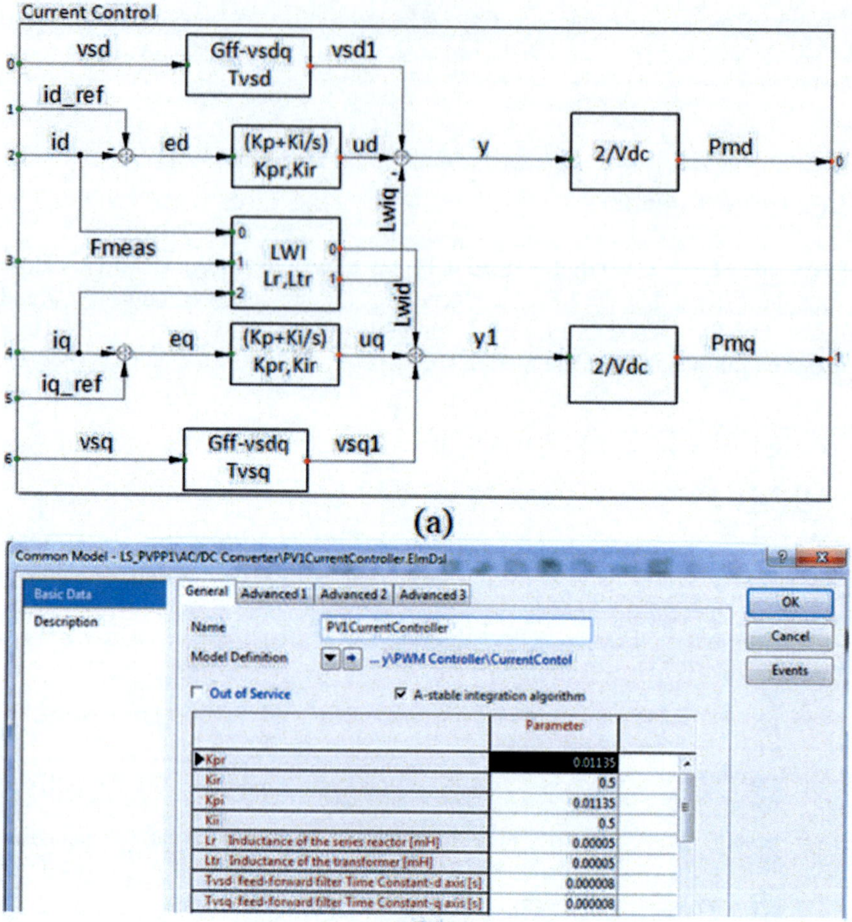

Fig. 2.6 **a**. *Current control* in DSL environment and **b** the parameters of *current control* used in an LS-PVPP in the test system

controller on the output voltage and current of PV arrays (V_{dc} and I_{dc} in Fig. 2.1) should be modelled. Therefore, in this section, the PV model has been implemented in DSL to simulate the dynamic behaviour of PV arrays. Figure 2.8 shows the composite frame and composite model of PV arrays, which will be explained in the following sub-subsection.

2.3.2.1 Solar Radiation and Temperature Models

Figures 2.9 and 2.10 show solar radiation and temperature variation models in which when *dE and dT* (*dtheta*) become positive/negative, the irradiance and temperature

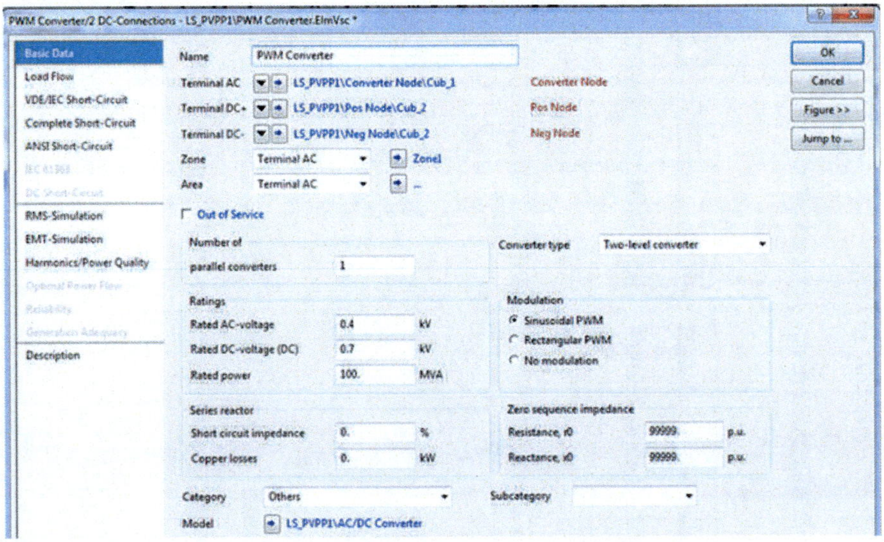

Fig. 2.7 PWM converter window for an LS-PVPP in the test system

increase/decrease. It should be noted that standard condition refers to $E = 1000\ \mathrm{W/m^2}$ and $T = 25\ °\mathrm{C}$. The initialisation of solar radiation and temperature models is:

```
! The initialisation of solar radiation
inc(xrmp)=E
inc(dE)=0
vardef(E)='W/m²';'Irradiance'
vardef(dE)='W/(m²*s)';'Change of Irradiance per s'
!The initialisation of temperature
inc(xrmp)=theta
inc(dtheta)=0
inc(theta)=25
vardef(theta)='°C';'Temperature'
vardef(dtheta)='°C/s';'Change of Temperature per s'
```

2.3.2.2 Photovoltaic Model

The structure of this model is shown in Fig. 2.11. The main component of this model is *PV module* which is available in the global library of PowerFactory. As shown, the output voltage of PV arrays (U_{array} which has been passed through a low-pass filter to remove its high-frequency components) is divided by the number of series modules (*nSerialModules*) to calculate the output voltage of each module (U). Then, this voltage along with E and T is applied to *PV* module as input to calculate the output current of each PV module (I) and the voltage of PV module at maximum power point (V_{MPP}). Figure 2.12 shows the I–V and P–V curves in which all operating points in the right-hand side of the maximum power point (MPP) are stable ones and PV may work in this section. Finally, the number of series modules (*nSerialModules*) and

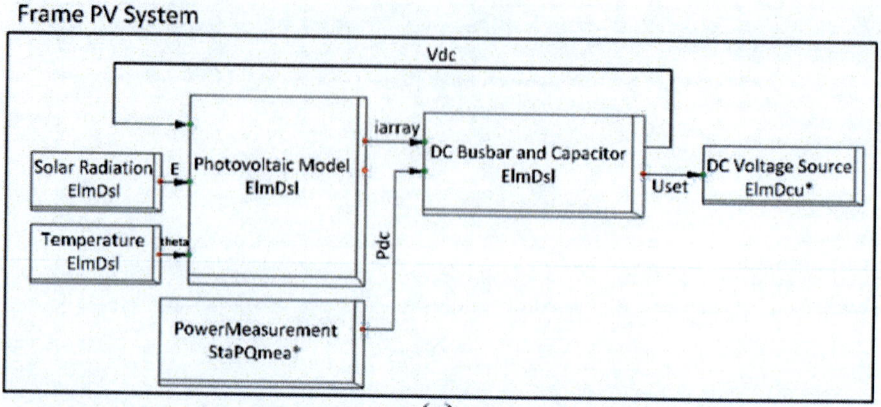

(a)

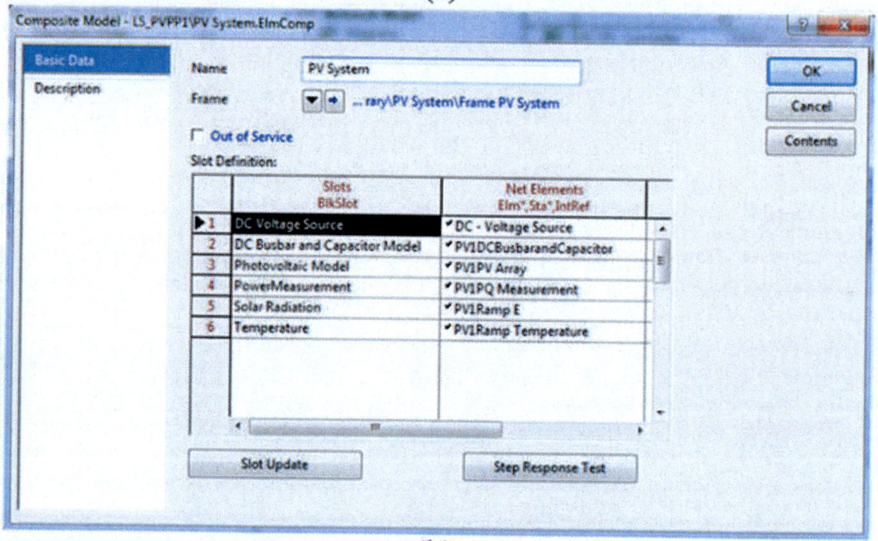

(b)

Fig. 2.8 Composite frame and composite model of PV arrays in DSL

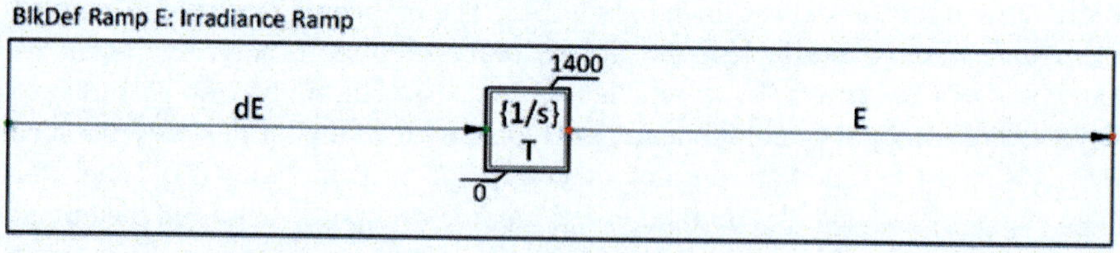

Fig. 2.9 Structure of the *solar radiation* slot

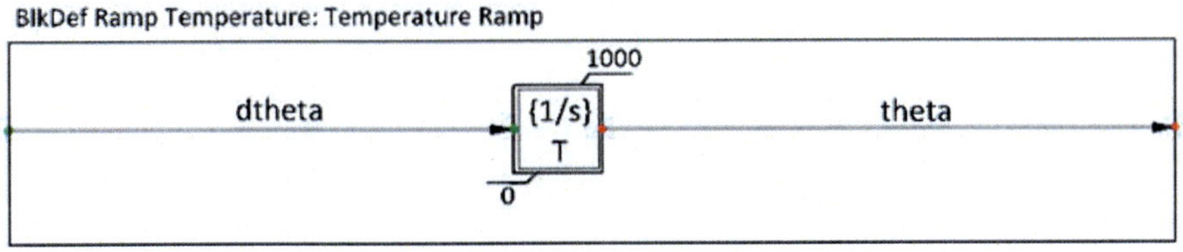

Fig. 2.10 Structure of the *temperature* slot

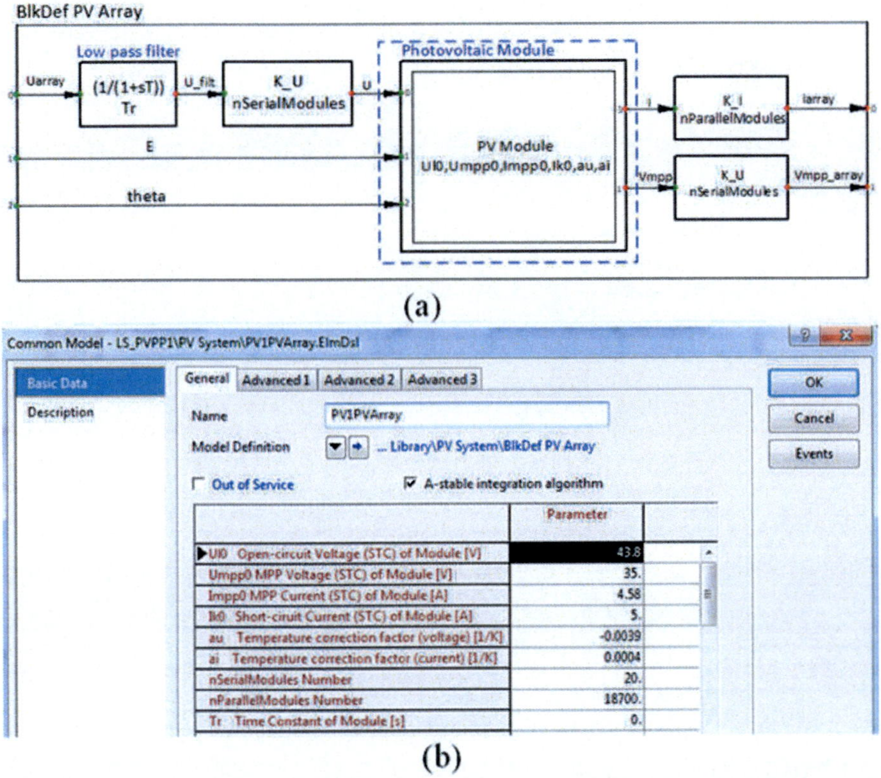

Fig. 2.11 **a** Photovoltaic array model in DSL environment and **b** the parameters of PV array of an LS-PVPP in the test system

parallel modules (*nParallelModules*) are used to calculate the output current (i_{array}) and V_{MPP} of the array. Also, the parameters of a PV array used to model an LS-PVPP in the test system are shown in Fig. 2.11b.

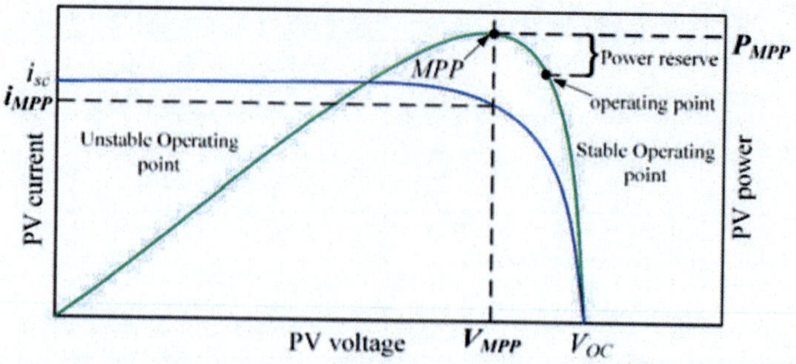

Fig. 2.12 I–V and P–V curves of a PV

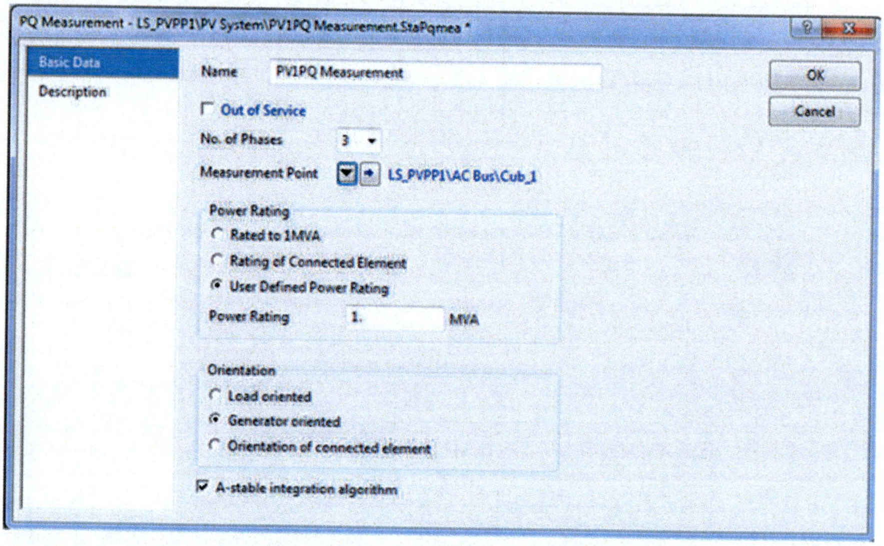

Fig. 2.13 *Power measurement* window for an LS-PVPP in the test system

2.3.2.3 Power Measurement

In Fig. 2.8, *power measurement* is used to measure the active power in DC side of the inverter and send it to the *DC bus and capacitor* model. In this regard, neglecting the inverter losses which is significantly less than the input power, *power measurement* model (shown in Fig. 2.13) measures the output power of the inverter (in AC side), P_{dc}.

2.3.2.4 DC Bus and Capacitor Model

According to Fig. 2.1, the voltage of DC capacitor changes as follows:

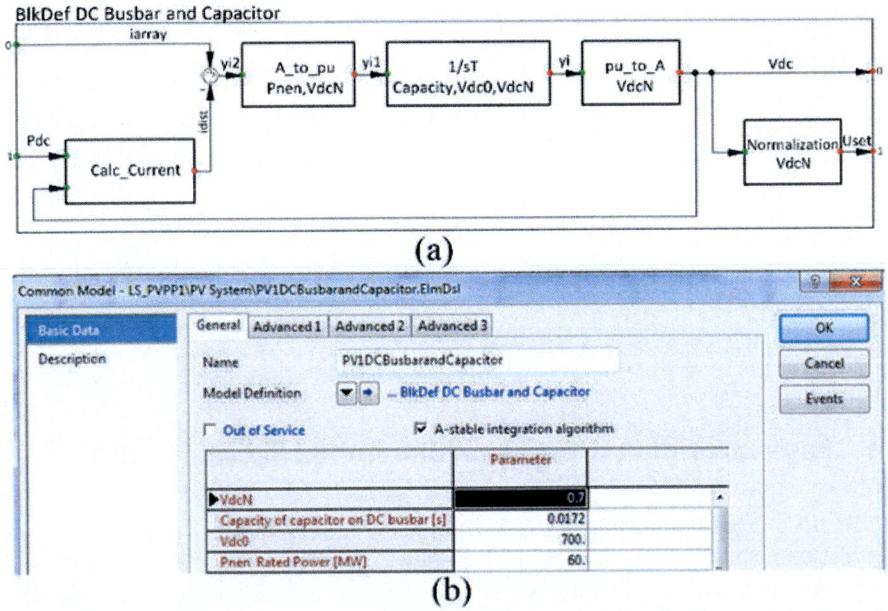

Fig. 2.14 **a**. *DC bus and capacitor* model and **b**. the parameters of *DC bus and capacitor* of an LS-PVPP in the test system

$$i_{\text{array}} - i_{dc} = C\frac{\mathrm{d}V_{dc}}{\mathrm{d}t}; \quad i_{dc} = \frac{P_{dc}}{V_{dc}} \tag{2.13}$$

Therefore, in the *DC bus and capacitor* model, V_{dc} is calculated based on the above equation, which is shown in Fig. 2.14a. It should be noted that in this model, U_{set} is the normalised DC-side voltage. Also, the parameters of *DC bus and capacitor* used to simulate an LS-PVPP in the test system are shown in Fig. 2.14b. The initialisation of this model is:

```
vardef(Capacity)='s';'of capacitor on DC busbar'
vardef(Vdc0)='V';'Initial DC-voltage'
vardef(VdcN)='kV';'Nominal DC Voltage'
vardef(Pnen)='MW';'Rated Power'
vardef(Vdc)='V';'DC busbar voltage'
inc(Vdc)=VdcN
inc(iv)=Pdc/Vdc0*1000000
```

2.3.2.5 DC Voltage Source

The last slot in Fig. 2.8a is the *DC voltage source* which receives U_{set} (the output voltage of DC voltage source in p.u.) from *DC bus and capacitor* model and generates V_{dc} (shown in Fig. 2.1). Figure 2.15 shows the *DC voltage source* window for an LS-PVPP in the test system.

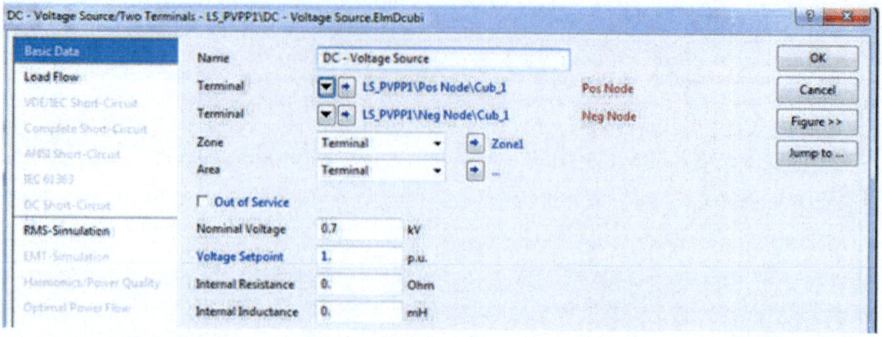

Fig. 2.15 *DC voltage source* window of an LS-PVPP in the test system

2.4 Implementation of the Remedial Action Scheme in DPL

DIgSILENT Programming Language is a proper tool to implement a remedial action algorithm and analyse its operation. To show the effectiveness of LS-PVPP in improving stability status, in this section, a remedial action scheme (or a Special Protection System, SPS) is proposed and implemented in DPL. The proposed method will be tested against different contingencies on a modified version of the IEEE 14-bus where four LS-PVPPs have been installed (Fig. 2.16). Of this subject, three events (an outage of *Lood3* in scenario 1, *Gen1* in scenario 2 and *Line2_3* in scenario 3) are considered to illustrate the effect of the LS-PVPPs. It should be noted that in the sequel, only the DPL of the proposed algorithm against load outage (an outage of *Lood3* in scenario 1) will be explained.

The flow chart of the proposed remedial action scheme and the hierarchical structure of DPL scripts prepared for this study is shown in Fig. 2.17, which will be described below. Also, Fig. 2.18 shows the main script of this algorithm, i.e. *Application of DPL in SPS Modelling.ComDPL* which is used for calling different subscripts explained in the following subsections.

2.4.1 *CreateResults.ComDPL*

This script is executed on line 10 of Fig. 2.18 to define desired results (voltage and frequency of buses, active and reactive power of loads, and active power generation of generators) which will be stored in *results.ElmRes* and used to assess the stability status [8]. It should be stated that, although simulation results are usually stored in *Allcalculations.ElmRes*, in this study, these results will be stored in *results.ElmRes*. Figure 2.19 shows *CreateResults.ComDPL* whose input is *Result* which points to *results.ElmRes*.

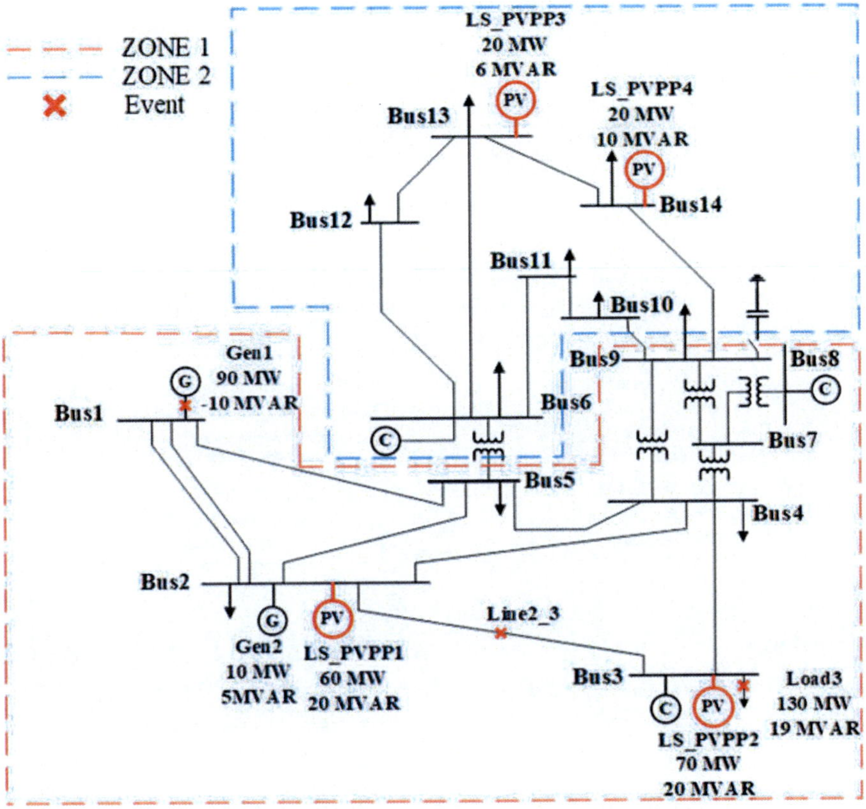

Fig. 2.16 Single-line diagram of the test system

2.4.2 *CreateEvent.ComDPL*

Considering those scenarios described in the following section, this script (executed on line 11 of Fig. 2.18) creates the intended events (an outage of *Lood3* in scenario 1, *Gen1* in scenario 2 and *Line2_3* in scenario 3) which are occurred at $t = 39.5$ s.

As shown in Fig. 2.18, *CntngnyLoc* (load outage in scenario 1) and *tevent* (= 39.5 s) are applied to *CreateEvent.ComDPL* as input. Figure 2.20 shows *CreateEvent.ComDPL* in which old events are deleted on line 10 and new ones created on lines 13–17.

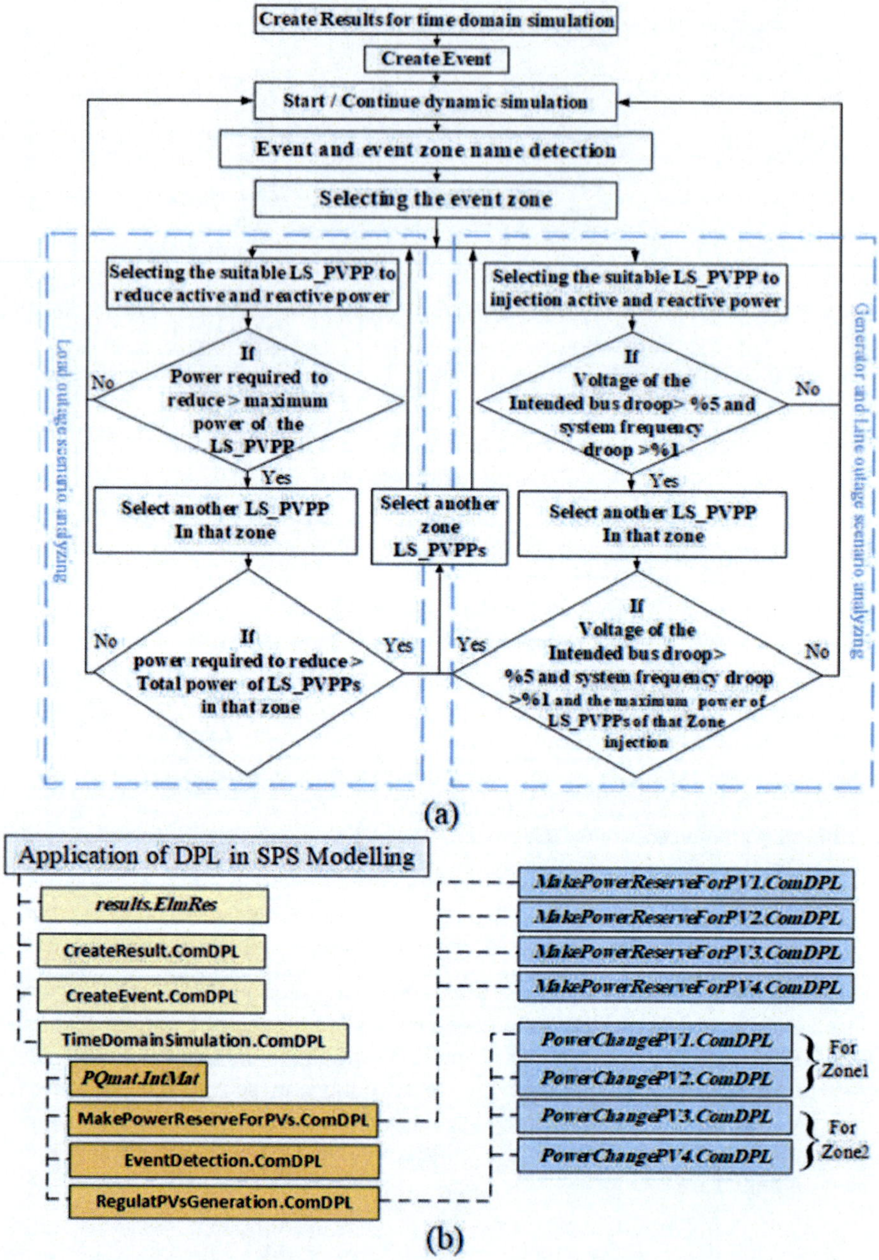

Fig. 2.17 **a**. Flow chart of the proposed algorithm for stability improvement and **b**. hierarchical structure of DPL scripts for the first scenario

```
string CntngnyLoc; ! contingency location
int tevent;
object Result;
double T;
ResetCalculation(); !Reset Calculations
ClearOutput(); !Clear output window
Result = results; !object "Result" points to results.ElmRes
CntngnyLoc='Load3'; ! Load outage (senario 1)
tevent=39.5; !At t=tevent, 'Load'will be disconnected
CreateResults.Execute(Result); !Create results of simulation
CreateEvent.Execute(CntngnyLoc,tevent);!Create Event for simulation
TimeDomainSimulation.Execute(Result,CntngnyLoc);!Execute dynamic simulation
```

Fig. 2.18 DPL script of the *Application of DPL in SPS Modelling.ComDPL* for load outage event (scenario 1)

```
!variable declaration
int EndCnd,cmp,i;
object mon,Olod,OSym,Inc,EventDir,oAllBars;
string str;
set oldmon,sLod,sAllBars,sSym;
!Create a set of all Busbars, loads and generators and sort them to their
! names (from A to Z)
sAllBars = AllRelevant('*.StaBar,*.ElmTerm'); sAllBars.SortToName(0);
sLod =     AllRelevant('*.ElmLod'); sLod.SortToName(0);
sSym =     AllRelevant('*.ElmSym'); sSym.SortToName(0);
Inc = GetCaseCommand('ComInc');
EventDir= Inc:p_event;
! Get old contents of results (from previous simulations) and Delete them
oldmon = Result.GetContents();
Delete(oldmon);
!Voltage and frequency of buses add to results
for(oAllBars=sAllBars.First();oAllBars;oAllBars=sAllBars.Next()){
  str=oAllBars:loc_name;
  mon=Result.CreateObject('IntMon', str);
  mon:obj_id = oAllBars;
  mon.AddVar('m:u1');!Add voltage  of buses to results
  mon.AddVar('m:fe');!Add frequency of buses to results
}
!Active power of loads add to results
for(Olod=sLod.First();Olod;Olod=sLod.Next()){
  str=Olod:loc_name;
  mon=Result.CreateObject('IntMon', str);
  mon:obj_id = Olod;
  mon.AddVar('m:P:bus1'); ! Add the Active Power (P) of load to results
  mon.AddVar('m:Q:bus1'); ! Add the Rective Power (Q) of load to results.
}
!Active power of Generators are added to results
for(OSym=sSym.First();OSym;OSym=sSym.Next()){
   str=OSym:loc_name;
   mon=Result.CreateObject('IntMon', str);
   mon:obj_id = OSym;
   mon.AddVar('m:P:bus1');!Add active power of Generators are added to results
}
```

Fig. 2.19 DPL script of the *CreateResults.ComDPL*

```
set sLod,sEvent;
object OLod,Inc,EventDir,event;
string str;
!Create a set of all load and sort them to their names (from A to Z)
sLod  = AllRelevant('*.ElmLod');     sLod.SortToName(0);
! Delete Existing Event Set
Inc        = GetCaseCommand('ComInc');
EventDir  = Inc:p_event;
sEvent     = EventDir.GetContents();
Delete(sEvent);
EchoOff ();
! Creat New Switch Event for 'Load3.ElmLod':
str =sprintf('%s%s',CntngnyLoc,'.ElmLod');     !str='Load3.ElmLod'
OLod = sLod.FirstFilt(str);
event           = EventDir.CreateObject('EvtSwitch','Switch Event');
event:p_target = OLod;
event:time      = tevent;
```

Fig. 2.20 DPL script of the *CreateEvent.ComDPL* for load outage (in scenario 1)

2.4.3 TimeDomainSimulation.ComDPL *for Load Outage (Scenario 1)*

After creating intended results and event, *TimeDomainSimulation.ComDPL* is executed on line 12 of the main script (see Fig. 2.18). As shown in this figure, the inputs of *TimeDomainSimulation.ComDPL* are:

- *Result*: This object points to *results.ElmRes* where the results are stored.
- *CntngnyLoc*: In load outage scenario, this string contains *load 3*.

The main goal of *TimeDomainSimulation.ComDPL* is to simulate the outage of system components (*Lood3, Gen1,* and *Line2_3*) at $t = 39.5$ s and the operation of the proposed control scheme. In this script, at the beginning of the simulation, *MakePowerReserveForPVs.ComDPL* will be executed to regulate the generation of PV resources and increase their power reserve (shown in Fig. 2.12) to enhance the capability of these resources in maintaining system stability [9].

According to Fig. 2.21, in *TimeDomainSimulation.ComDPL* which includes three subscripts, a *for* loop (lines 19–32) is used to perform the following commands, iteratively (in scenario 1):

- Simulation time is set to T on line 19 (initially $T = 0$ s). In the next iterations, T increases by 1 s and time-domain simulation will continue.
- Time-domain simulation is executed until T (line 21).
- At the beginning of the dynamic simulation (i.e. $T = 0$ s), the amount of the power reserve of the PV resources is regulated in the line 22.
- *EventDetection.ComDPL* is executed every $dT = 1$ s (line 23) to identify whether the event is happened or not. Obviously, this means that the procedure assesses event occurs every second. If this script identifies an event occurrence, *EventHappened* will be set to 1 (line 24), and the area where the disturbance happened along with the active and reactive power of the outaged load is determined in the line

```
int    EventHappened,GiveMatValue,mRow;                                      !1
double T,dT,PpowerToShed,QpowerToShed,Teventdetect;                          !2
object Ldf,Inc,Sim,PQmat;                                                    !3
string EvtZneName,PV1,PV2,PV3,PV4;                                           !4
Ldf = GetCaseCommand('ComLdf');!Call power flow analyses,                    !5
Inc = GetCaseCommand('ComInc');!Call Initialising analyses,                  !6
Sim = GetCaseCommand('ComSim');!Call time domain simulation analyses,        !7
Echo.Off();                                                                  !8
Ldf.Execute(); !Calculate Load FLow Analysis                                 !9
Inc.Execute(); !Calculate initialize Condition                               !10
PV1='PV1';PV2='PV2';!Name of LS_PVPPs at Zone1                               !11
PV3='PV3';PV4='PV4';!Name of LS_PVPPs at Zone2                               !12
dT=1;              !'dT' is step time to check event                         !13
GiveMatValue=0;    !Give PQmat Value                                         !14
EventHappened=0;   !"EventHappen =1": means that the load3 is disconnected from grid!15
PQmat=PQMat;                                                                 !16
PQmat.Init(0,3);   !Initialises the PQmat with given size                    !17
mRow=0;            !The number of PQmat row.                                 !18
for(T=0;T<=100;T=T+dT){                                                      !19
 Sim:tstop = T; !Stop time of time domain simulation set to Tstop            !20
 Sim.Execute(); !Run Time Domain Simulation until Tstop                      !21
 if(T=0){  MakePowerReserveForPVs.Execute(T,PV1,PV2,PV3,PV4);}               !22
 EventDetection.Execute(T,CntngnyLoc,GiveMatValue,mRow,PQmat);               !23
 EventHappened = EventDetection:EventHappened;                               !24
 EvtZneName    = EventDetection:EvtZneName;                                  !25
 PpowerToShed  = EventDetection:PpowerToShed;                                !26
 QpowerToShed  = EventDetection:QpowerToShed;                                !27
 Teventdetect  = EventDetection:Teventdetect;                                !28
 if({EventHappened=1}.and.{T=Teventdetect}){                                 !29
 RegulatPVsGeneration.Execute(PV1,PV2,PV3,PV4,EvtZneName,PpowerToShed,QpowerToShed,T);
  }                                                                          !31
}                                                                            !32
```

Fig. 2.21 DPL script of the *TimeDomainSimulation.ComDPL* for load outage (in scenario 1)

25 and lines 26–27, respectively. Also, on line 28, the moment at which the event occurrence is detected is assigned to *Teventdetect*.

- Just after the event detection, *RegulatPVsGeneration.ComDPL* is executed to regulate the output power of PV resources to prevent system instability.

2.4.3.1 MakePowerReserveForPVs.ComDPL

As mentioned, *MakePowerReserveForPVs.ComDPL* is executed on line 22 of Fig. 2.21 (at $T = 0$ s) to increase the power reserve of PV resources. According to this figure, the inputs to this script are T and the name of four LS-PVPPs available in the test system. Figure 2.22 shows the script of *RegulatPVsGeneration.ComDPL* where the active and reactive powers of LS-PVPPs are regulated in

```
MakePowerReserveForPV1.Execute(T,PV1);                                       !1
MakePowerReserveForPV2.Execute(T,PV2);                                       !2
MakePowerReserveForPV3.Execute(T,PV3);                                       !3
MakePowerReserveForPV4.Execute(T,PV4);                                       !4
```

Fig. 2.22 DPL script of the *MakePowerReserveForPVs.ComDPL*

```
set sPV,sEvent;                                                              !1
object oPV,Inc,EventDir,event;                                               !2
double Ppv,Qpv,Ppvnew,Qpvnew,Reserve;                                        !3
string val,str;                                                              !4
!Create a set of all LS_PVPP and sort them to their names (from A to Z)      !5
sPV = AllRelevant('*.ElmDsl'); sPV.SortToName(0);                            !6
Reserve=0.2; !means 20% reservation                                          !7
Inc = GetCaseCommand('ComInc');                                              !8
EventDir = Inc:p_event;                                                      !9
sEvent = EventDir.GetContents();                                             !10
str=sprintf('%s%s',PV1,'PowerController.ElmDsl');                            !11
oPV = sPV.FirstFilt(str);                                                    !12
Ppv=oPV:s:Pref; !get value of PV1 Active power                               !13
Ppvnew=Ppv*(1-Reserve);  !Calculate active power reservation value of PV1    !14
val=sprintf('%f',Ppvnew);                                                    !15
event = EventDir.CreateObject('EvtParam','Parameter Event');                 !16
event:variable='Pref';                                                       !17
event:p_target=oPV;                                                          !18
event:value=val;                                                             !19
event:mtime=0;                                                               !20
event:time=T;                                                                !21
event.Execute();  !execute parameter event to regulate active power reserve of PV1.!22
!for reactive power                                                          !23
Qpv=oPV:s:Qref; !get value of PV1 Reactive power                             !24
Qpvnew=Qpv*(1-Reserve); !Calculate reactive power reservation value of PV1   !25
val=sprintf('%f',Qpvnew);                                                    !26
event=EventDir.CreateObject('EvtParam','Parameter Event');                   !27
event:variable='Qref';                                                       !28
event:p_target=oPV;                                                          !29
event:value=val;                                                             !30
event:mtime=0;                                                               !31
event:time=T;                                                                !32
event.Execute();!execute parameter event to regulate reactive power reserve of PV1 !33
```

Fig. 2.23 DPL script of the *MakePowerReserveForPV1.ComDPL*

lines 1–4. Since the procedure of increasing power reserve of all PVs is the same, only *MakePowerReserveForPV1.ComDPL* will be described in the sequel.

MakePowerReserveForPV1.ComDPL

In this script, the active and reactive power reserve of the first LS-PVPP will be decreased by 20%. For this purpose, on line 7 of Fig. 2.23, *reserve* is set to 0.2 and then, in lines 13–22, a parameter event is executed to decrease the active power of the LS-PVPP (P_{ref} in Fig. 2.2a or Fig. 2.5) and increase the active power reserve. Similarly, in lines 24–33, a parameter event is executed to regulate the reactive power reserve of the LS-PVPP.

2.4.3.2 EventDetection.ComDPL

On line 23 of Fig. 2.21, *EventDetection.ComDPL* is executed every second to detect whether the disturbance has occurred or not. The script of *EventDetection.ComDPL* is shown in Fig. 2.24 in which the algorithm checks the connectivity of the loads to

```
string str;                                                              !1
object oLod,OCub,OSwitch;                                                !2
int iCount,iCub,opn,PreviousData;                                        !3
set sLod,sCntnt;                                                         !4
double Pload,Qload;                                                      !5
sLod = AllRelevant('*.ElmLod'); sLod.SortToName(0);                      !6
str = sprintf('%s%s',CntngnyLoc,'.ElmLod');        !str='Load3.ElmLod'   !7
oLod = sLod.FirstFilt(str);              !OLod points to Load3.ElmLod    !8
iCount = oLod.GetConnectionCount();!the number of electrical connections !9
for (iCub=0; iCub<iCount; iCub=iCub+1){                                  !10
  OCub=oLod.GetCubicle(iCub);           !OCub is a cubicle of OLod       !11
  sCntnt=OCub.GetContents();       !sCntnt includes contents of OCub     !12
  OSwitch=sCntnt.First();         !OSwitch points to switch in OCub      !13
  opn = OSwitch.IsOpen();       !opn=1: means that the load switch is open !14
   mRow=mRow+1;          !PQmat stores Pload3 and Qload3 every second    !15
  PQmat.Resize(mRow,3);                                                  !16
  Pload=oLod:m:P:bus1;   !active power of load3                          !17
  Qload=oLod:m:Q:bus1;   !reactive power of load3                        !18
  PQmat.Set(mRow,1,T);   !set time in the first column of the PQmat      !19
  PQmat.Set(mRow,2,Pload); !set active power in The second column of the PQmat   !20
  PQmat.Set(mRow,3,Qload);  !set reactive power in The Third column of the PQmat !21
  if({opn=1}.and.{GiveMatValue=0}){                                      !22
    Teventdetect= PQmat.Get(mRow,1);!get T at the moment of event detection !23
    PreviousData =mRow-1;            !Previous Time data                 !24
    !get active and reactive power at 1 sec before the event detection   !25
    PpowerToShed = PQmat.Get(PreviousData,2);                            !26
    QpowerToShed = PQmat.Get(PreviousData,3);                            !27
    EventHappened=1;                                                     !28
    EvtZneName=oLod:r:cpZone:b:loc_name; !event zone name                !29
    GiveMatValue=1;                                                      !30
  }                                                                      !31
}                                                                        !32
```

Fig. 2.24 DPL script of the *EventDetection.ComDPL* for load outage (scenario 1)

the network and $opn = 1$ when a load outage is occurred (line 14). Also, in lines 17–21, the current time T and the active and reactive power of load 3 are stored in *PQmat.IntMat.* Also, if the load outage has happened, in lines 22–31, the active and reactive power consumption of this load (before load outage) is obtained and assigned to *PpowerToShed* and *QpowerToShed*, respectively. Furthermore, in this condition, the area where the load outage has happened is assigned to *EvtZneName*, and *EventHappened* is set to 1.

2.4.3.3 RegulatPVsGeneration.ComDPL

As shown in lines 29–31 of Fig. 2.21, right after disturbance detection (i.e. at $T =$ *Teventdetect*), the algorithm executes *RegulatPVsGeneration.ComDPL* to change the active and reactive power reference values of LS-PVPPs to control the bus voltages and the system frequency. For this purpose, if the load outage occurs in a zone (which is assigned to *EvtZneName*), firstly, those PV resources located in that area decrease their active and reactive power generation by *PpowerToShed* and *QpowerToShed*, respectively, to shed extra power generation and balance the system. If these resources cannot reduce their generation sufficiently, the LS-PVPPs located in the other zone should change their generation to maintain system stability. However, due to the local nature of volt/var control, those resources in the other zone only change their active power generation.

Figure 2.25 shows the script of *RegulatPVsGeneration.ComDPL* in which, if the load outage occurred in zone 1 (*EventZone* = 1 on line 5), at the first step, the active and reactive power generation of PV2 and PV1 would be decreased (lines 6 and 8, respectively). Then, if there is excess power in the system, the active power generated from PVs in zone 2 will be decreased on lines 11 and 13. A similar procedure is performed on lines 15–24 to regulate the power generation of PV resources when a load outage occurs in zone 2.

In this script, the regulation of LS-PVPP generations is performed in *PowerChangePV2.ComDPL* (lines 6 and 23), *PowerChangePV1.ComDPL* (lines 8 and 21), *PowerChangePV3.ComDPL* (lines 11 and 16)*and PowerChangePV4.ComDPL* (lines 13 and 18) which use similar algorithm. Therefore, in the following sub-subsection, only *PowerChangePV2.ComDPL* will be described.

PowerChangePV2.ComDPL

As mentioned before, this script is used to decrease the active and reactive power generation of PV2 by *PpowerToShed* and *QpowerToShed*, respectively, which are applied to this script as input.

According to the script shown in Fig. 2.26, if the amount of active power scheduled for shedding (i.e. *PpowerToShed*) becomes less than the existing output power of PV2, a parameter event should be defined on lines 14–23 to decrease the active power generation of this resource from *Ppv* (existing active power generation of PV2) to

```
1  string  zone1;
2  int     EventZone;
3
4  EventZone=strcmp(EvtZneName,zone1);!get event zone and zone1 name to detect zone
5  if(EventZone=1){  !means that the event is occurred at zone1
6  PowerChangePV2.Execute(PV2,PpowerToShed,QpowerToShed,T);
7  PpowerToShed=PowerChangePV2:PpowerToShed; QpowerToShed=PowerChangePV2:QpowerToShed;
8  PowerChangePV1.Execute(PV1,PpowerToShed,QpowerToShed,T);
9  PpowerToShed=PowerChangePV1:PpowerToShed; QpowerToShed=PowerChangePV1:QpowerToShed;
10  ! if the system is not stable, select LS_pvpp of zone2
11 PowerChangePV3.Execute(PV3,PpowerToShed,0,T);
12 PpowerToShed=PowerChangePV3:PpowerToShed; QpowerToShed=PowerChangePV3:QpowerToShed;
13 PowerChangePV4.Execute(PV4,PpowerToShed,0,T);
14 }
15 else{
16 PowerChangePV3.Execute(PV3,PpowerToShed,QpowerToShed,T);
17 PpowerToShed=PowerChangePV3:PpowerToShed; QpowerToShed=PowerChangePV3:QpowerToShed;
18 PowerChangePV4.Execute(PV4,PpowerToShed,QpowerToShed,T);
19 PpowerToShed=PowerChangePV4:PpowerToShed;QpowerToShed=PowerChangePV4:QpowerToShed;
20  ! if the system is not stable, select LS_pvpp of zone2
21 PowerChangePV1.Execute(PV1,PpowerToShed,0,T);
22 PpowerToShed=PowerChangePV1:PpowerToShed; QpowerToShed=PowerChangePV1:QpowerToShed;
23 PowerChangePV2.Execute(PV2,PpowerToShed,0,T);
24 }
```

Fig. 2.25 DPL script of *RegulatPVsGeneration.ComDPL* for load outage (scenario 1)

```
set    sPV,sEvent;                                                    !1
object oPV,Inc,EventDir,event;                                        !2
double Ppv,Qpv,dp,dq,Ppvnew,Qpvnew;                                   !3
string val,str;                                                       !4
sPV = AllRelevant('*.ElmDsl'); sPV.SortToName(0);                     !5
Inc = GetCaseCommand('ComInc');                                       !6
EventDir = Inc:p_event;                                               !7
sEvent = EventDir.GetContents();                                      !8
str=sprintf('%s%s',PV2,'PowerController.ElmDsl');                     !9
oPV = sPV.FirstFilt(str);                                             !10
Ppv=oPV:s:Pref;          !Existing active power generation of PV2    !11
dp=PpowerToShed-Ppv+1;                                                !12
!Calculate the rate of reduction of LS_PVPP2 active  power            !13
if(Ppv<PpowerToShed){ Ppvnew=1;   PpowerToShed=dp; }                  !14
else{Ppvnew=Ppv-PpowerToShed;     PpowerToShed=0; }                   !15
val=sprintf('%f',Ppvnew);                                             !16
event = EventDir.CreateObject('EvtParam','Parameter Event');          !17
event:variable='Pref';                                                !18
event:p_target=oPV;                                                   !19
event:value=val;                                                      !20
event:mtime=0;                                                        !21
event:time=T;                                                         !22
event.Execute();                                                      !23
!for reactive power                                                   !24
Qpv=oPV:s:Qref;  !Existing reactive power generation of PV2          !25
dq=QpowerToShed-Qpv+1;                                                !26
if(Qpv<QpowerToShed){Qpvnew=1;    QpowerToShed=dq;}                   !27
else{ Qpvnew=Qpv-QpowerToShed;    QpowerToShed=0;}                    !28
val=sprintf('%f',Qpvnew);                                             !29
event=EventDir.CreateObject('EvtParam','Parameter Event');            !30
event:variable='Qref';                                                !31
event:p_target=oPV;                                                   !32
event:value=val;                                                      !33
event:mtime=0;                                                        !34
event:time=T;                                                         !35
event.Execute();                                                      !36
```

Fig. 2.26 DPL script of *PowerChangePV2.ComDPL* for load outage (scenario 1)

Ppvnew (= *Ppv-PpowerToShed*). Otherwise, the output power of PV2 is decreased to 1 MW and *PpowerToShed* is set to *PpowerToShed-Ppv* + *1* which indicates the extra active power generation that should be shed by other LS-PVPPs (as described in previous sub-subsection). Similarly, the same procedure is performed on lines 25–36 to shed extra reactive power generation.

2.5 Simulation Results

In this section, a modified version of the IEEE 14-bus with four LS-PVPPs is used to illustrate the impact of LS-PVPP on the power system stability. As shown in Fig. 2.16, this test system is divided into two voltage control areas (zones) using the electrical distance criterion [10] where PV1 and PV2 are located in zone 1, and PV3 and PV4 are located in zone 2.

In this test system, as mentioned in Sect. 2.3, LS-PVPPs have been implemented in DSL environment to accurately simulate the dynamic behaviour of PV arrays as well

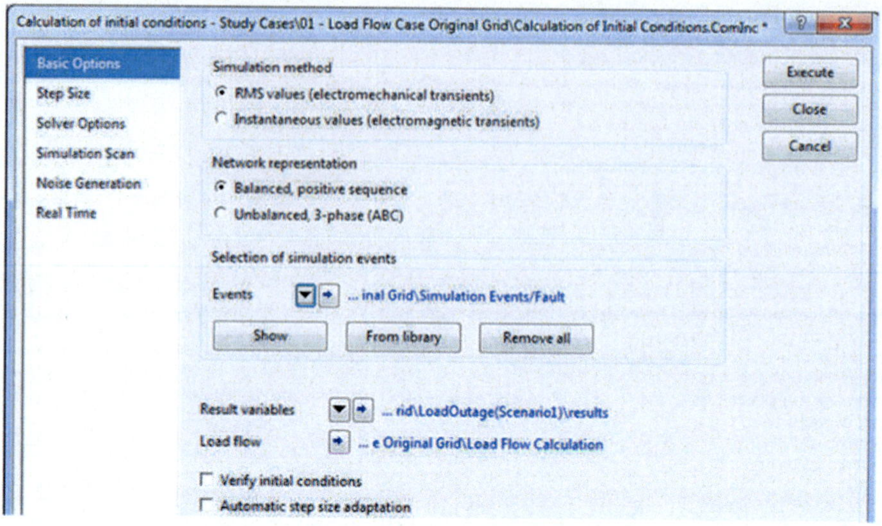

Fig. 2.27 The *results.ElmRes* should be assigned to result variables

as the DC/AC converter controllers and analyse their impacts on the system stability. It is worth mentioning that although usually, each LS-PVPP includes several PV arrays as well as AC-DC converters connected to the network, implementation of this structure in PowerFactory significantly reduces the simulation run-time. Therefore, in this research work, to increase the simulation speed, the number of PV modules connected in parallel and series is increased to generate the required active and reactive power of each LS-PVPP. The parameters used to simulate the first LS-PVPP (i.e. LS-PVPP1 connected to bus 2) have been shown in Sects. 2.3.1 and 2.3.2. Also, in this test system, all generators and their controllers (AVRs and governors) have been accurately modelled.

In the following section, three different disturbances are used to demonstrate the LS-PVPP capabilities in improving the stability status of the power systems. It should be noted that in these simulations, the influence of different penetration levels of PV sources on the stability of the test system will be investigated. Also, before the execution of these scripts, in the calculation of initial condition page, *results.ElmRes* should be assigned to result variables (Fig. 2.27).

2.5.1 Scenario 1: Load Outage

In this scenario, the impact of an outage of load 3 located in zone 1 (which consumes 130 MW and 19 Mvar) with/without LS-PVPP is investigated. According to the results of simulation given in Fig. 2.28a, when there is no LS-PVPP, disconnection of the load 3 at $t = 39.5$ s causes about 5% frequency overshoot (i.e. 1.0757 pu) and

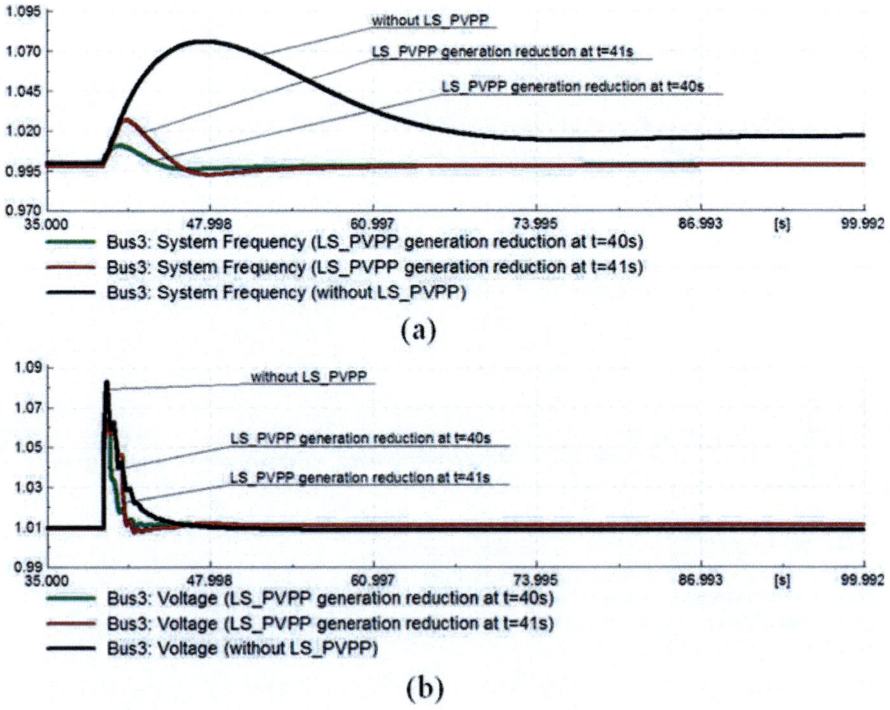

Fig. 2.28 Impact of LS-PVPP in scenario 1: **a** the system frequency and **b** the voltage of bus 5

finally, the system frequency reaches to $f = 1.0165$ pu (in steady-state condition) which indicates that synchronous generators cannot regulate the system frequency accurately. However, considering the proposed remedial action algorithm, when the LS-PVPPs exist and change their generation at $t = 40$ s (the reserves of LS-PVPPs are 20%) the frequency overshoot reaches about 1.011 pu. Also, according to this figure, when the reaction time decreases, the frequency overshoot reduces significantly. Similarly, Fig. 2.28b shows that the proposed remedial action slightly improves the bus voltage.

2.5.2 *Scenario 2: Generator Outage*

In this scenario, at $t = 39.5$ s, Gen1 (which generates 90 MW) suddenly disconnects from the grid and causes a significant drop of the system frequency. Clearly, such a condition may result in the operation of the frequency relays that may disconnect some loads and generation units and make the system unstable. However, as shown in Fig. 2.29, the fast reaction of LS-PVPPs (with 10, 15 and 20% penetration levels) can improve system stability significantly. In this condition, after the generator outage,

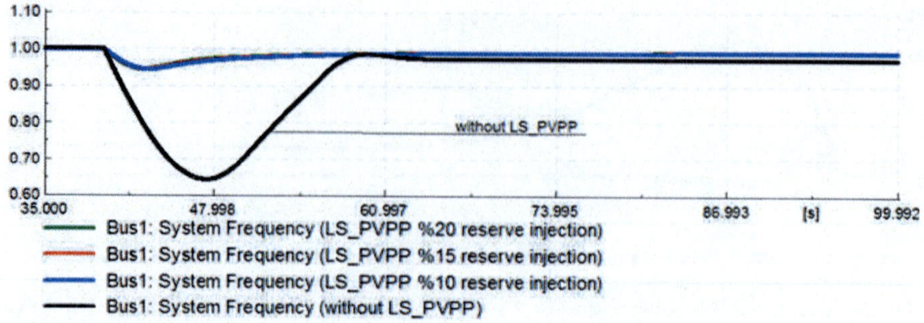

(a)

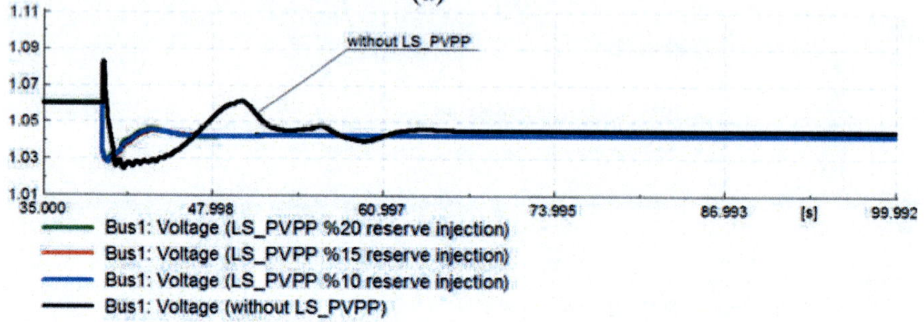

(b)

Fig. 2.29 Impact of LS-PVPP in scenario 2: **a** the system frequency and **b** the voltage of bus 1

the active and reactive powers of the LS-PVPPs are increased quickly to compensate for the power shortage. It should be stated that before the event occurrence, the operating point of LS-PVPPs is in the right-hand side of the MPP and the active power reserves of them are 20% (see Fig. 2.12).

2.5.3 Scenario 3: Line Outage

In this scenario, lines 2–3 outage occurs at $t = 39.5$ s which causes a severe voltage drop in bus 3 (in zone 1). In this condition, LS-PVPP2 quickly injects its active and reactive power reserve, and then, after $\Delta T_w = 1.0$ s, the voltage at bus 3 reaches the acceptable level. Thus, there is no need to inject further active and reactive powers by other LS-PVPPs. Also, as shown in Fig. 2.30, the fast response of LS-PVPP2 prevents a severe voltage drop and improves the dynamic behaviour of the network.

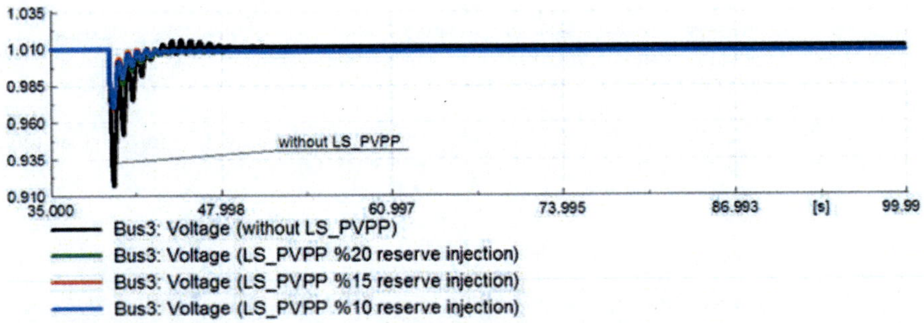

Fig. 2.30 Impact of LS-PVPP in scenario 3

2.6 Conclusion

Recent advancements in the power electronics industry made large-scale photovoltaic resources more accessible, and consequently, the integration of VSC-based generation units into the transmission system has been increased. The fast dynamic response of power electronic switches provides the solar PV generators with a quick response that enables these resources to contribute to system stability. In this study, using the capabilities of DIgSILENT Simulation Language (DSL) and DIgSILENT Programming Language (DPL) environments, large-scale PVs are modelled in PowerFactory software to validate the VSC ability in improving the dynamic response of the system. For this purpose, a modified version of IEEE 14-bus test system has been used to evaluate the impact of LS-PVPPs and their controllers in maintaining the system operating point in a stable region. The simulation results have clarified the fact that VSCs can decay the deteriorating impacts of disturbances by performing timely remedial actions.

References

1. S. Eftekharnejad, V. Vittal, G.T. Heydt, B. Keel, J. Loehr, Impact of increased penetration of photovoltaic generation on power systems. IEEE Trans. Power Syst. **28**(2), 893–901 (2012)
2. R. Shah, N. Mithulananthan, R.C. Bansal, V.K. Ramachandaramurthy, A review of key power system stability challenges for large-scale PV integration. Renew. Sustain. Energy Rev. **41**, 1423–1436 (2015)
3. G. Lammert, D. Premm, L.D.P. Ospina, J.C. Boemer, M. Braun, T.V. Cutsem, Control of photovoltaic systems for enhanced short-term voltage stabilility and recovery. IEEE Trans. Energy Convers. **34**(1), 243–254 (2018)
4. D. Remon, A.M. Cantarellas, J.M. Mauricio, P. Rodriguez, Power system stability analysis under increasing penetration of photovoltaic power plants with synchronous power controllers. IET Renew. Power Gener. **11**(6), 733–741 (2017)
5. Q. Li, M. Baran, A novel frequency support control method for PV plants using tracking LQR. IEEE Trans. Sustain. Energy (2019)

6. R. Shah, N. Mithulananthan, R.C. Bansal, oscillatory stability with high penetrations of large-scale photovoltaic generation. Energy Convers. Manage. **65**, 420–429 (2013)
7. A. Yazdani, R. Iravani, *Voltage-Sourced Converters in Power Systems*, vol. 34 (Wiley, Hoboken NJ, USA, 2010)
8. H. Khoshkhoo, A. Parizad, Dynamic stability improvement of Islanded power plant by smart power management system: Implementation of PMS Logic, in *Advanced Smart Grid Functionalities Based on PowerFactory* (Springer, Cham, 2018), pp. 125–151.
9. A. Cabrera-Tobar, E. Bullich-Massagué, M. Aragüés-Peñalba, O. Gomis-Bellmunt, Review of advanced grid requirements for the integration of large scale photovoltaic power plants in the transmission system. Renew. Sustain. Energy Rev. **62**, 971–987 (2016)
10. Sk. Razibul Islam, D. Sutanto, K.M. Muttaqi, Coordinated decentralized emergency voltage and reactive power control to prevent long-term voltage instability in a power system. IEEE Trans. Power*Syst.* **30.5**: 2591–2603 (2014)

Chapter 3
Dynamic Modelling and Simulation of Power Electronic Converter in DIgSILENT Simulation Language (DSL): Islanding Operation of Microgrid System with Multi-energy Sources

Piyadanai Pachanapan

Abstract In this chapter, the dynamic performances of a microgrid system under the islanding operation are examined based on RMS transient simulation in DIgSILENT *PowerFactory*. The dynamic models of energy sources are implemented in the DIgSILENT Simulation Language (DSL) including battery storage, photovoltaic (PV), small hydropower plant and diesel generator. The converter-connected energy sources, such as battery and PV systems, are the main controllable devices to support frequency and voltage controls in the islanded microgrid system, with the fast control response. The dynamic models of grid-tied inverter in battery and PV systems are presented which the control structure is based on the synchronous reference frame with PI and PID controllers. On the other hand, the models of governors and excitation systems of hydrogenerator and diesel generator are built from the standard models established in the DIgSILENT library. An example project proposes the coordinated control manner among PV, battery, small hydrogenerator and diesel generator based on the centralised control scheme. The results from RMS transient simulations illustrate the control performances of those controllable devices to deal with the significant changes in system frequency and voltage after the planed islanding is applied into the microgrid system.

Keywords DIgSILENT Simulation Language · DSL · Islanding operation · Microgrid · RMS transient simulation

Electronic supplementary material The online version of this chapter (https://doi.org/10.1007/978-3-030-54124-8_3) contains supplementary material, which is available to authorised users.

P. Pachanapan (✉)
Department of Electrical and Computer Engineering, Faculty of Engineering, Naresuan University, Phitsanulok, Thailand
e-mail: piyadanip@nu.ac.th

F. M. Gonzalez-Longatt and J. L. Rueda Torres (eds.), *Modelling and Simulation of Power Electronic Converter Dominated Power Systems in PowerFactory*, Power Systems,
https://doi.org/10.1007/978-3-030-54124-8_3

3.1 Introduction

The growth of *distributed generation* (DG), both conventional and renewable energy sources, can improve power quality, reliability and security of supply to existed distribution networks in the form of a microgrid system . Also, the microgrid system is an interconnected network of loads and DG units that can function whether they are connected to or separated from the main utility system. The microgrid system can enhance power quality, such as voltage improvement and loss reduction, to the network during the grid-connecting mode, while it increases stability and reliability to the network in the islanding mode.

Islanding operation is an efficient control technique applied in the microgrid system when isolating from the main power system. The islanding condition is generally classified in two types: (1) intentional or planed islanding which is a process of opening the circuit breaker/switch at the *point of common coupling* (PCC) to disconnect the microgrid and utility, and (2) unintentional islanding which occurs due to a fault in the system and may lead to severe instability. During the islanding condition, there are significant changes in frequency and voltage. Therefore, the control strategy should secure the system frequency and voltage to stay within the statutory limits, as provided by some international standards such as IEEE 929–2000 or IEC 62116 [1] (Table 3.1).

Although the increase of DG can enhance power generation and stability of microgrid system, the mix of several types of energy sources can cause the complexity to the microgrid control system during islanding operation. The control operation should be quick enough to deal with the large rapid changes of frequency and voltage, especially during the islanding transition. Therefore, the converter-connected DG and energy storage system, such as PV and battery, are necessary to deal with the fast fluctuations of frequency and voltage level in the system, while the other types of DG, such as a conventional synchronous machine, can support either frequency or voltage control, with the slower control response, depending on their availability and control functions.

The typical microgrid system adapted from [2] is shown in Fig. 3.1. The hierarchical control system architecture comprised the following two control levels: (1) device level at DG and energy storage units and (2) *microgrid central control*

Table 3.1 Different standards for microgrid islanding

Standards	Nominal frequency range	Nominal voltage range	Islanding detection time
UL 1741	$59.3 \le f \le 60.5$	$88\% \le V \le 110\%$	$t < 2$ s
IEEE 929–2000	$59.3 \le f \le 60.5$	$88\% \le V \le 110\%$	$t < 2$ s
VDE 0126-1-1	$(f - 1.5\text{ Hz}) \le f \le (f + 1.5\text{ Hz})$	$88\% \le V \le 100\%$	$t < 0.2$ s
IEC 62116	$59.3 \le f \le 60.5$	$85\% \le V \le 115\%$	$t < 2$ s
IEEE 1547	$59.3 \le f \le 60.5$	$88\% \le V \le 110\%$	$t < 2$ s

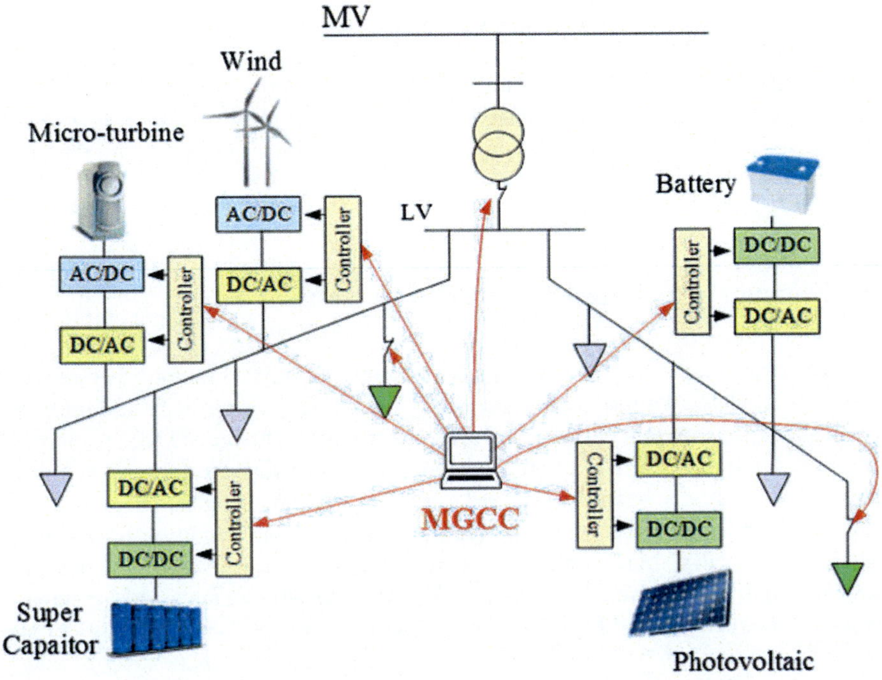

Fig. 3.1 Typical microgrid architecture

(MGCC). At the device level, DG and energy storage units try to generate power to the microgrid system that meets the grid requirement, such as system frequency and voltage level. The localised control methods are usually applied to this level using the droop control method with the supervision from the MGCC.

To operate the microgrid system in both grid-connecting and islanding modes, the MGCC is the key control device in the typical centralised control scheme of the microgrid system. The MGCC can command and exchange information with other controllable devices in the microgrid system via communication links, such as optical fibre cable. In the islanding mode, the MGCC will manage the power production from DG units and battery to match with the load demand in the microgrid system. Moreover, the MGCC is able to disconnect the uncontrollable DG and specific loads with the aim to maintain the stability of the system.

In this chapter, the test system is the microgrid system which consists of small hydrogenerator and diesel generator, PV and battery systems. The study is based on transient simulations aiming to demonstrate the power balancing mechanism and volt-var control from DG and energy storage systems in the microgrid system during the islanding condition. The balanced RMS simulation in DIgSILENT *PowerFactory* software is used to examine the fast dynamic performances of microgrid system during the islanding operation. Additionally, the development of dynamic models of those energy sources, using *DIgSILENT Simulation Language* (DSL), is presented in detail. The simulation results illustrate that, in the islanding mode, the cooperated control among PV, battery, small hydrogenerator and diesel generator can deal with the deviations of frequency and voltage during the islanding condition effectively. Those energy sources can prevent the interruption event from occurring in the microgrid system, while the system frequency and voltage are maintained within the grid requirements.

3.2 Microgrid System Operation and Control

The study aims to demonstrate the operation and control of the microgrid system in the islanding mode, with satisfying frequency and voltage level, while the number of load shedding should be a few. However, the uncertainty of renewable resources causes the difficulty of the microgrid system when operating the power balancing process. The battery is generally selected as the main storage device to deal with the sudden changes of frequency and voltage after the islanding condition is detected. Renewable energy sources remain to connect to the system if they can provide frequency and voltage control abilities. The PV system, which is a converter-connected DG, can be used for supporting power and voltage controls to the system if the solar irradiance is available. The power curtailment control may be required in case of using a PV system to prevent excessive power supplied into the system. The droop control methods are applied to share active and reactive power among those PV and battery units.

Alternatively, the small hydrogenerator and diesel generator, based on synchronous machine, can be used as frequency and voltage controllers if turbine-governing and voltage excitation systems are included. The small hydrogenerator is mostly a run-of-river design which the output power, decided by the water level, is quite steady. The diesel generation aims to be employed when support from the battery energy storage is insufficient.

3.2.1 *Photovoltaic and Battery Storage Systems*

The PV and battery storage systems are typically based on back-to-back *voltage source converters* (VSCs) which consist of DC/DC converter cascading with DC/AC inverter, as illustrated in Fig. 3.2. The energy source-side converter establishes the maximum power extraction function, while the grid-side converter performs the grid interface control. The active and reactive powers from the grid-side converter can be controlled following the supervision from the MGCC. Even though both PV and battery systems have the same converter configuration, the PV system is using unidirectional power converter while the bidirectional power converter is employed for the battery system.

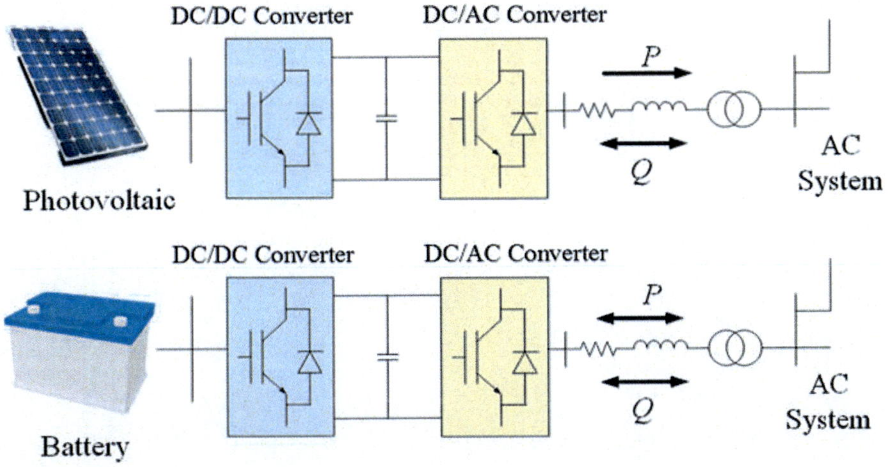

Fig. 3.2 Configurations of grid-connected PV and battery

During the grid-connecting condition, the power converter of PV system can operate in either *PQ* or *PV* mode [2]. In *PQ* mode, the reactive power of PV system is controlled at the fixed value or to be generated by the MGCC demand while the active power is still produced according to a maximum power extraction rule. Alternatively, the *PV* mode will arrow the PV system providing voltage control, to maintaining the AC voltage of inverter at the desired value. The voltage controllability is limited by the allowable reactive power (Q_{PV}) of PV system, which depends on the output active power (P_{pv}) and the power factor of a grid-tied inverter (PF_{PV}). This reactive power is calculated by

$$Q_{PV} = P_{PV} \times \tan(\cos^{-1} PF_{PV}) \tag{3.1}$$

In the case of a battery storage system, the direction of the active power of the battery is depending on the battery conditions. The battery injects the active power to the microgrid system during the discharge mode, while, in the charge mode, the battery creates negative active power. The battery will be operated in charge or discharge mode depending on the level of *state of charge* (SOC). If SOC is high, the battery has been filled with electricity and is ready to discharge power to the system. On the contrary, the battery should be recharged immediately when SOC is lower than the permission value. The battery system aims to keep the system frequency to stay within the statutory limit in both on- and off-grid conditions, while the reactive power is fixed or depending on the change of voltage level (*FQ* or *FV* mode).

In the islanding mode, the grid-side inverter of either PV or battery system, especially the one chosen as the master device, will operate as a grid-forming inverter aiming to emulate the behaviour of a synchronous machine by providing a reference for voltage and frequency [3]. If two or more converter-based PV and energy storage

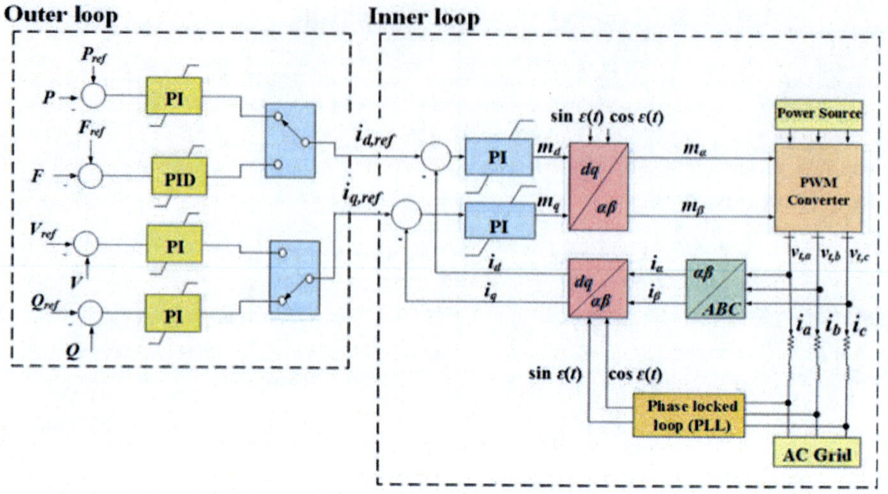

Fig. 3.3 Grid-side converter controller of PV and battery systems

units participate in grid frequency and voltage controls, the droop control methods are applied to share active and reactive powers among those PV and battery units.

The control scheme of grid-side inverter is developed based on the synchronous rotating reference frame which consists of two cascaded control loops. The output of converter is generated by using the *pulse width modulation* (PWM) technique. The inner control loop decouples the control for the real and imaginary current components [4]. This control method consists of two main classes of two-dimensional frames, the $\alpha\beta$-frame and the *dq*-frame. Therefore, the two current components, i_d and i_q, are controlled independently, as shown in Fig. 3.3, where the reference values for the i_d and i_q controls are provided by the outer control loop. The $i_{d,\ ref}$ is from the active power (*P*) or system frequency (*F*) controller on the *d*-axis, while the $i_{q,\ ref}$ is from the reactive power (*Q*) or bus voltage (*V*) controller on the *q*-axis. In both control loops, the error signals are compensated using PID and PI controllers.

Inside the inner control loop, the $\alpha\beta$-frame is the transformation from *abc*-frame into the Cartesian coordinate system where real and imaginary parts of current from the transformation are function of time. If *i*(*t*) is the current-controlled signal, that is, time-varying function, the transformation from three-phase *abc*-frame into $\alpha\beta$-frame using Clark's transformer can be written as:

$$\begin{bmatrix} i_\alpha(t) \\ i_\beta(t) \end{bmatrix} = \frac{2}{3}\begin{bmatrix} 1 & -1/2 & -1/2 \\ 0 & \sqrt{3}/2 & -\sqrt{3}/2 \end{bmatrix}\begin{bmatrix} i_a(t) \\ i_b(t) \\ i_c(t) \end{bmatrix} \tag{3.2}$$

Park's transformation from $\alpha\beta$-frame to *dq*-frame can be represented as:

$$\begin{bmatrix} i_d(t) \\ i_q(t) \end{bmatrix} = \begin{bmatrix} \cos\varepsilon(t) & \sin\varepsilon(t) \\ -\sin\varepsilon(t) & \cos\varepsilon(t) \end{bmatrix} \begin{bmatrix} i_\alpha(t) \\ i_\beta(t) \end{bmatrix} \tag{3.3}$$

In case of a constant frequency system, $\varepsilon = \varepsilon_0 + \omega_0 t$, where ω_0 is the AC system operating frequency and ε_0 is an initial phase angle between the real axes of the $\alpha\beta$- and dq-frames (constant value). The value of ε can be obtained by using a phase-locked loop (PLL).

The real and imaginary parts of terminal voltages are calculated using (3.4). Thus, the dq voltages of the PMW converter can be transformed into three-phase voltages by transforming from $\alpha\beta$-frame to abc-frame.

$$\begin{aligned} v_{t,\alpha}(t) &= k_o m_\alpha(t) V_{dc}(t) \\ v_{t,\beta}(t) &= k_o m_\beta(t) V_{dc}(t) \end{aligned} \tag{3.4}$$

where $v_{t,\alpha}$ and $v_{t,\beta}$ are the real and imaginary parts of the RMS terminal voltage of PWM converter, respectively. Similarly, m_α and m_β are real and imaginary parts of the modulation index, correspondingly. k_o is a constant factor that depends on the modulation method, and V_{dc} is the DC link voltage.

3.2.2 *Small Hydropower Generator*

The small-scale hydrogenerators are widely used in the rural areas. The power plant is commonly based on run-of-river design, which is lacking of significant water storage. The power output is decided by the water level which varies with rainfall. Typically, there is no automatic power and voltage controllers in the small hydropower plant. This means that the hydroturbine will produce the power as much as possible to the microgrid system. Therefore, it may cause the surplus of power generation during islanding condition. To prevent this situation, the hydrogenerator will be disconnected or be asked for the power curtailment if the active power controller is available.

The power production of hydrogenerator can be reduced by adjusting the position of penstock gate to dropping the water flow rate. This gate is adjusted by the servo actuator which is controlled by the governor. In addition, the speed governor will monitor the variation of system frequency or turbine speed, and then command the change of gate position. Moreover, the excitation system which consists of DC exciter and voltage regulator is used to regulate the field voltage inside the synchronous generator, in order to supply the satisfied voltage level and smooth power into the grid system. Figure 3.4 demonstrates the simplified relationship between the basic elements of power generation and voltage regulation processes in a small hydropower plant [5].

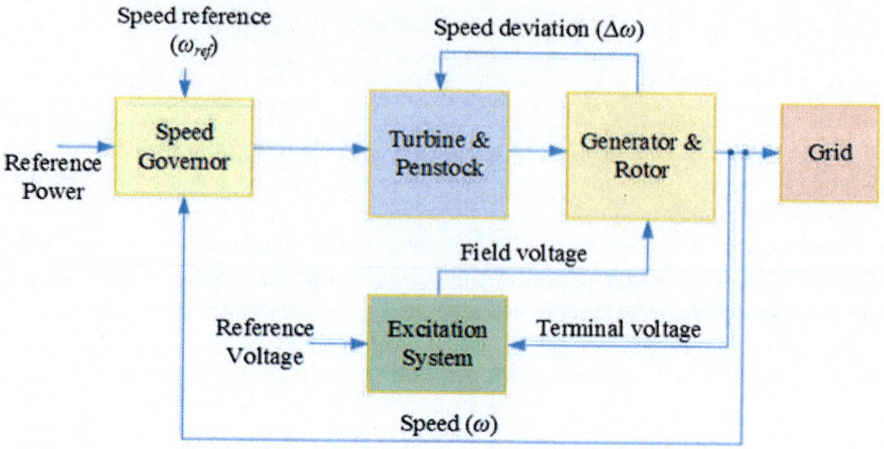

Fig. 3.4 Functional block diagram of hydropower plant

3.2.3 *Diesel Power Generator*

A small diesel generator is commonly used as the emergency power supply if the main grid fails. It is the combination of a reciprocating four-stroke engine, run on diesel fuel, driving a three-phase synchronous generator. Similar to the hydrogenerator, the frequency and voltage outputs of diesel generator are regulated by using speed governor and excitation, respectively, as seen in Fig. 3.5 [6].

The speed of generator depends on the amount of injected fuel and the load applied into the engine crankshaft. The speed governor keeps the engine speed and the speed of the generator constant by adjusting the flow of fuel injection. Additionally, the output of the speed governor is the throttle signal that controls the fuel going into the engine. The clutch may be required to decouple the synchronous machine from the diesel engine, in a case of high-speed combustion engine.

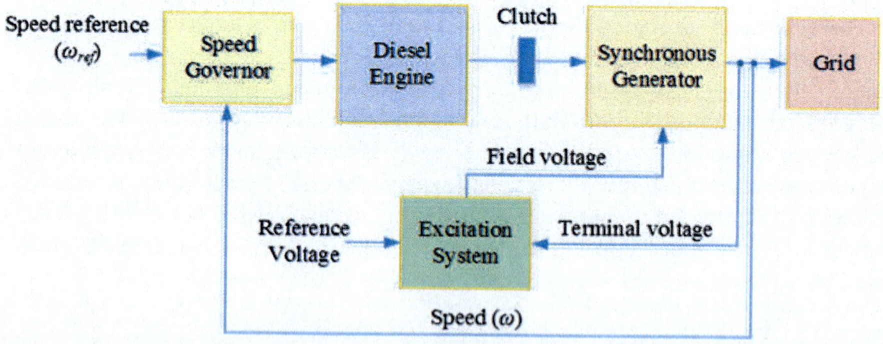

Fig. 3.5 Functional block diagram of diesel generator

Although a standby diesel generator is a popular option for supplying electricity to specific locations during the loss of supply from the main utility, a 1–2 gallon fuel tank making the diesel generation can only last from 8 to 10 h. It also needs to be stopped for around 15 min for refuelling, and then it continues working. Hence, this makes it difficult to use only diesel generation supporting the system especially in a case of long-term interruptions.

3.3 Preparing for RMS Simulations

The balanced RMS simulation is considered as dynamics in electromechanical, control and thermal devices in a time domain. This function uses a symmetrical, steady-state representation of the passive electrical networks, which only the fundamental components of voltages and currents are taken into account, for midterm and long-term transients under balanced network conditions.

To implement the microgrid system for RMS transient studies in DIgSILENT *PowerFactory* software, there are seven steps as follows:

Step 1: Building a test system, which consists of power system equipment such as generators, transformers, lines, converters and loads. Additionally, objecting type and input parameters of each network element need to be assigned.
Step 2: Creating the dynamic models, which include composite models and common models, into the controllable devices such as grid-tied inverters, hydrogenerator and diesel generator.
Step 3: Adding network elements, common models, measurement devices, etc., into the dynamic models.
Step 4: Calculating the initial conditions of test system, to identify the initial values of all internal variables and the internal operating status of connected machines, controllers and other transient models.
Step 5: Defining the simulation events.
Step 6: Defining the result variables.
Step 7: Executing the simulation and creating simulation plots.

The further information of building a test system in **step 1** can be found in Chap. 9 (Networks graphics) and Chap. 12 (Building networks) of the PowerFactory 2018: User Manual [7], while Chapter 29 (RMS/EMT simulations) gives more detail about the time-domain simulation processes in **steps 4** to **7**.

To demonstrate the dynamic responses of controllable energy sources in the microgrid system, the frequency and voltage controllers in Sect. 3.2 must be implemented in the form of dynamic modelling, as in **steps 2** and **3**. The dynamic modelling approach in DIgSILENT *PowerFactory* is designed to be flexible, and then any model can be constructed using the same set of tools. The basic building blocks are the model equations that are written in the DSL, which the nested block structures

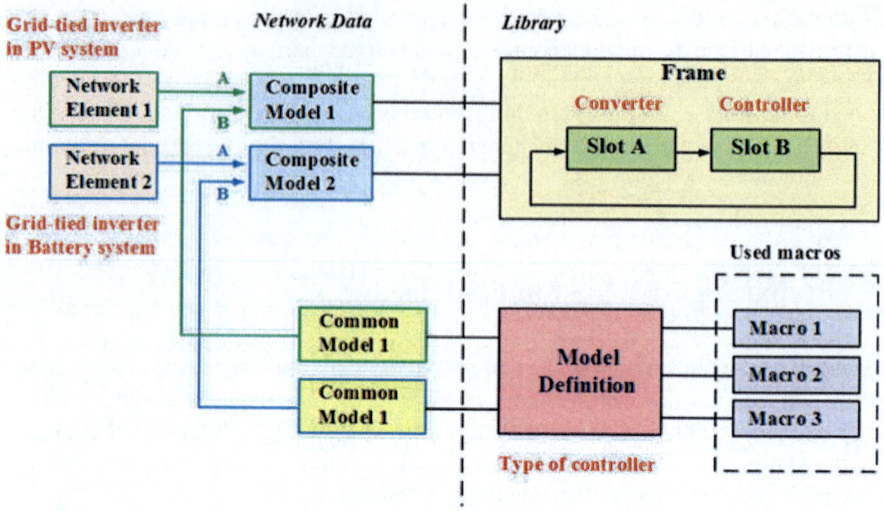

Fig. 3.6 Block diagram of dynamic modelling (one-to-many scheme)

are automatically converted to a set of DSL equations. The DSL equations are then parsed and interpreted during the RMS simulation.

The dynamic modelling uses an object-oriented approach that encourages reuse by strictly enforcing the distinction between library (type) and grid (element) objects. In addition, there are type objects in the library (i.e. composite frame and model definitions) that are referenced by element objects in the network (i.e. composite models and common models). The block diagram of dynamic modelling is illustrated in Fig. 3.6 [8].

The composite model is a grid object that represents a complete dynamic system. It links the composite frame to actual network elements, measurement devices and DSL common models in the system. The composite frame is an overview diagram showing the interconnection between slots. Each slot indicates the type of object that should be assigned to the slot. The model definition is a block that defines the transfer function of a dynamic model, in the form of equations and/or graphical block diagram. In addition, both equations and graphical block diagram can be built in the form of object, called as the block definition (*BlkDef*). Furthermore, the DSL common model will link a model definition to an actual piece of equipment with specific parameter settings.

For example, the frame in Fig. 3.6 represents the connection of a PWM converter (Slot A) to a controller (Slot B). This frame is used for two converters that can be grid-tied inverters in PV and battery systems (e.g. network elements 1 and 2). If both converters use the same type of controller, this controller is defined once as a type (i.e. model definition) and then two different common models are created for each of the converters which each has its own local parameter settings.

From the DSL advance tutorial in [8], the suggested procedure for creating a dynamic model in *PowerFactory* is as follows:

(1) Consider the structure of the system to be modelled and how it can be broken down into discrete blocks that can be modelled separately.
(2) Construct a composite frame illustrating how the slots are interconnected.
(3) Create each of the model definitions and set appropriate initial conditions.
(4) Crate a composite model and fill the slots with the relevant grid elements, such as common models, built-in models and measurement device.
(5) Test the complete model.

This chapter focuses on the development of dynamic models of grid-tied inverter control system that is used in PV and battery systems, whereas the composite frames and model definitions of hydrogenerator and diesel generator are brought from the dynamic model library (ver. 2018: *DIgSILENT Library* → *Dynamic Models*).

3.3.1 Grid-Tied Inverter Control System

The built-in PWM converter/1 DC connection model (*ElmVscmono*) is used as grid-tied inverter in PV and battery systems. This model includes the built-in *dq* current controller (see more detail in [9]), which is the same as the inner control loop as presented in Fig. 3.3. The PWM converter's currents (i_d and i_q) are adjusted inside this built-in current controller using PI regulators. In addition, the input signals of this converter model are including the *cosphi* and *sinphi* signals from the built-in frequency measurement model (*ElmPhi__pll*) and the current references (*id_ref* and *iq_ref*) from the outer control loop in Fig. 3.3.

The outer control loop consists of *P/F* controller and *Q/V* controller. The *P/F* controller is used for controlling either active power (*P*) or system frequency (*F*), while the reactive power (*Q*) or bus voltage (*V*) is controlled by the *Q/V* controller. Under the normal condition, the PV and battery systems are in *PQ* mode and *FQ* mode, respectively. Besides, they can be switched to frequency and voltage control mode (*FV* mode) after the islanding condition is detected.

The PWM converter control system can be summarised by a block diagram as shown in Fig. 3.7. The output signal of *P/F* controller is the *d*-axis reference current (*id_ref*), and the input signals are including the measured frequency (*fmea*) from the built-in frequency measurement model (*ElmPhi__pll*) and output active power (*p*) from the built-in power measurement model (*StaPqmea*). In a case of *Q/V* controller, the output signal is the *q*-axis reference current (*iq_ref*) and the input signals consist of the output reactive power (*q*) from the built-in power measurement model (*StaPqmea*) and the magnitude of positive-sequence voltage (*u*) from the built-in voltage measurement model (*StaVmea*). Additionally, frequency, power and voltage measurements are located at the PCC of grid-tied inverter.

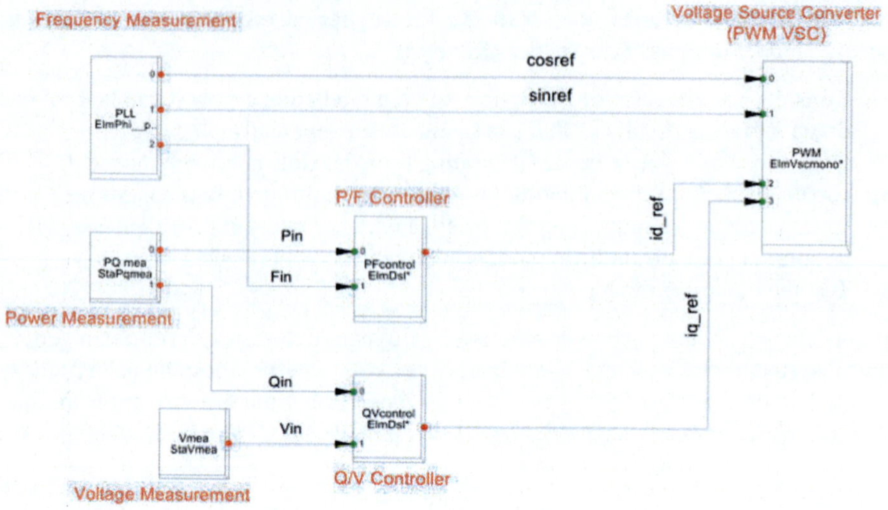

Fig. 3.7 Composite frame for grid-tied PWM converter

3.3.1.1 Model Definitions

The model definition of *P/F* controller (*PFsupport.BlkDef*) is based on a simple PI and PID controllers, as shown in Fig. 3.8. The reference active power (*Pref*) and

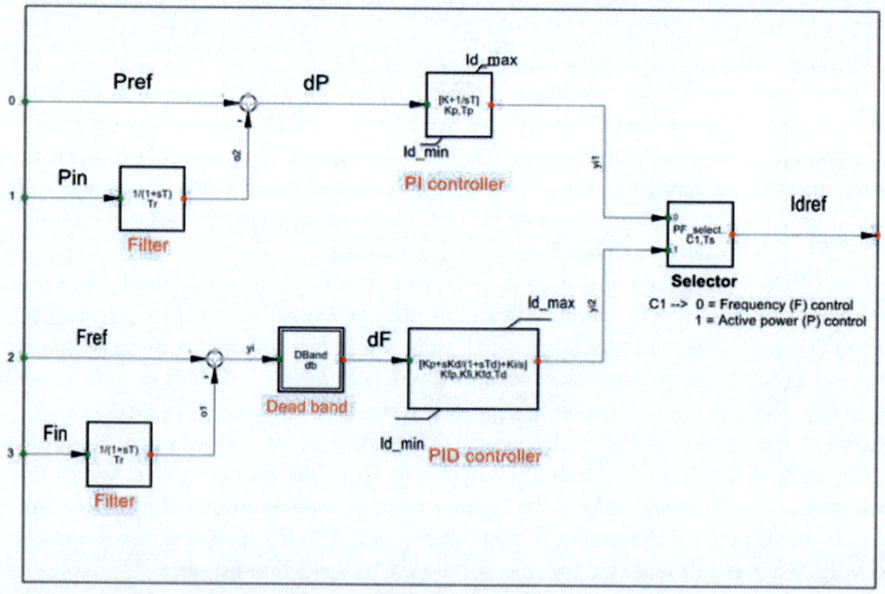

Fig. 3.8 Model definition for *P/F* controller

reference frequency (*Fref*) are defined from the initial condition calculation. The active power and frequency inputs that are *Pin* and *Fin* are filtered by using the first-order delay (*1/(1 + sT).BlkDef*). At the *P* controller, the active power is regulated by using a PI controller with int./ext. non-windup limiter (*[K + 1/sT].BlkDef*), while the frequency in the *F* controller is controlled by using a PID controller with saturation (*[Kp + sKd/(1 + sTd) + Ki/s].BlkDef*). Moreover, the continuous dead band (*DBand.BlkDef*) is added into the *F* controller, to avoid the frequency fluctuation when the grid-tied converter is operating under the grid-connected condition.

This *P/F* controller will be operating in *P* or *F* control mode, by defining the value of C_1 in the selector block definition (*PF_selection.BlkDef*). Additionally, it is *P* control mode if C_1 is 1, while it is switched to *F* control mode if C_1 is 0.

In a case of *Q/V* controller, the model definition (*QVsupport.BlkDef*) is implemented similar to the *P/F* controller as demonstrated in Fig. 3.9. The reference reactive power (*Qref*) and reference voltage (*Vref*) are defined from the initialisation process. The first-order delay (*1/(1 + sT).BlkDef*) is used to filter the inputs from reactive power and voltage measurements. The PI controller with int./ext. non-windup limiter (*[K + 1/sT].BlkDef*) is employed as the regulator for both *Q* and *V* controllers. Furthermore, this *Q/V* controller will be in *Q* control mode if the value of C_2 in the selector block definition (*QV_selection.BlkDef*) is set to 1. Otherwise, it will be operating in the *V* control mode.

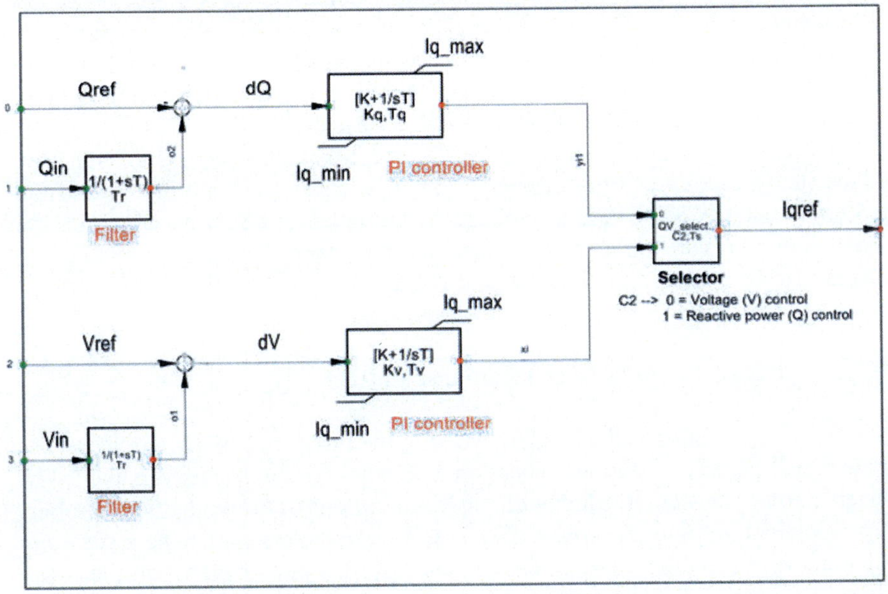

Fig. 3.9 Model definition for *Q/V* controller

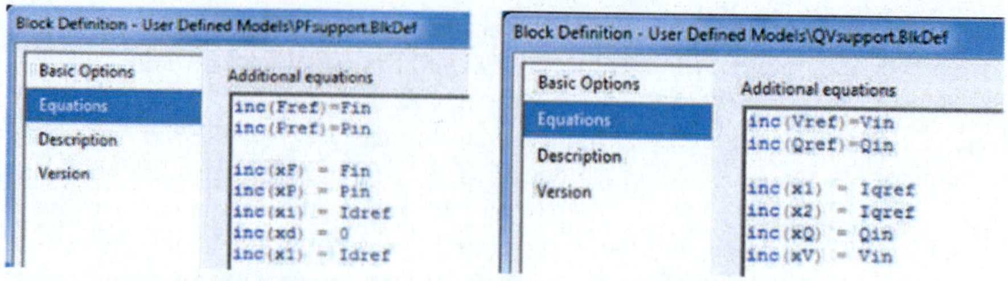

(a) *P/F* controller (b) *Q/V* controller

Fig. 3.10 Equation windows of block definition of *P/F* and *Q/V* controllers

3.3.1.2 Model Initialisation

All dynamic models must be initialised according to a steady-state load flow calculation before the start of a time-domain simulation. The manual initialisations are requested to enter in code form inside the equation window of the model definitions of *P/F* controller (*PFsupport.BlkDef*) and *Q/V* controller (*QVsupport.BlkDef*), as presented in Fig. 3.10.

The state variables of the *P/F* controller in Fig. 3.8 need to be manually initialised. The unknown inputs, *Pref* and *Fref*, and the state variables of first-order lag filters (i.e. *xP* and *xF*) are initialised from the known outputs of power and frequency measurements, which are *Pin* and *Fin*, respectively. The state variable of integrator in PI controller (*x1*) is initialised from the steady-state output which is *Id_ref*. Similarly, the state variables of integrator and differentiator (i.e. *xi* and *xd*) in the PID controller are initialised from the values of *Id_ref* and 0, respectively.

To initialise the *V/Q* controller in Fig. 3.9, the unknown inputs, *Qref* and *Vref*, and the state variables of first-order lag filters (i.e. *xQ* and *xV*) are initialised from the known outputs of power and voltage measurements, which are *Qin* and *Vin*, respectively. The state variables of integrators in PI controllers (i.e. *x1 and x2*) are initialised from the known steady-state output, *Iq_ref*.

3.3.1.3 Composite Models and Common Models

A common model inherits the block diagram of the linked model definition, but has its own local parameter settings. There are two common models used in the PWM converter control system that link with model definitions of *P/F* and *Q/V* controllers (*PFsupport.BlkDef* and *QVsupport.BlkDef*). All parameters need to be assigned in each common model including gain parameters of filter and PI/PID controllers, and values of C_1 and C_2 (for selecting the control mode). For example, the grid-tied inverter in battery system is operating in *FQ* mode, for providing the frequency control with constant power factor during the normal condition, hence $C_1 = 0$ and $C_2 = 1$. Figure 3.11 demonstrates the common models and their parameters of battery system. Alternatively, the common models of grid-tied inverter in PV system, which

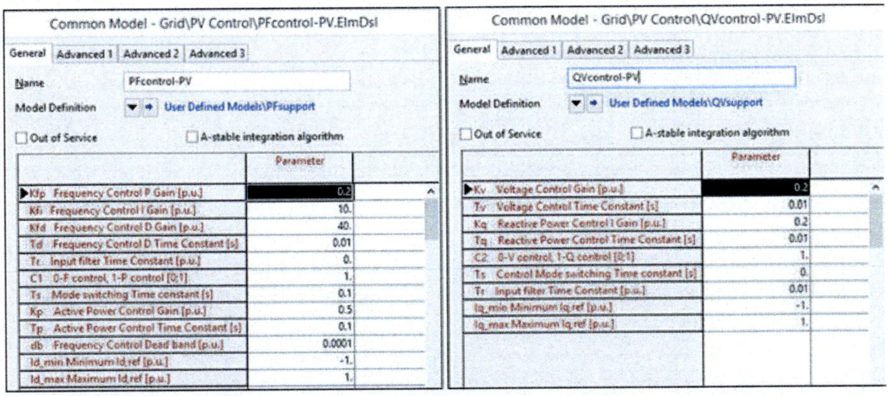

Fig. 3.11 Common models for grid-tied inverter in battery system

operates in *PQ* mode, use the same model definitions as employed in the battery system, but their parameters are different ($C_1 = 1$ and $C_2 = 1$).

The composite model is a grid object that represents a complete dynamic system. It is found that the composite models of grid-tied inverter in both PV and battery systems employ the same composite frame (*PWM Converter Frame.BlkDef*). However, the relevant system components, such as network elements, common models and measurement devices, are assigned differently in each composite model. Figure 3.12 shows the composite model of battery system in the data manager and the elements assigned to slots in the composite model.

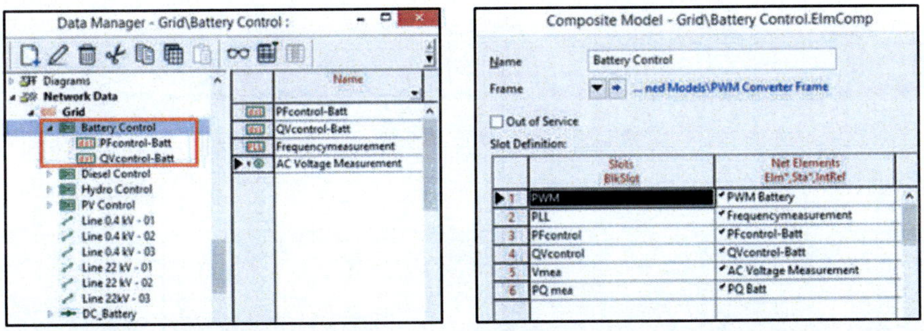

(a) In the data manager

(b) With elements assigned to slots

Fig. 3.12 Composite model for grid-tied inverter in battery system.
Note that the composite frame, model and block definitions of PWM converter used in this project are located in *Project Library* → *User Defined Models*

3.3.2 *The Control Systems of Hydrogenerator and Diesel Generator*

Both hydrogenerator and diesel generator use the built-in synchronous machine model (*ElmSym*) with speed governor and excitation controllers. Therefore, the turbine power (*pt*) and excitation voltage (*ve*), which are the synchronous generator inputs, are controlled throughout the duration of the simulation. In this work, the composite frame of synchronous machine and model definitions of governor and excitation system are employed from the typical models in *DIgSILENT Library* (ver. 2018).

These two types of generators use the same composite frame which is *SYM Frame_no droop.BlkDef* that is located in "*DIgSILENT Library* → *Dynamic Model* → *PSS\E Compatible* → *Composite Model Frames*". Additionally, this frame provides the synchronous machine signal interconnections which includes the slots of governor, voltage exciter and stabiliser. The governor and excitation system of hydrogenerator are based on the models of HYGOV (*gov_HYGOV.BlkDef*) and IEEE Type AC1 Excitation System (*avr_EXAC1.BlkDef*), respectively, whereas the diesel generator uses the models of DEGOV1 (*gov_DEGOV1.BlkDef*) and IEEE Type ST2 Excitation System (*avr_EXST2A.BlkDef*), as speed regulator and voltage controller, respectively. These typical definition models can be found in "*DIgSILENT Library* → *Dynamic Model* → *PSS\E Compatible*".

The block diagrams of governor and excitation system used in this work are presented in [10], while the values of controller's parameters are selected based on the guideline in [10] and [11]. The common models of governors for hydrogenerator and diesel generator are illustrated in Fig. 3.13.

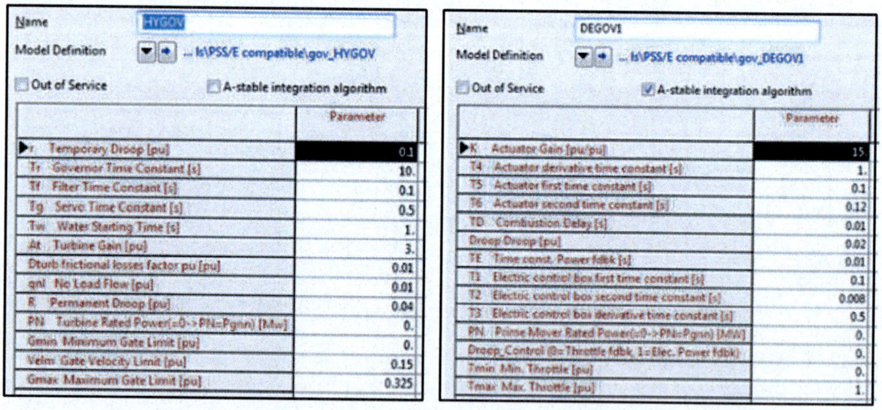

(a) Hydro governor (b) Diesel governor

Fig. 3.13 Common models of governors for hydrogenerator and diesel generator

3.3.3 *Simulation Events*

During time-domain simulations, the controller's settings or breaker status (i.e. close or open) can be changed at the certain time by defining in simulation event objects. When the islanding is detected, some breakers in the microgrid system need to be switched to the new status. Therefore, the switching events (*EvtSwitch*) are applied to those breakers for triggering the switch action to be closed or opened.

The MGCC, which is the central controller, is able to change the control mode or set point of each controllable energy source in the microgrid system. To adjust the settings in energy source controllers, the parameter events (*EvtParam*) are employed. With this type of event, an input parameter of any elements or DSL common models can be set or changed at a specific time.

The simulation events for operating the islanded microgrid system are illustrated in Fig. 3.14. It can be explained that the islanding condition is defined by tripping the breaker at the PCC at time = 10 s. After the islanding is detected, the MGCC is able to adjust the status of automatic transfer switches (ATSs) and breakers within 0.1 s. In the meantime, the MGCC orders the grid-tied inverters to support voltage control by changing C_2 to 0 in the common model of *Q/V* controller. Moreover, the minimum q-axis current in p.u. (*iq_min*) is limited based on the allowance reactive power that can be injected into the grid by each inverter, as determined in (3.1)

3.4 Example Project

The following project demonstrates the performances of microgrid operation with multi-energy sources during the islanding operation. The RMS transient simulation is used to present the dynamic response of each energy source when the planed islanding has occurred. The dynamic models proposed in Sects. 3.3.1 and 3.3.2 are employed into the test system. Moreover, the scenarios in a case study are implemented by defining in simulation events, as described in Sect. 3.3.3.

3.4.1 *Test System*

The test system is a balanced three-phase 22/0.4 kV, 50 Hz distribution network which includes 90 kW hydrogenerator, 20 kW PV system and 100 kW/150 kWh battery storage and 419 kW diesel generator, as shown in Fig. 3.15. There are three load connections which load 1 and load 2 are small industrial plants and load 3 is the accommodation. Assume that the power factor of this system is 0.9 lagging. Moreover, the power output of hydrogenerator is assumed as constant for the whole day, while the PV generation has varied depending on the sunlight intensity. Figure 3.16

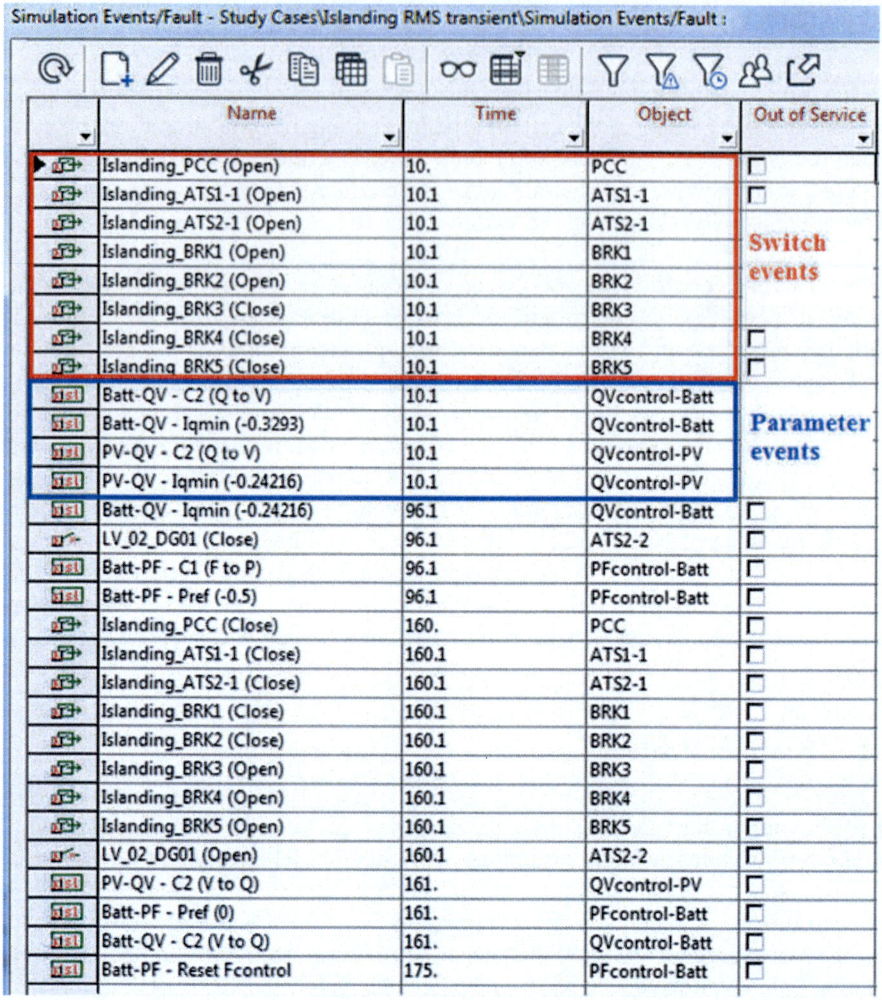

Name	Time	Object	Out of Service
Islanding_PCC (Open)	10.	PCC	
Islanding_ATS1-1 (Open)	10.1	ATS1-1	
Islanding_ATS2-1 (Open)	10.1	ATS2-1	
Islanding_BRK1 (Open)	10.1	BRK1	
Islanding_BRK2 (Open)	10.1	BRK2	
Islanding_BRK3 (Close)	10.1	BRK3	
Islanding_BRK4 (Close)	10.1	BRK4	
Islanding_BRK5 (Close)	10.1	BRK5	
Batt-QV - C2 (Q to V)	10.1	QVcontrol-Batt	
Batt-QV - Iqmin (-0.3293)	10.1	QVcontrol-Batt	
PV-QV - C2 (Q to V)	10.1	QVcontrol-PV	
PV-QV - Iqmin (-0.24216)	10.1	QVcontrol-PV	
Batt-QV - Iqmin (-0.24216)	96.1	QVcontrol-Batt	
LV_02_DG01 (Close)	96.1	ATS2-2	
Batt-PF - C1 (F to P)	96.1	PFcontrol-Batt	
Batt-PF - Pref (-0.5)	96.1	PFcontrol-Batt	
Islanding_PCC (Close)	160.	PCC	
Islanding_ATS1-1 (Close)	160.1	ATS1-1	
Islanding_ATS2-1 (Close)	160.1	ATS2-1	
Islanding_BRK1 (Close)	160.1	BRK1	
Islanding_BRK2 (Close)	160.1	BRK2	
Islanding_BRK3 (Open)	160.1	BRK3	
Islanding_BRK4 (Open)	160.1	BRK4	
Islanding_BRK5 (Open)	160.1	BRK5	
LV_02_DG01 (Open)	160.1	ATS2-2	
PV-QV - C2 (V to Q)	161.	QVcontrol-PV	
Batt-PF - Pref (0)	161.	PFcontrol-Batt	
Batt-QV - C2 (V to Q)	161.	QVcontrol-Batt	
Batt-PF - Reset Fcontrol	175.	PFcontrol-Batt	

Fig. 3.14 List of simulation events defined in this work

illustrates the power output characteristic of 20 kW PV system and the daily aggregated demand which the minimum and maximum demands are approximately 61 kW (at 3.45 a.m.) and 127 kW (at 14.30 p.m.), correspondingly.

The PV and battery systems can adjust the power factor in a range of 0.9 (leading) to 0.9 (lagging), while they are operating at unity power during the grid-connected condition. Before the islanding is detected, the PV system is operating in *PQ* mode, which active and reactive powers are controlled to maintain the maximum power extraction and constant power factor, respectively, whereas the battery system is assigned as the master control devices and working in *FQ* mode to keep the system frequency and reactive power output constant.

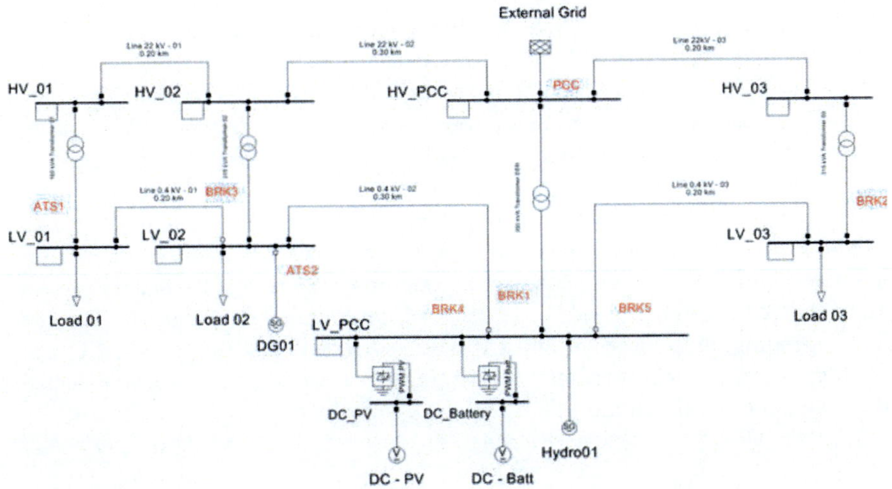

Fig. 3.15 Test system

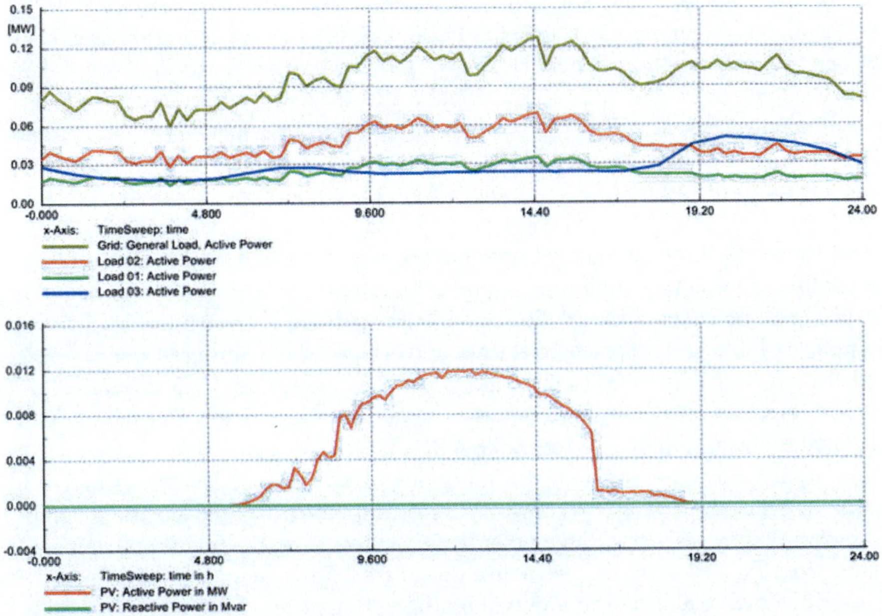

Fig. 3.16 Total load demand and PV generation

The MGCC will detect the islanding condition, and the controllable switches/breakers can be connected or disconnected by the decision of MGCC to keep the stability of microgrid in the islanding mode. Furthermore, MGCC is able to adjust the controller functions and power dispatches of battery and PV systems, for managing the islanded microgrid system to meet the grid code requirements.

The critical situation for operating this microgrid system occurs in the low water season which the power supplied from hydrogeneration is quite low and requires a lot of support from PV and battery systems. Hence, the power output of hydrogenerator is assumed as 40 kW constant. At the initial condition, assume that the battery system is fully charged (SOC = 90%). The battery will be operated in charge or discharge mode depending on the level of SOC. If SOC is lower than 10%, the MGCC will suddenly turn on the diesel generator to charging the battery and supplying connected loads in the microgrid system.

3.4.2 *Transient Performance of Islanded Microgrid System*

Assume that the islanding is occurred at 14.30 p.m. which the microgrid system has the peak load consumption, at about 127 kW. At this time, the PV is supplying 10 kW and hydro remains supplying at 40 kW. Meanwhile, the battery is required to support the active power about 70 kW for the power balancing mechanism in the system. In addition, the load demand in the islanded microgrid system is slightly decreased due to the drops of system frequency and voltage (see more detail about the dynamic part of load in [12]). The islanded microgrid system also requests battery, PV and hydrosystems to inject the reactive power at about 61.5 kVar to maintain the voltage level within the statutory limit.

The simulation events are applied into the test system, as shown in Fig. 3.14. In addition, the main grid is separated at time = 10 s and then resuming at time = 160 s. Furthermore, it also assumes as the battery's SOC is lower than 10% at time = 96 s and then it requires the charging support from the diesel generator. The sequence of simulated events is summarised in Table 3.2.

The results from RMS simulation in Fig. 3.17 show that the battery system will support active and reactive powers with the fast response when the islanding condition is detected by MGCC. On the other hand, the hydro system and PV system will change from *PQ* to *PV* mode. The active power from PV system is injected into the microgrid system according to a maximum power extracted from solar irradiance, while the supporting reactive power is limited by (3.1). It is found that the coordinated controllers among battery, PV and small hydro systems can maintain system frequency and voltage level in islanding microgrid effectively.

At time = 96 s, the SOC is lower than 10% which the battery needs to be charged. Hence, the diesel generator starts to support active and reactive power into the microgrid system for both supplying load (~70 kW) and charging the battery (~50 kW). It is found that the diesel generator can support this battery's transition, from discharging to charging condition, successfully.

Table 3.2 Sequence of simulation events

Time (s)	Events
10	Switch PCC is opened
10.1	MGCC orders to change the status of selective switches/breakers
	MGCC orders PV system to switch from *PQ* mode to *PV* mode [$C_2 = 0$] • PV system keeps supplying at 10 kW • *Q* support is limited as 4.84 kVar [*iq_min* = −0.24216 p.u.]
	MGCC orders battery system to switch from *FQ* mode to *FV* mode [$C_2 = 0$] • Associated power from the battery is about 70 kW • *Q* support is limited as 32.93 kVar [*iq_min* = −0.3293 p.u.]
96	The SOC of battery is lower than 10%
96.1	MGCC turns on the diesel generator [switch ATS2 is closed]
	MGCC orders battery system to switch from *FV* mode to *PV* mode [$C_1 = 1$] • Defined charged power is 50 kW [*Pref* = 0.5 p.u.] • *Q* support is limited as 24.216 kVar [*iq_min* = −0.24216 p.u.]
160	Switch PCC is reconnected
160.1	MGCC orders to change the status of selective switches/breakers
	MGCC disconnects the diesel generator [switch ATS2 is opened]
161	MGCC orders PV system to switch from *PV* mode to *PQ* mode [$C_2 = 1$]
	MGCC orders battery system to switch from *PV* mode to *PQ* mode [$C_2 = 1$] • The charging process is stopped [*Pref* = 0]
175	MGCC orders battery system to switch from *PQ* mode to *FQ* mode [$C_1 = 0$]

Note that the base VA of PV system is 20 kVA, while the base VA of battery system is 100 kVA

3.5 Conclusions

The dynamic models of grid-tied inverter and synchronous generator, used in the microgrid system, are presented in this chapter. The dynamic modelling is based on the DSL aiming to examine the control responses of PV and battery systems, small hydropower plant and diesel generator during the islanding operation. The dynamic performances of those energy sources are demonstrated in time domain by using RMS transient simulations. In addition, the case scenarios in the example project are created by applying simulation events. The simulation results show that the proposed dynamic models can deal with the deviations of frequency and voltage, during islanding transition effectively. The grid-tied inverters in PV and battery system can provide the fast voltage control by supplying the reactive power into the microgrid system.

The grid-tied inverter in battery system will act as the master device by providing the reference voltage and frequency with the fast response. It is used as a frequency responsive device to maintain the stability of the islanded microgrid system. The surpassing energy can be absorbed by charging the battery until the SOC reaches the upper limit, while the battery will stop to dispatch further energy if the SOC

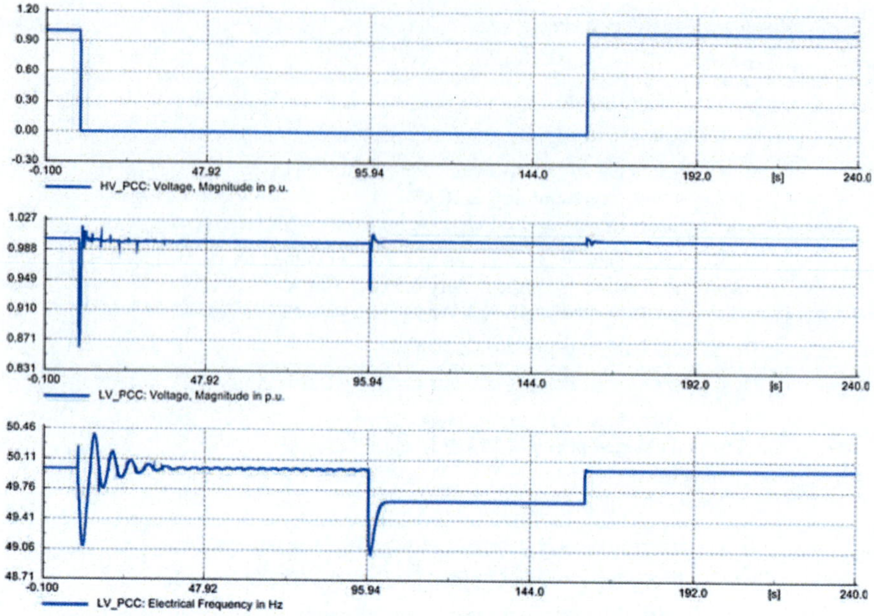

(a) Voltage and frequency at the PCC bus

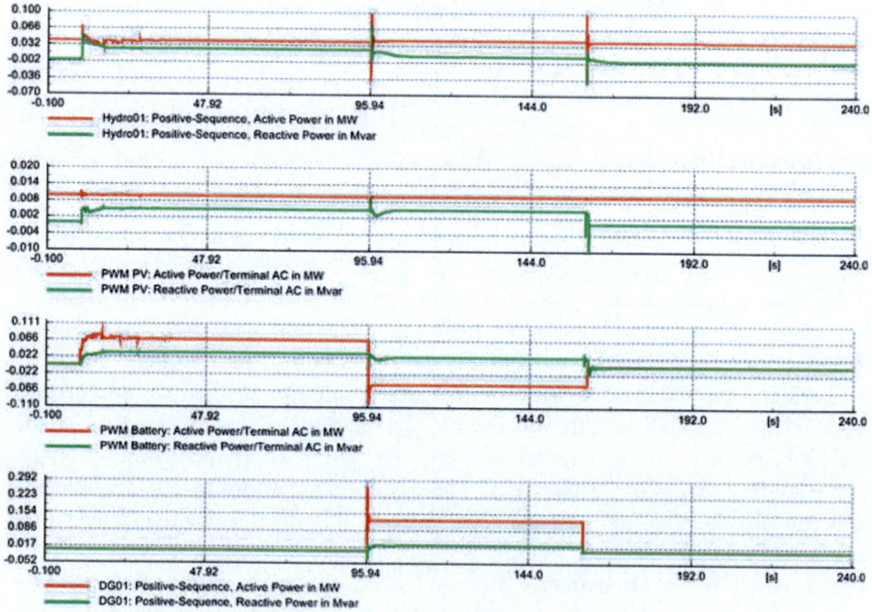

(b) P and Q of hydro, PV, battery and diesel systems (from top to bottom)

Fig. 3.17 Results of RMS transient simulation during islanding operation is too low. Meanwhile, the standby diesel generator is required to assist the battery charging and to prevent the interruption during the islanding operation.

Appendix

The development of test system with dynamic modelling for RMS transient simulation is explained, step by step, in this section. The process of adding dynamic models, as created in Sect. 3.3, into the test system is described. Furthermore, the procedures for creating simulation events and plots are presented.

Step 1: Creating Composite Model and Common Models

The composite frame and block/model definitions of PWM converter are built in *Project Library* → *User Defined Models.* Before creating composite model, the voltage, power and frequency measurement devices should be built first. All dynamic models and measurement devices can be created within a data manager.

Voltage Measurement

- At the data manager → pressing "*New Object* " → *Elements\Other\Station Element (Sta*)\Voltage measurement (StaVmea).*
- Defining "*Name*" and "*Measurement Point*" such as at PCC bus.
- Enable *Rating of connected Terminal* in "*Voltage Rating*".

Power Measurement

- At the data manager → pressing "*New Object* " → *Elements\Other\Station Element (Sta*)\ PQ Measurement (StaPqmea).*
- Defining "*Name*" and "*Measurement Point*" such as at the AC side of PWM converter.
- Enable *Rating of connected Terminal* in "*Power Rating*".
- Enable *Generator Oriented* in "*Orientation*".

Frequency Measurement

- At the data manager → pressing "*New Object* " → *Elements\Other\ Net Elements* (*Elm**)\ *Phase Measurement Device PLL-Type* (*ElmPhi__pll*).
- Defining "*Name*" and "*Measurement Point*" such as at the PCC bus, and then select the model version.
- Defining the controller gains in "*RMS simulation*" menu.

Common Model

- At the data manager → pressing "*New Object* " → *Elements\Common Model.*
- Defining "*Name*" and then assigning the model definition in "*Model Definition* ".
- Assigning all parameters related to the controller (see Fig. 3.11).

Composite Model

- At the data manager → pressing "*New Object* " → *Elements\Composite Model.*
- Defining "*Name*" and assigning the composite frame into "*Frame* ".
- At the data manager, the relevant common models and measurement devices should be put inside the composite model (see Fig. 3.12).
- Assigning those relevant network element, common models and measurement devices into slots in the composite model (see Fig. 3.12).

Step 2: Creating the Simulation Events

The event objects can be accessed from the Simulation RMS/EMT toolbar, by pressing the *Edit Simulation Events* icon. Then, a list of the currently defined events will be displayed including the set of simulation time, when the event will happen, and the related object.

Two event objects are employed in this work, which are switch event (*EvtSwitch*) and parameter event (*EvtParam*). These two types of events can be created as follows.

Switch Event

- At Simulation RMS/EMT toolbar, pressing the *Edit Simulation Events* icon.
- Pressing "*New Object* " → *Elements\ Switch Event (EvtSwitch).*

- Setting the execution time.
- Adding the chosen switch or breaker in "*Breaker or Element* ".
- Defining the action of switch/breaker that to be opened or closed.

Parameter Event

- At the Simulation RMS/EMT toolbar, pressing the *Edit Simulation Events* icon.
- Pressing "*New Object* " → *Elements\ Parameter Event (EvtParam)*.
- Setting the execution time.
- Adding the network element or DSL common model in "*Element* ".
- Inserting the name and new value of the element parameter in "*Name of variable*" and "*New value*", respectively.

Step 3: Creating the Simulation Plots

Time-dependent variables which can be monitored within dynamic simulation fall into the following categories: current/voltage/power, bus results, signals (i.e. input signals, output signals, state variables and derivatives of state variables), calculation parameter, element parameter and reference parameter.

An easier way to add the variables of different elements into the result object for RMS simulation is done by right-click on the chosen element, and then

- Selecting *Define* → *Results for Simulation RMS/EMT*.
- Double-click on the chosen element in the list of result objects. Hence, a variable selection window is automatically opened.
- The variables of interest to be recorded can be selected.

To edit or inspect the result object, it can be done by pressing the *Edit Results Variables* icon on the Simulation RMS/EMT toolbar. This will enable the user to edit the contents of the currently selected result object.

After defining the result objects, the results can be visualised in a plot by using the *Insert Plot* icon from the main menu ( ), which will open the insert plot dialog. Therefore,

- Selecting the plot design. In this work, the curve plots are chosen for demonstrating the results from the RMS simulation.
- Adding the chosen elements and variables to plots. After that, the plot page is automatically created.

Step 4: Executing the RMS Simulation

Before running the RMS simulation, the initial conditions of test system must be calculated. In addition, all state variable of controllers and dynamic models, and any other device which is active and will affect the time-domain simulation are also calculated.

Calculation of Initial Conditions

- At the Simulation RMS/EMT toolbar, pressing the *Calculate Initial Conditions* icon.
- *Simulation method → RMS value (electromechanical transients).*
- *Network representation → Balanced, positive sequence.*
- Then, pressing *"Execute"*.

After the initial conditions are calculated successfully, the *Run Simulation* icon on the Simulation RMS/EMT toolbar will be activated.

Executing the RMS Simulation

- Pressing the icon.
- Defining the stop time (in seconds).
- Then, pressing *"Execute"* for running the simulation.

References

1. R. Singh, M. Kirar, Transient stability analysis and improvement in microgrid, in *2016 International Conference on Electrical Power and Energy Systems (ICEPES)* (Bhopal, 2016), pp. 239–245
2. S. Li, J. Proano, D. Zhang, Microgrid power flow study in grid-connected and islanding modes under different converter control strategies, in *2012 IEEE Power and Energy Society General Meeting* (San Diego, CA, 2012)
3. F. Katiraei, R. Iravani, N. Hatziargyriou, A. Dimeas, Microgrids management. IEEE Power Energy Mag. **6**(3), 54 (2008)
4. A. Yazdani, R. Iravani, *Voltage-Sourced Converters in Power Systems*, 1st ed. (Wiley, 2010)
5. W. Ali, H. Farooq, A. ur Rehman, M.E. Farrag, Modeling and performance analysis of micro-hydro generation controls considering power system stability, in *2017 First International Conference on Latest trends in Electrical Engineering and Computing Technologies (INTELLECT)* (Karachi, 2017), pp. 1–7

6. A.S. Kini, U.R. Yaragatti, Modelling and simulation of a wind/diesel hybrid power system, in *2006 IEEE International Conference on Industrial Technology*, (Mumbai, 2006), pp. 1670–1675
7. DIgSILENT PowerFactory, *User Manual* (2018)
8. DIgSILENT PowerFactory, *Advance Tutorial: Dynamic Modelling (DSL)* (2015)
9. DIgSILENT PowerFactory, *Technical Reference Documentation: PWM Converter* (2018)
10. TURBINE-GOVERNOR MODELS, *Standard Dynamic Turbine-Governor Systems in NEPLAN Power System Analysis Tool*. NEPLAN AG, p. 91
11. P. Kundur, *Power System Stability and Control* (McGraw-Hill, New York, 1994)
12. DIgSILENT PowerFactory, *Technical Reference Documentation: General Load* (2018)

Chapter 4
Dynamic Modelling and Co-simulation Between MATLAB–Simulink and DIgSILENT PowerFactory of Electric Railway Traction Systems

Luis Chiza, Jaime Cepeda, Jonathan Riofrio, Santiago Chamba, and Marcelo Pozo

Abstract The inclusion of electric transportation systems generates more complex interactions among multiple grid components. The action of these elements actively affects the state of power distribution grids and their complexity for operation analyses, which requires proper studies in order to avoid potentially bad situations. A feasible answer to this constraint can be the usage of digital co-simulation that is a well-developed technique for the performance assessment of power systems. Therefore, this chapter presents a co-simulation tool for assessing the impact of electric railway traction systems into the grid. The co-simulation tool is applied to mass electric mobility interacting with an electric power system by means of OLE for process control (OPC), which allows controlling and supervising the communication between DIgSILENT PowerFactory and MATLAB–Simulink. DIgSILENT PowerFactory is used for railway and utility power systems simulation; meanwhile,

Electronic supplementary material The online version of this chapter (https://doi.org/10.1007/978-3-030-54124-8_4) contains supplementary material, which is available to authorised users.

L. Chiza (✉) · S. Chamba
CELEC EP Coca Codo Sinclair, Quito, Ecuador
e-mail: luis.chizas@gmail.com

S. Chamba
e-mail: santiagochamba@gmail.com

J. Cepeda
Operador Nacional de Electricidad CENACE, Av. Atacazo and Panamericana Sur km 0, Quito, Ecuador
e-mail: cepedajaime@ieee.org

J. Riofrio
Budapest University of Technology and Economics (BME), Budapest, Hungary
e-mail: ariofriotrujillo@edu.bme.hu

M. Pozo
Escuela Politécnica Nacional EPN, Ladrón de Guevara E11-253, Quito, Ecuador
e-mail: marcelo.pozo@epn.edu.ec

F. M. Gonzalez-Longatt and J. L. Rueda Torres (eds.), *Modelling and Simulation of Power Electronic Converter Dominated Power Systems in PowerFactory*, Power Systems,
https://doi.org/10.1007/978-3-030-54124-8_4

MATLAB–Simulink simulates electrical drives as well as control and operation of asynchronous machines (i.e. the power electronic converters). The combination of both computer programs through OPC Simulation Server sets a powerful platform up to test complex control systems applied in traction systems of electric trains.

Keywords Co-simulation · Control system · DIgSILENT PowerFactory · FOC · MATLAB–Simulink · MatrikonOPC Server · Subway systems · Traction system

4.1 Introduction

Nowadays, due to the global market, the continuous and generalised use of private vehicles has conducted to big cities to face significant environmental and traffic congestion problems. As a result, electric massive transportation systems have assumed a vital role in the transition to a low-carbon society process because of their high efficiency and low environmental impact [1, 2].

Electrical traction systems design is a complex task because they are composed of multiple complex subsystems (e.g. power supply, power electronics, traction drives, train mechanic, among others), all of which have to interact during operation stages. Moreover, these subsystems are designed separately instead of as a whole, so there is a real need for system operators and utility companies to understand how they could affect the entire grid [3]. Consequently, railway system modelling appears as a feasible way to assess its performance and possible operational risks, which could lead to not fulfil quality and safety requirements [4].

Train movement on a defined pathway is the foremost feature during railway traction systems modelling stage. Actually, this motion constantly produces variations of physical couple points along the route. Also, trains' load tends to change regularly, and it depends on (1) operation conditions (acceleration, constant speed, braking), (2) railway track features (slope and cant of a railway) and (3) the total weight (passenger number and operation timetable) [5]. Hence, the majority of these characteristics are managed by control systems of electric trains.

Control systems applied in railway can be broadly partitioned into two categories. One is concerned with the vehicle dynamics control, which deals with issues such as vehicle stability, vehicle running behaviour and passenger ride comfort. The other is in charge of controlling braking and tracking during railway vehicles operation [6].

Railway vehicles development and vehicle dynamics control have progressively brought greater comfort and convenience for passengers and technical–economic benefits to the railway companies [7]. For these reasons, not only physical but also electrical features have to be included at the modelling stage of traction equipment and electronic control units.

In this sense, previous studies were oriented to the design of railway control systems. For instance, a speed control model for an electric vehicle driving motor using MATLAB–Simulink software has built-in [8]. This research intends to verify the suitable strategy control option between fuzzy PID and conventional PID controllers. Also, using MATLAB–Simulink simulations of control systems, the double closed-loop speed regulation system implemented in DC motors is analysed in [9]. However, this research is not focused on the possible effects of electrical grids because of that electric mobility systems commissioning.

Moreover, a vector control drive system to determine traction load, drive, and train movement for British Rail Class 350/1 in the UK is proposed in [10]. Nevertheless, the simulations are performed considering only a short section of railway track (between two passenger stations), which limits its application scope. At the same time, a new traction supply system based on single-phase transformer and active power controller is modelled in [11]. Its improved topology helps to minimise power quality issues; nonetheless, this method has not yet been tested in a more complex system. Besides that, [12] and [13] present modelling methodologies to deal with problems in power quality and AC feeding systems, respectively. Nonetheless, these models only include one type of technology for inverters and analyse power quality on connection points instead of the whole system. On the other hand, a new methodology for probabilistic assessment of underground railway systems impact over distribution grids has been presented in [5]; however, this proposal is focused on steady-state analysis.

In order to develop new automation and control systems and analyse their impact over electrical grids operation, appropriate methods and tools are necessary. In this context, recent research has started adopting coordinated simulations or better known as co-simulation environment, which covers electrical grids and control of power systems, including supervisory tasks and communication issues. The main co-simulation approach is characterised by using a complete power system modelling and simulation environment where detailed control systems are implemented in a dedicated tool. For example, the open-source co-simulation framework has been successfully used to analyse coupled transmission and distribution systems in [14].

Likewise, another co-simulation environment, designed to exchange information between different institutions, is presented in [15]. The environment is able to couple an arbitrary number of instances from DIgSILENT PowerFactory running on different virtual servers, at different rates, each represents a different area. On the other hand, [16] presents the concept of co-simulation as an alternative to model migration. This work compares the pros and cons of two approaches: (1) the model migration to a domain-independent modelling and simulation tool (OpenModelica) and (2) the co-simulation between instances of DIgSILENT PowerFactory.

Furthermore, among other researchers, in [17], a simple TCP/IP interface between DIgSILENT PowerFactory and MATLAB–Simulink is proposed. It can be used for development and validation of smart grid automation concepts. In the same way, a co-simulation framework developed to test optimal control methods for root mean square (RMS) simulations on DIgSILENT PowerFactory is presented in [18]. To do this, the *digexfun* interface is applied where DIgSILENT PowerFactory can send and receive data from other mathematical software APIs such as MATLAB.

Summarising, most of the above-mentioned studies focus their analyses on either power systems or simulation/emulation of automation and control systems. In some researches, coupling different programs in a co-simulation environment has already been covered for power systems and smart grids. However, the integration of railway control system, supervisory tasks and effects over distribution grids, which are also very important for future power systems, has not been discussed in detail so far.

In response to this lack of studies, this chapter presents the usage of OLE for process control (OPC, which is a well-used communication standard for controlling and supervising industrial processes) for communicating DIgSILENT PowerFactory and MATLAB–Simulink, based on the fact that OPC enables data exchange among devices or software from different manufacturers. Indeed, MatrikonOPC Server was chosen for the developed practical applications. In this sense, taking advantage of OPC, a proper interaction between DIgSILENT PowerFactory and MATLAB environments is reached during simulations in every timestamp [19].

Additionally, this research presents a simulation tool that is applied to assess the impact of mass electric mobility systems in which traction motors operation could generate significant disturbances on electric grids. For this purpose, DIgSILENT PowerFactory is used for railway and utility power systems simulation. Meanwhile, MATLAB–Simulink simulates electrical drives as well as control and operation of asynchronous machines (i.e. the power electronic converters). In the same way, MatrikonOPC Server is responsible for communicating DIgSILENT PowerFactory and MATLAB–Simulink using their settings for data exchange within the OPC client/OPC server architecture.

4.2 Co-simulation Between DIgSILENT PowerFactory and MATLAB–Simulink

DIgSILENT PowerFactory offers useful functionalities for modelling, simulation and analysis of power systems, especially for smart grids or new tendencies in power systems [20]. Among these functionalities, communication flexibility and interfaces for co-simulation are the most valuable in the development of this research. In this chapter, DIgSILENT PowerFactory was coupled with MATLAB–Simulink for continuous interaction as a single model by means of OLE for process control (OPC) [20,21]. Next, an explanation for MATLAB and OPC interfaces provided by DIgSILENT PowerFactory is presented, and both definitions are illustrated in Fig. 4.1.

OPC server is an industry-standard interface for asynchronous communication and data exchange in process interaction. It connects client processes with applications that run in real time such as automation and control components, or electronic drives. That means, if any data value varies at OPC server, this value will be propagated to OPC clients (DIgSILENT PowerFactory) from an external application (MATLAB–Simulink).

DIgSILENT PowerFactory represents a client OPC with which external applications sync and transfer RMS data. For this purpose, a DSL model must be defined in DIgSILENT PowerFactory programming environment in which global variables

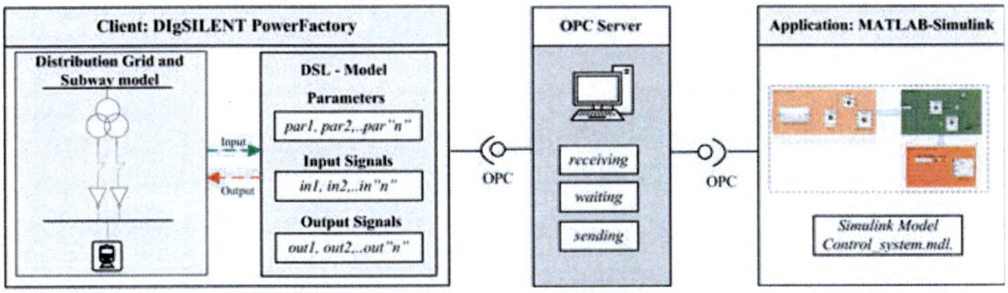

Fig. 4.1 Co-simulation interfaces of DIgSILENT PowerFactory with MATLAB–Simulink and OPC

are declared. In the same way, this DSL receives the output parameters generated from MATLAB, which are usually parameters (timestamps, state variables), and input/output signals.

MATLAB–Simulink represents an external application that has a direct interface with DIgSILENT PowerFactory, which lets to execute specific models or different instances of algorithms programmed in MATLAB code (m-file).

4.3 Proposed Co-simulation Environment Interaction

In this section, a comprehensive overview of the proposed co-simulation environment is provided. Additionally, the different software packages used for specific simulations and railway power system modelling are also described. Figure 4.2 summarises how every software interacts with each other and its main functions.

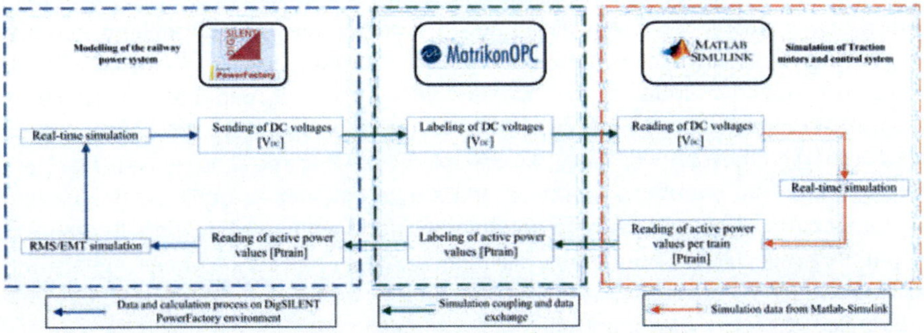

Fig. 4.2 Proposed co-simulation environment interaction

4.3.1 OLE for Process Control: MatrikonOPC

MatrikonOPC is a tool developed by OPC Foundation, and its main goal is to make easier industrial communications as well as equipment integration and management [22]. Specifically, MatrikonOPC Simulation Server was chosen for this study because it is free of the payment application and lets to create a simulation test environment in which data can be exchanged with multiple clients, in the current case: DIgSILENT PowerFactory and MATLAB–Simulink. Moreover, the fundamental objective of MatrikonOPC, in this application, is to label voltage signals from DIgSILENT PowerFactory (electric grid topology) and active power values from MATLAB–Simulink (traction system), which are indispensable information for coupling independent simulations from both environments. In this context, MatrikonOPC makes possible the interaction between DIgSILENT PowerFactory and MATLAB–Simulink, which is based on the RMS/EMT simulation of the railway power system in DIgSILENT PowerFactory and traction control system simulation in MATLAB–Simulink, respectively.

Nevertheless, when OPC is applied in co-simulation applications, a synchronisation mechanism should be used to sync the simulation time step between simulation components. In this case, MatrikonOPC is responsible for managing simulation time in both environments, and each software fits its own simulation time stamp to the update time provided by MatrikonOPC that is a crucial feature in co-simulation applications [23]. Update frequency is responsible for it in RMS/EMT simulation mode, and it depends on the real-time PC-clock synchronisation.

4.3.2 DIgSILENT PowerFactory: Subway System Model

DIgSILENT PowerFactory is used to model the railway power system topology and to analyse power disturbances on distribution systems. Here, a distribution grid, as it is better explained in [5], must be modelled besides including traction stations, catenary systems, trains, and coupling points between utility company and mass transportation system (in this chapter, the IEEE 14-bus test system is used for representing the grid). Similarly, DIgSILENT PowerFactory allows reading and sending information regarding electrical parameters while uses its own calculation tools. In this case, power flow calculation and RMS/EMT simulation tools are used as trigger components in the proposed co-simulation environment. Next, real-time voltage signals in direct current (V_{DC}) are sent to MATLAB–Simulink throughout MatrikonOPC.

Taking advantage of the modelling tools available in DIgSILENT PowerFactory, a power grid and an electric subway system were implemented. Specifically, it is based on the 14-bus system model of the PowerFactory example library. It is composed of one interconnection point between a utility company's grid (Bus_0014) and a subway system at 33 AC kV, one traction station which converts 33 AC kV to 1.5 DC kV and two electric trains represented by general loads, as shown in Fig. 4.3. This

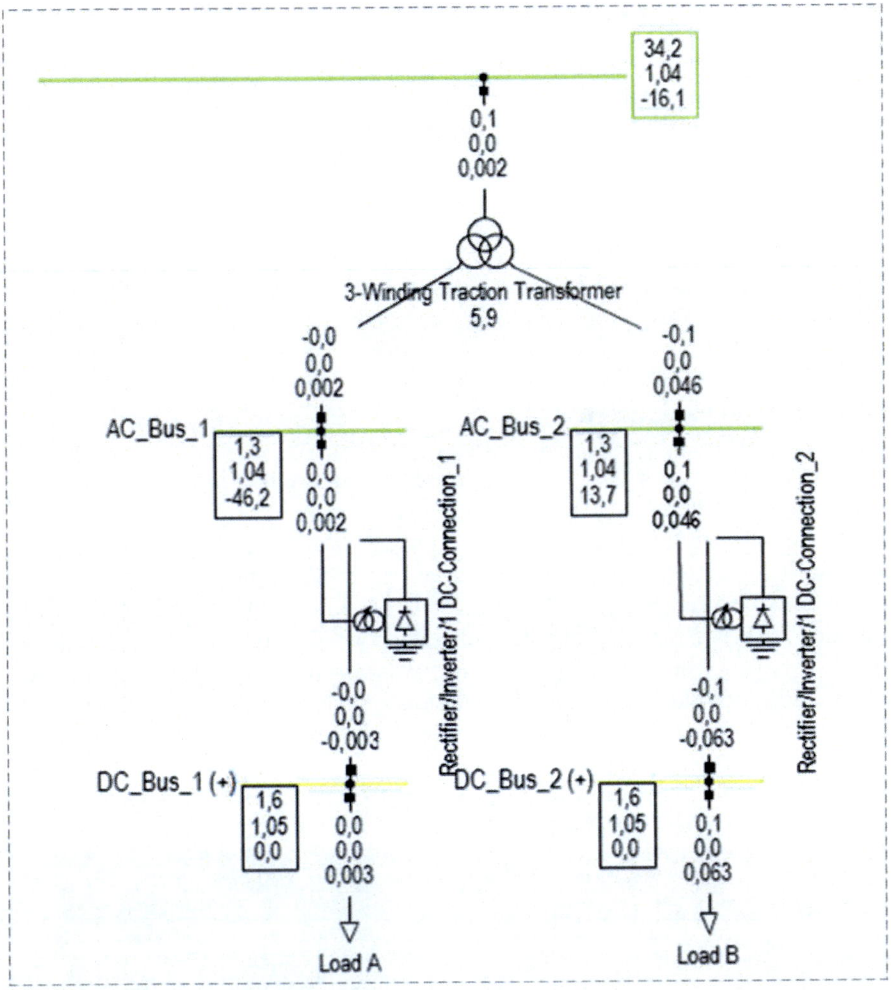

Fig. 4.3 Traction substation scheme

test power system is used in order to identify the possible dynamic impacts caused by the penetration of massive electric transportation systems.

The trains operate at 1.5 kV DC voltage, and they are provided with traction motors of 92 kW power rate. A more detailed explanation of the power systems' features appears in [5,24].

It is important to highlight that electric trains were modelled as general load (*ElmLod*) in DIgSILENT PowerFactory due to the fact that this element lets to receive external signals from MATLAB–Simulink. However, converters, traction motors and electrical drives were modelled with detail in Simulink.

4.3.3 *MATLAB–Simulink Model*

MATLAB–Simulink is used for modelling and executing the control system for traction motors and electrical drives (i.e. the power electronic converters), as shown in Fig. 4.4, so that the train load can be replicated. For this purpose, the option for sending and reading signals must be active not only in MATLAB but also in MatrikonOPC. Next, active power values (P_{train}) are sent to DIgSILENT PowerFactory by means of external measurement devices. It is important to highlight that data writing process between MATLAB–Simulink and MatrikonOPC is carried out in asynchronous mode since processing times are higher in the simulation environment.

4.3.3.1 Three-Phase Inverter

A three-line diagram of a bi-level inverter used for controlling induction motors is shown in Fig. 4.5. This control is also known as *voltage source inverter* (VSI) because it is able to regulate the output voltage according to operating conditions [25]. It comprises indispensable equipment for controlling and operating induction motors. Its main function is to transform DC voltage in the catenary system to AC voltage that is supplied to traction motors. This AC voltage has a variable frequency and amplitude [26, 32, 33]. This feature lets the inverter to work properly and efficiently against load variations of induction motors. However, applying some control strategies and modulation techniques is required in this sense. As a result, low noise levels and load losses can be reached as well.

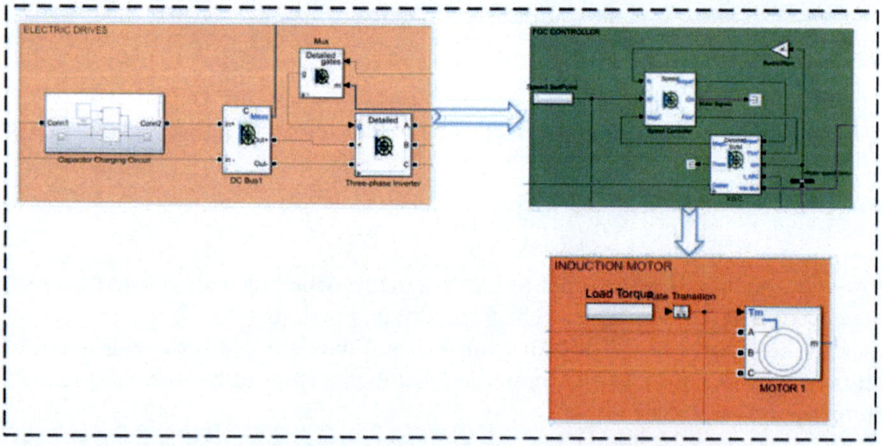

Fig. 4.4 Electrical drives of the railway traction system in MATLAB–Simulink

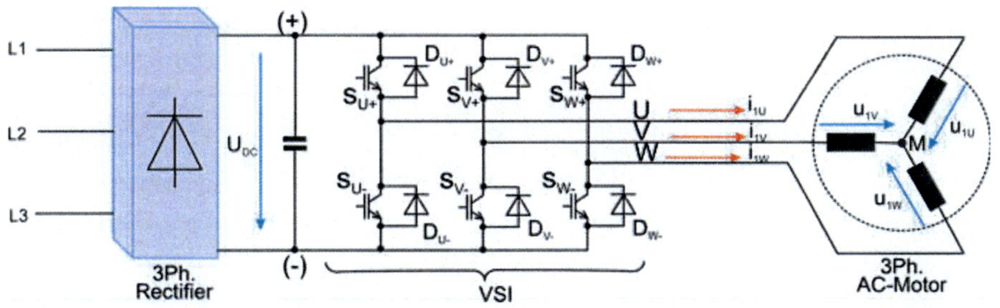

Fig. 4.5 Three-line diagram of a bi-level inverter [32]

4.3.3.2 Modulation

The modulation in electric drives is needed since it constitutes the interface between the power side and the digital control side in an electrical control drive. Although the modulator is programmed in a digital embedded system, it works as an interface from the control output to the gate drive signals of the power inverter, as shown in Fig. 4.6 [32,33].

There are a lot of modulation techniques, but the most used are: the sinusoidal pulse width modulation (SPWM) and the space vector modulation (SVM). The SPWM technique (whose block diagram is depicted in Fig. 4.7) compares three sinusoidal

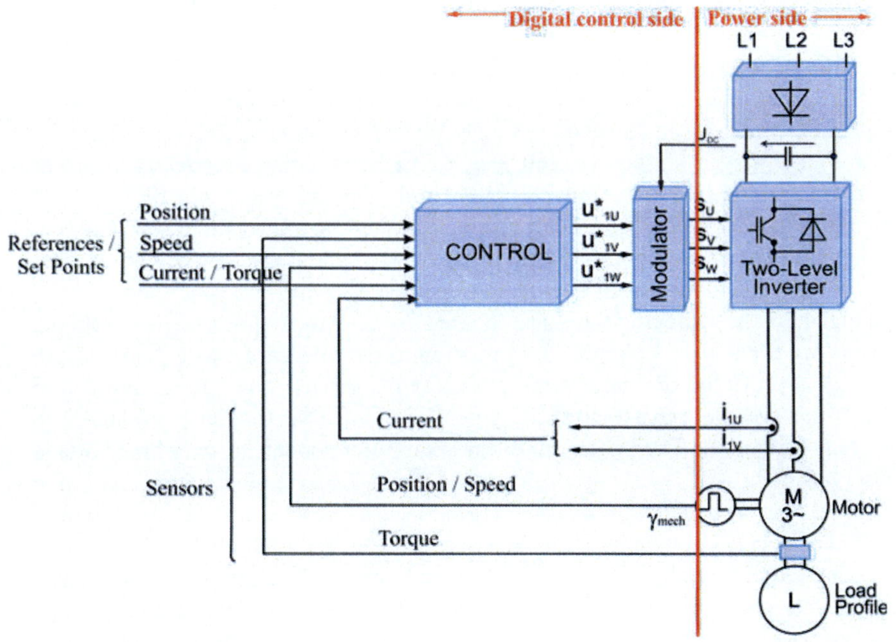

Fig. 4.6 Electrical control drive [32]

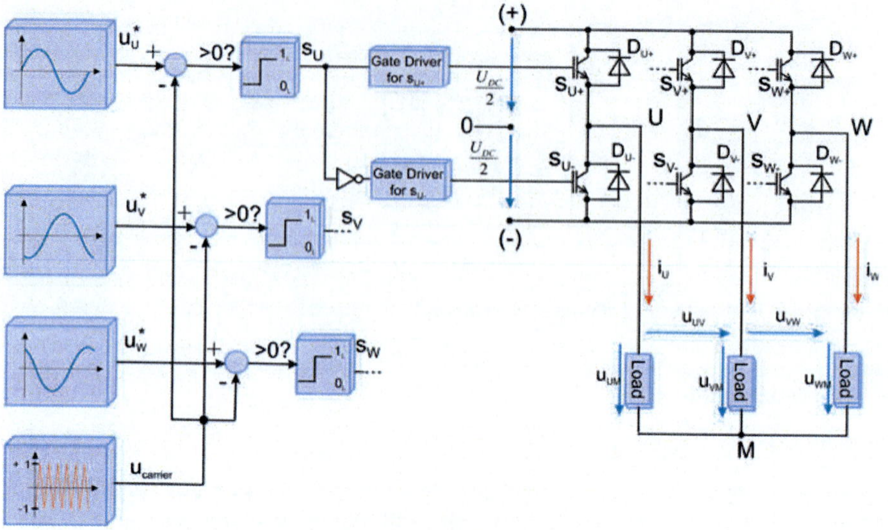

Fig. 4.7 Sinusoidal pulse width modulation-SPWM scheme

signals (which come from the control) and a triangular signal with high frequency (about 5–10 kHz) in order to generate the high-side firing pulses of the inverter. The low-side firing pulses of the inverter, instead, are the negation of the high side in order to prevent short circuits in each leg. Figures 4.8 and 4.9 show the results of the comparison, and the amplified window in one period, respectively, of the triangle signal and the three sinusoidal signals [33].

The SVM technique is based on space vector diagrams of inverters, as in Fig. 4.10. Its operating principle is based on the eight possible switching states presented by the inverter. Using a voltage reference $\underline{u}_s$ of each switching inverter position, as shown in Fig. 4.10, sequences are computed and modulated, and thus the length of commutation states is too [27]. Each vector is plotted on a hexagonal star diagram, based on the two-axis stationary system (α,β). With this strategy, the switching losses of the power semiconductor in the inverter become minimal [31, 32].

In Fig. 4.10, the switching states are detailed in one of the regions of the hexagonal diagram, specifically in the region I, in relation to the reference axis α. The voltage vector $\underline{u}_s^*$ is projected on the switching vectors $\underline{u}_1$ and $\underline{u}_2$. The voltage projection presents the average value that each switching voltage vector must produce in a given interval of time. The behaviour of the vector components of switching voltage is observed in Fig. 4.11, where according to the switching states, there is the respective voltage value in each of the supply phases (u_{U0}, u_{V0}, u_{W0}), in addition of its switching periods $\frac{T_n}{2}$ [31].

Among the main advantages of the SVM technique is the characteristic of generating low-harmonic distortion levels in output voltages and currents, that is why this is the most useful technique for power systems, especially for utilities' grids [27]. For this reason, the SVM is implemented in MATLAB–Simulink in this chapter.

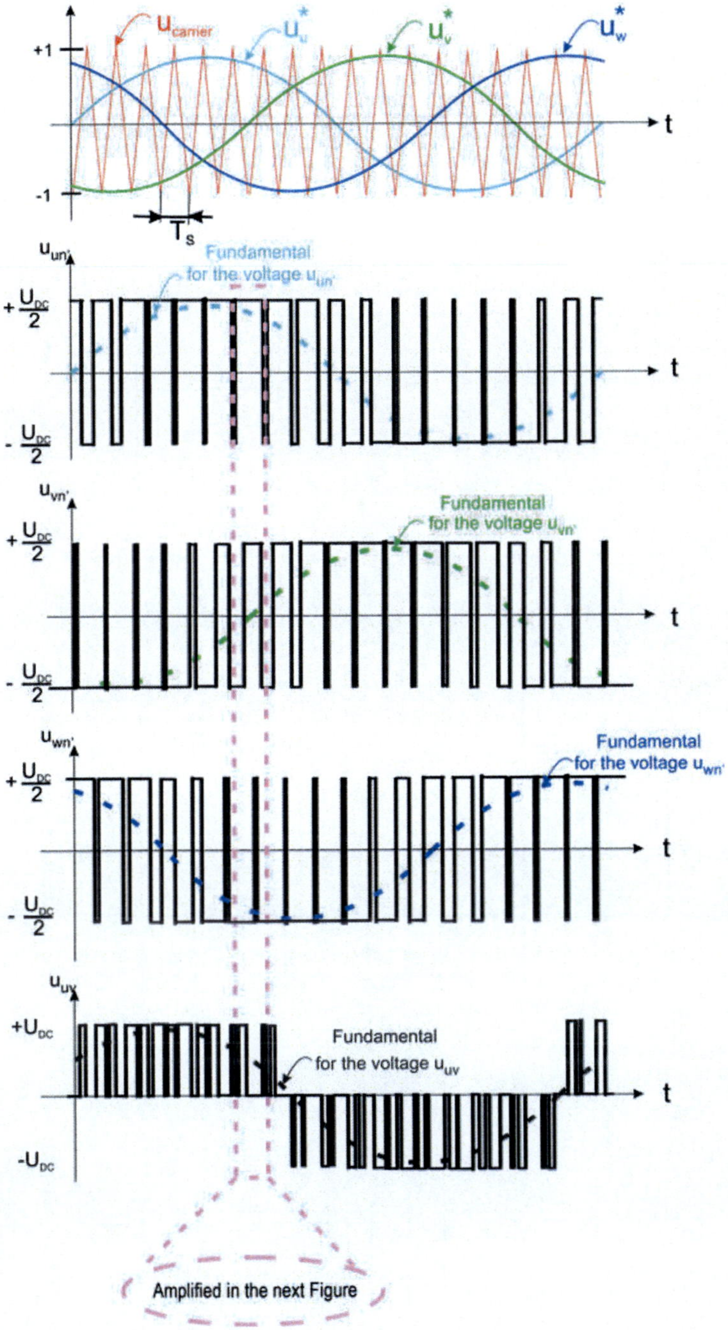

Fig. 4.8 Space vector modulation. Waveforms [33]

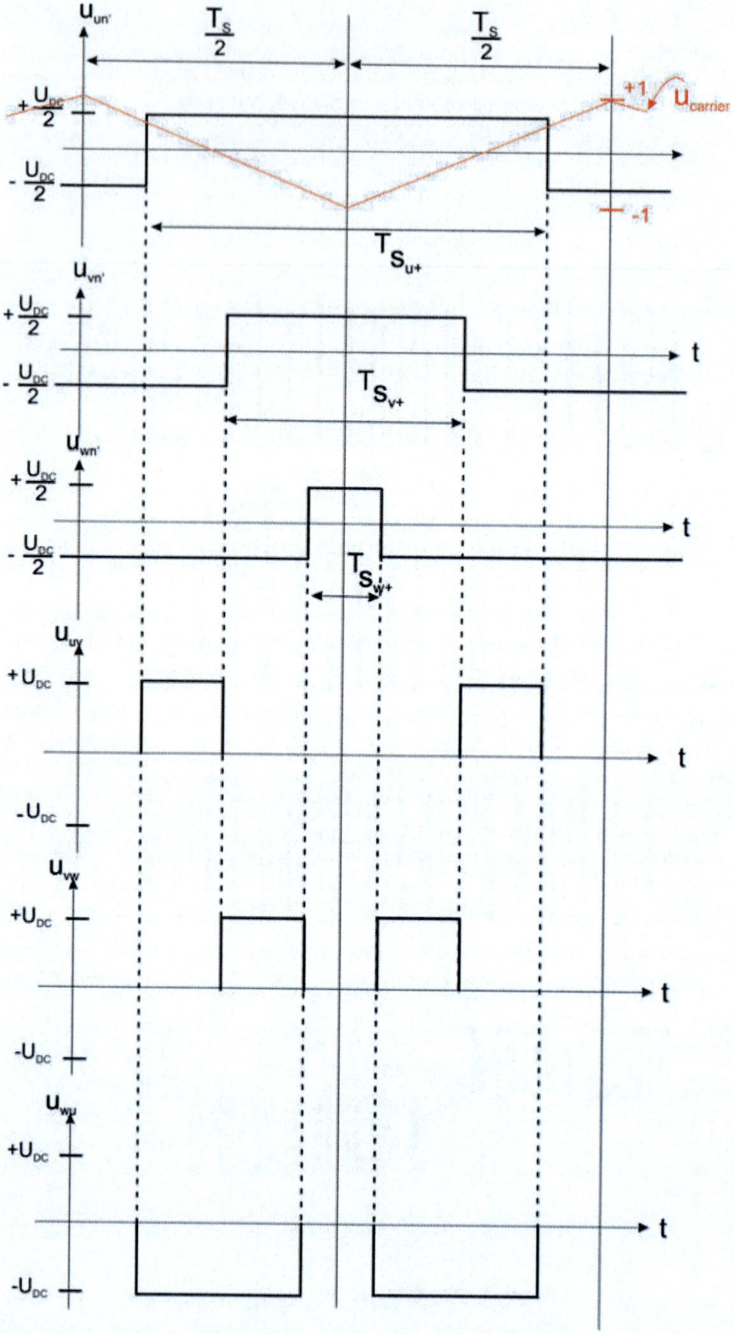

Fig. 4.9 Space vector modulation. Waveforms magnification in an instant of time [33]

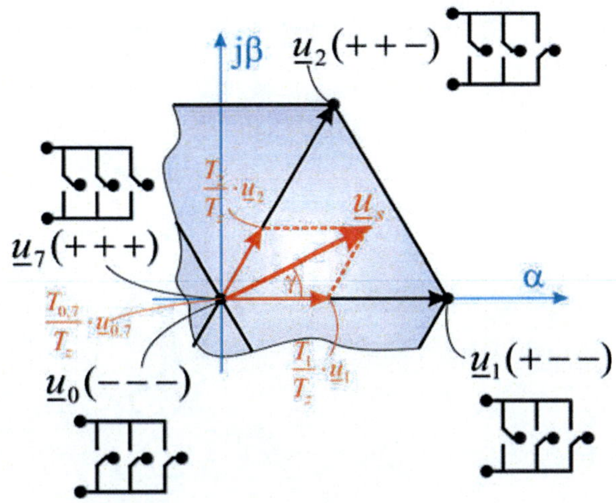

Fig. 4.10 Space vector modulation gate switching description [33]

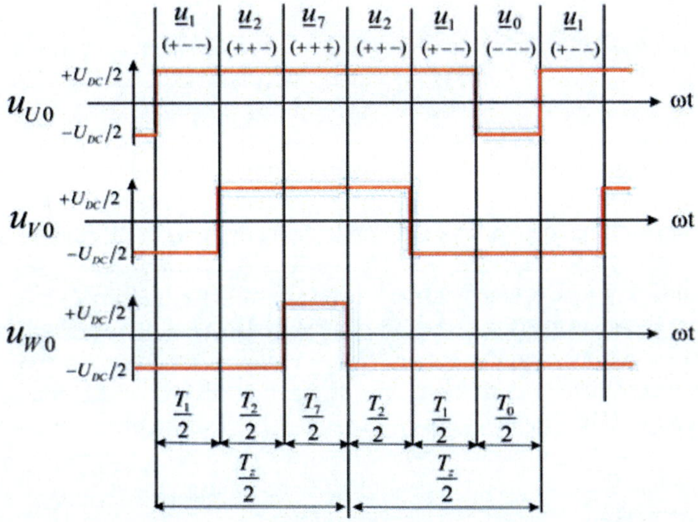

Fig. 4.11 Space vector modulation signals [33]

4.3.3.3 Braking Chopper

At brake functioning, a current flow recirculation appears that has an opposite direction than in normal operation. This current generates large quantities of energy, which can affect nearby equipment. Thus, protective devices or mechanisms should be included in this stage to minimise or dispel the excess of energy. The working principle of these mechanisms is to convert the kinetic energy at braking into a form

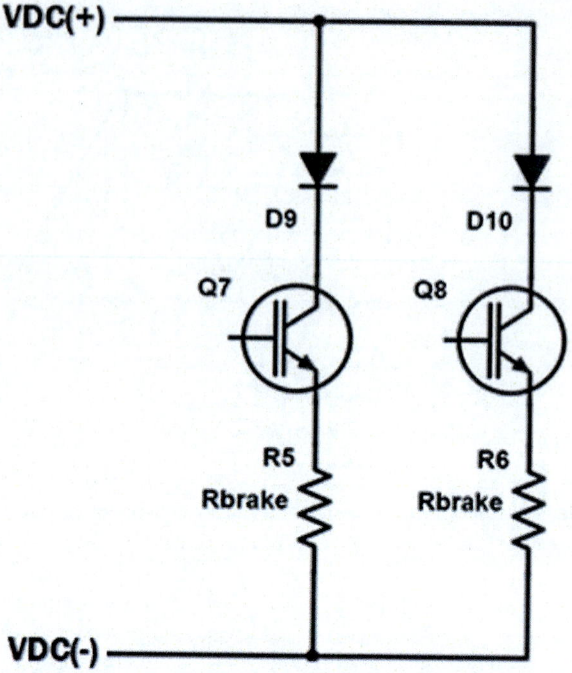

Fig. 4.12 Scheme of commutation circuit for braking chopper [32]

which can be either used immediately or stored in some cases (regenerative braking). In the case of subway systems, this energy can be directly injected to the supply grid. However, it depends on loadability levels of catenary systems or the impossibility of working in any of the four quadrants in the previous rectification stage [33].

To solve this problem, braking chopper is responsible for limiting DC voltage by switching that energy to a resistor where it is converted to heat. Also, braking chopper is also activated in overvoltage cases in DC buses or even inside traction converters. Finally, this mechanism is composed of a dissipation resistor and a commutation circuit ruled by an IGBT diode as appears in Fig. 4.12. The circuit is complemented with a diode connected in parallel, which controls recirculating currents. What is more, IGBTs has thermal protection through temperature probes located in their module, while braking resistors use natural cooling.

4.3.3.4 Line Contactor and Preload Circuit

This equipment limits the charge current of the capacitor during its load time. Before the switch is closed, the capacitor is completely discharged. Therefore, an initial current pick occurs, so diodes from the rectifier bridge and capacitor bank are protected by a start circuit or initial load. The main goal of this circuit is that the capacitor starts charging slowly and softly until reaching the rated voltage of DC buses [32].

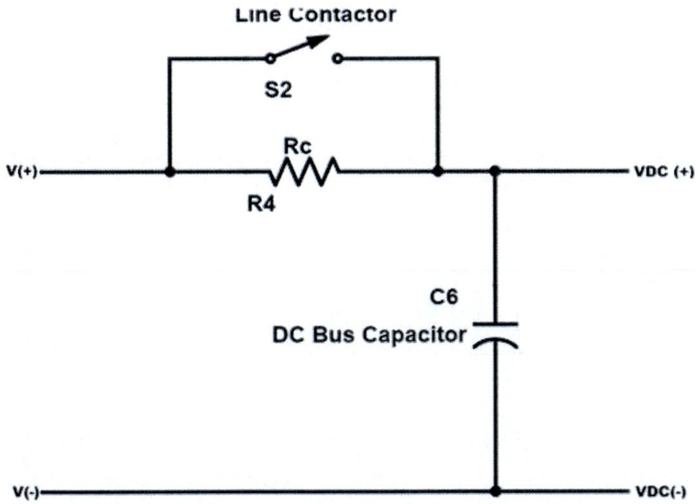

Fig. 4.13 Operation of preload circuit for capacitors [32]

In the case of Fig. 4.13, the capacitor (C6) is charged through the resistor (Rc), so its rate value depends on the initial pick current. Next, at the time of the capacitor charge starts being constant, the switch shorts-circuit R and the DC bus is working.

4.3.3.5 Field-Oriented Control (FOC)

Figure 4.14 presents the control scheme of FOC, which is based on decoupling the components of stator current and relates the electrical torque and magnetic flux. These components are aligned to an orthogonal reference system of*q*-*d* axes in which electrical torque corresponds to coordinate "*q*", and flux is represented by coordinate "*d*" [28, 32, 33]. The three-phase system, variable in time and speed, is transformed into this new two-phase reference system. Subsequently, the operation of a three-phase machine is achieved similarly as the structure of DC machine.

A means of instantaneous control of torque and flux is based on the position of the rotor combined with two-phase currents. Control of both the magnitude and the phase of AC measures (voltage and current) is required. The FOC control allows obtaining results with a good dynamic response to torque variations against a wide speed range, and it can induce a high torque at zero speed [34].

4.4 MatrikonOPC Interface

The configuration of OPC interface in different simulation environments, including prerequisites and the configuration of OPC server and OPC Tags, is deeply explained

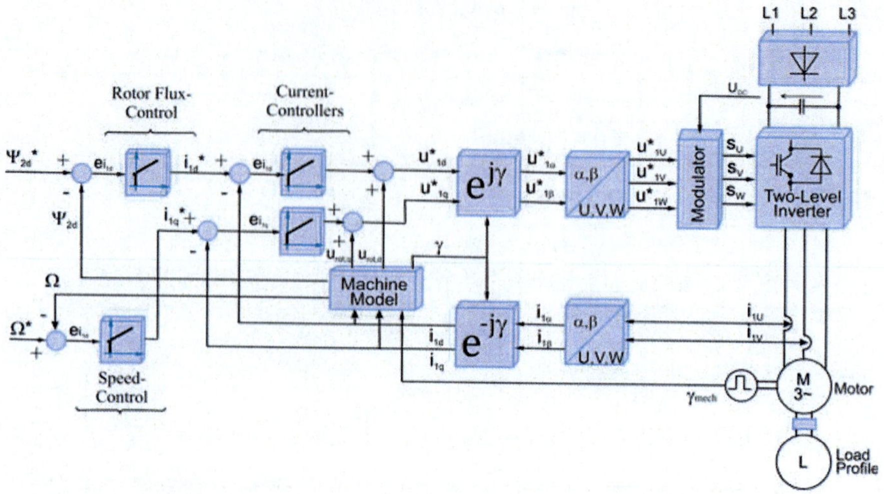

Fig. 4.14 Scheme of field-oriented control [33]

in [30]. This section will focus on the specific aspects required for configuring the MATLAB–Simulink—PowerFactory interface for simulating the railway system presented in this chapter. Interested readers must refer to [30] in order to be capable of complement the knowledge of OPC simulation and configuration based on PowerFactory.

4.4.1 *OPC Interface MATLAB—Simulink*

MATLAB includes a proprietary toolbox called OPC Toolbox for data exchange and communication by means of OPC. This tool helps to carry out OPC data recording, reading and writing processes in MATLAB as well as Simulink. Thus, the integration of multiple devices is possible, such as distributed control systems, data collection, PLCs, among others. Due to OPC Toolbox, MATLAB–Simulink is able to communicate with OPC servers and access to real-time information and historical data following the standards OPC DA, OPC HDA and OPC UA [29].

Within the MATLAB environment, a hierarchical structure for managing the connection between OPC servers and data should be employed. Here, an OPC DA client object is created to make the communication possible with the OPC servers, whereas in the Simulink environment, the vast majority of configurations for OPC clients are set using the OPC configuration block. These configurations let to define features, such as specifications for OPC DA clients applied in models, the behaviour of events, errors in OPC communication and real-time performance.

Furthermore, the processes of OPC data writing or reading in MATLAB can be carried out in synchronous or asynchronous mode via OPC servers. In this regard, OPC Read and OPC Write blocks are used for configuring read/write elements.

- **OPC interface configuration in MATLAB/Simulink**

The OPC configuration block is shown in Fig. 4.15, where OPC clients are set at the start. Inside the option "Configure OPC Clients", a complete list of all the available OPC clients appears. It lets that Simulink interacts as an OPC client.

Down below, all the information regarding the installation directories of "OPC Clients" can be found in the tab "OPC Client Manager", which is depicted in Fig. 4.16. In this case, the denomination "localhost" appears because the directories for the OPC server and MATLAB/Simulink are the same. Also, it is possible to find the

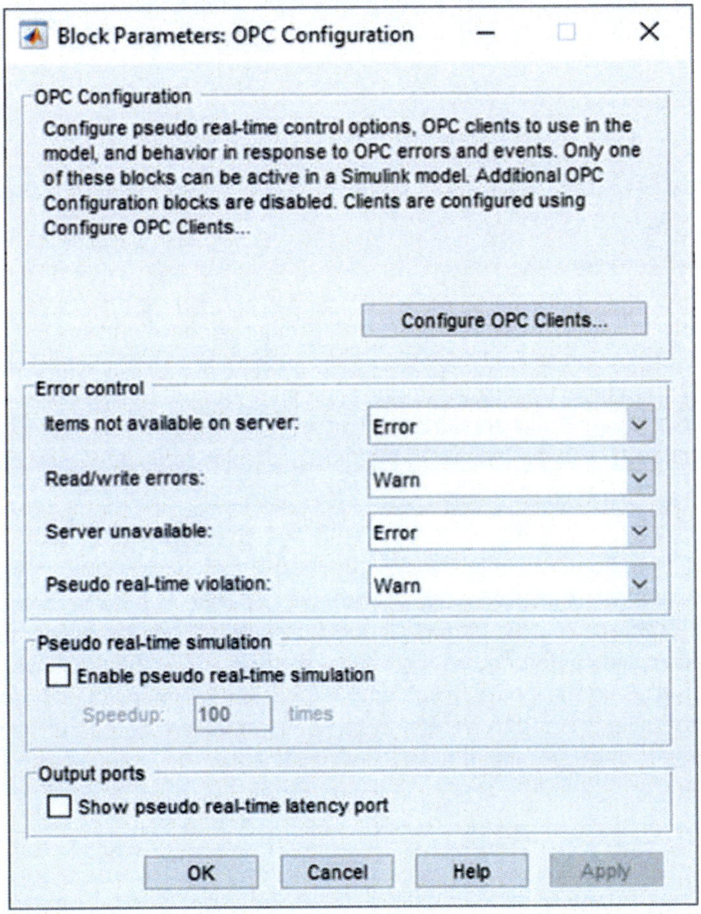

Fig. 4.15 Main parameters of OPC configuration block

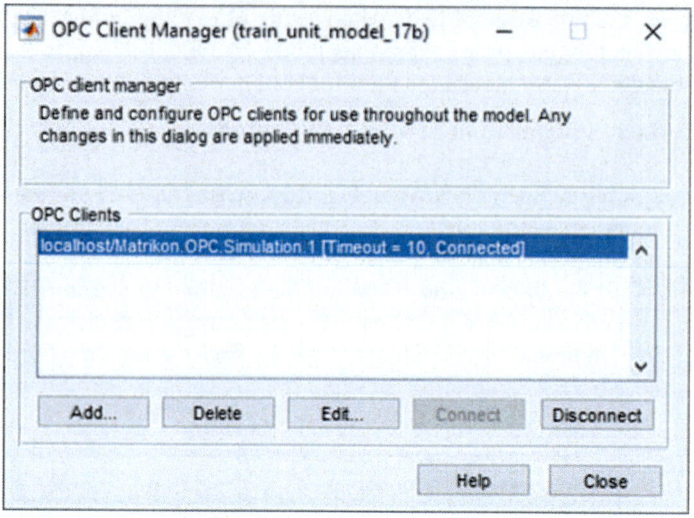

Fig. 4.16 Configuration of OPC clients

OPC server ID (Matrikon.OPC.Simulation.1), its connectivity status ("Connected") and the waiting time (Timeout = 10).

- OPC Read configuration

OPC Read block is used for this purpose, and its main window is shown in Fig. 4.17. Here, the configuration tab includes a drop list with all the OPC clients configured besides all the available options for it. The parameter "Item IDs" should be assigned in every data reading device from the OPC server. The assigned names of every item should be matched with the tag names previously defined in the OPC server.

- OPC Write configuration

Data writing from Simulink to DIgSILENT PowerFactory is carried out by means of the OPC Write block, Fig. 4.18. In this block, the OPC server links to its respective OPC client; Simulink for this purpose is set. In the same way, the synchronisation for data writing and configurations for writing devices are established here. To this end, the assigned names should match with the declared names in OPC server. On the other hand, data writing from Simulink is done in an asynchronous way due to processing times during simulations in its environment.

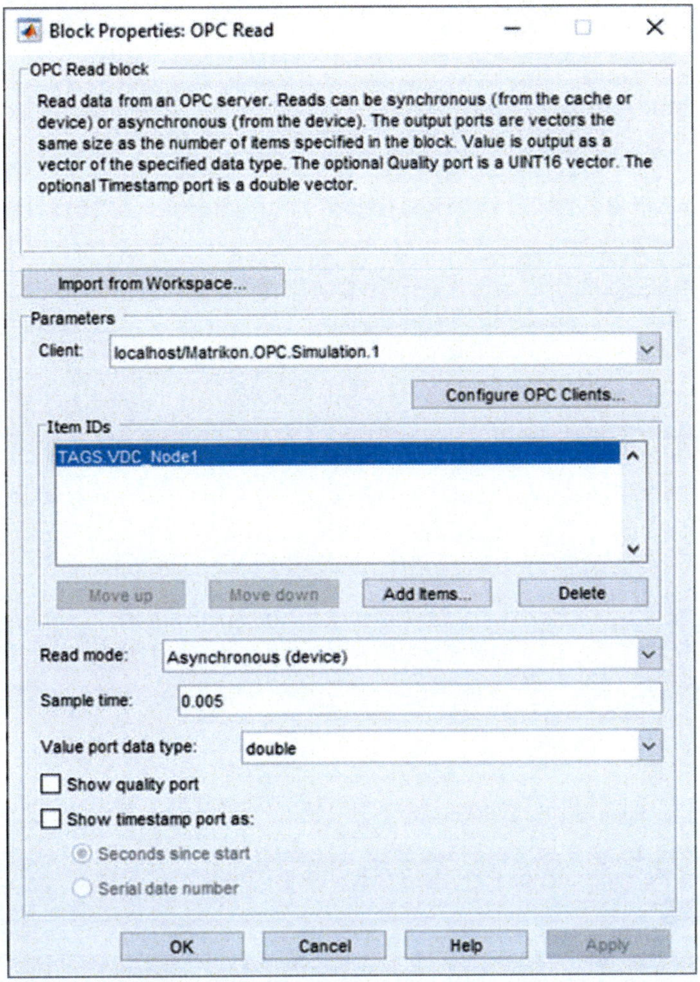

Fig. 4.17 Configuration of OPC Read block

4.4.2 OPC Interface in DIgSILENT PowerFactory

The first step is to configure an OPC client in DIgSILENT PowerFactory through the communication object denominated "External Data Link" (*ComLink*). This object lets to set up all the necessary OPC parameters for PowerFactory as a client. Figure 4.19 shows the configuration window of the External Data Link.

The parameters that must be set up at this point are:

- Type of OPC clients: OPC TDS.
- Prog ID: Matrikon.OPC.Simulation.1.

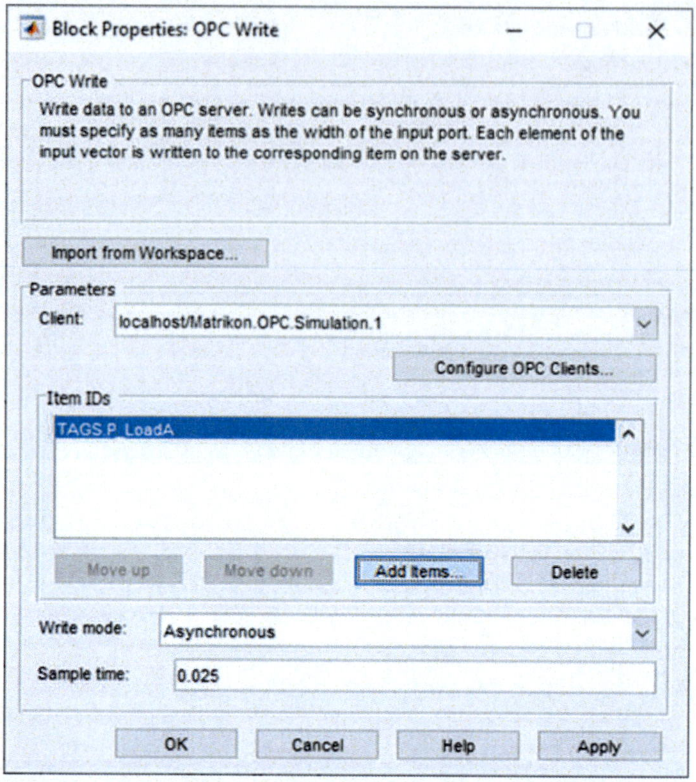

Fig. 4.18 Configuration of OPC Write block

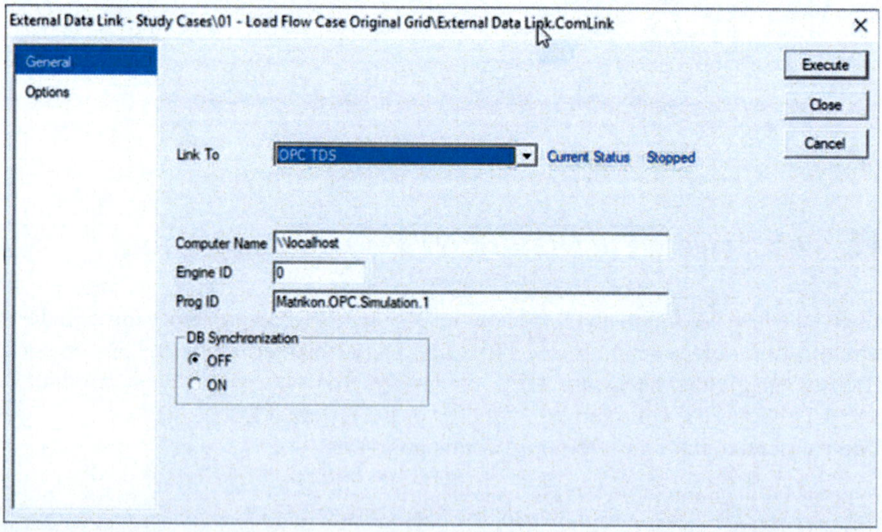

Fig. 4.19 Configuration window of External Data Link

- Directory address of OPC server (Computer ID): localhost (installed inside the same PC).
- Optionally, a dead band for data update can be established. Its default value is equal to 0.1.
- Engine ID can also keep its default value (0).

A communication object (*ComLink*) has to be executed to initialise the communication process with the server. Thus, the client PowerFactory is connected to the defined OPC server. After the initialisation, all the communication devices configured are added. A more detailed explanation of how to configure all the OPC configuration, including the OPC Tags, can be found in [30].

4.4.3 Configuration of Communication Objects and Tags in DIgSILENT PowerFactory

Communication objects are linked to each variable utilising an external measurement device that is responsible for data reading/writing in the OPC server. These devices can have a two-way configuration; in other words, they can read and write data at the same moment. As part of the configuration process, data type should be pointed, and it has to match with the data type declared in the tags of OPC server. After that, the main parameters of external measurement are listed:

- Tag ID: Identification of tag

  Figure 4.20 depicts the given name for a DC voltage node in the catenary system. Additionally, it is advisable to use as ID the name assigned to the external measurement device for a better reference. How to set up ID parameters within "Basic Data" tab is clarified in Fig. 4.20.Besides, the "Basic Data" tab has two extra parameters for configuration:

  - Multiplicator: establishes a multiplying factor for variables from the OPC server.
  - DAT type: lets to choose the type of data for variables.

- Status: data writing and reading

Data writing and reading of external measurement devices are set in this option. Figure 4.21 presents the main window for "Status" configuration.

- Post-processing: external measurement device control

The type of external measurement device is controlled in this option. To illustrate, variables for DC voltage and active power use an "External DAT Measurement" device type. Indeed, Fig. 4.22 illustrates how a DC voltage variable is associated with the electric model in DIgSILENT PowerFactory using the name "m:U11".

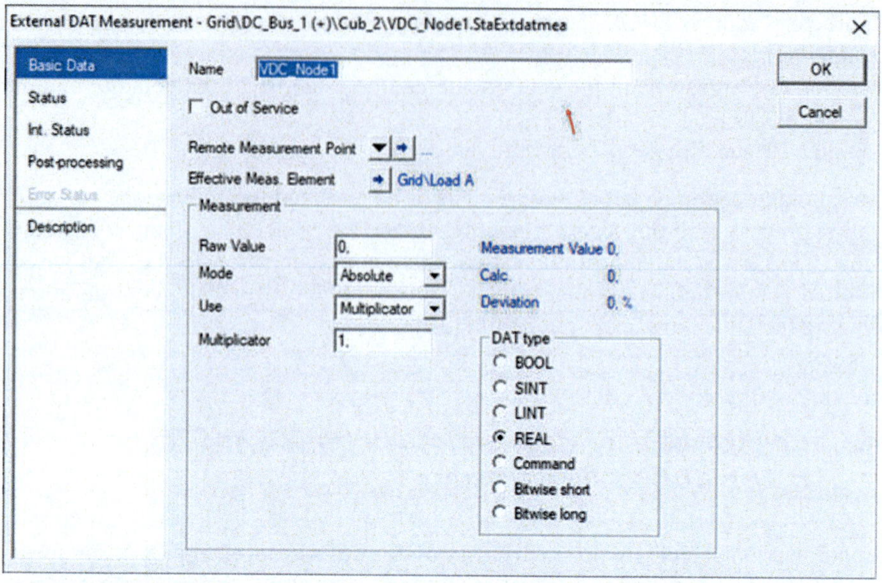

Fig. 4.20 Configuration of OPC Tag ID name of an external data measurement

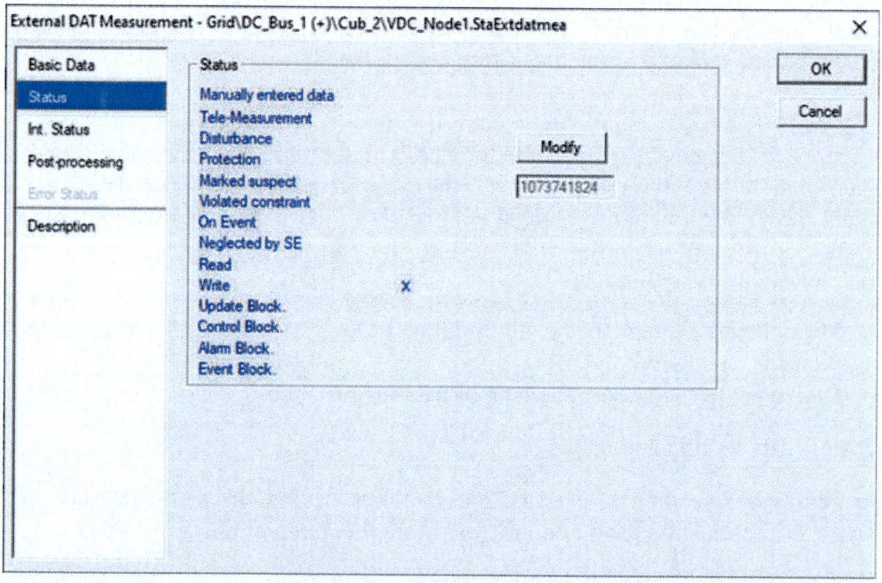

Fig. 4.21 Configuration of data writing and reading status for External Data Measurement devices

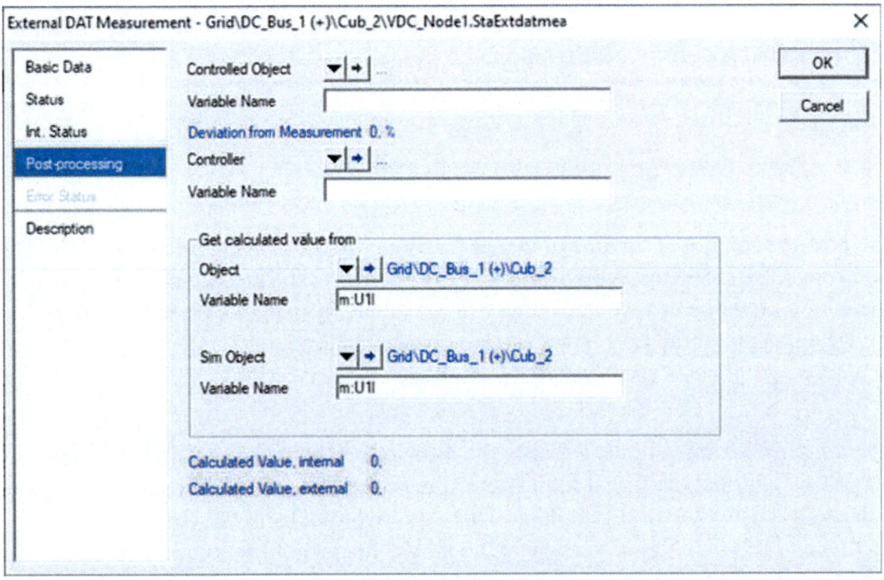

Fig. 4.22 Post-processing tab for "External DAT Measurement" object associated with a DC voltage variable

Contrary, active power variables are set as appears in Fig. 4.23. These variables are

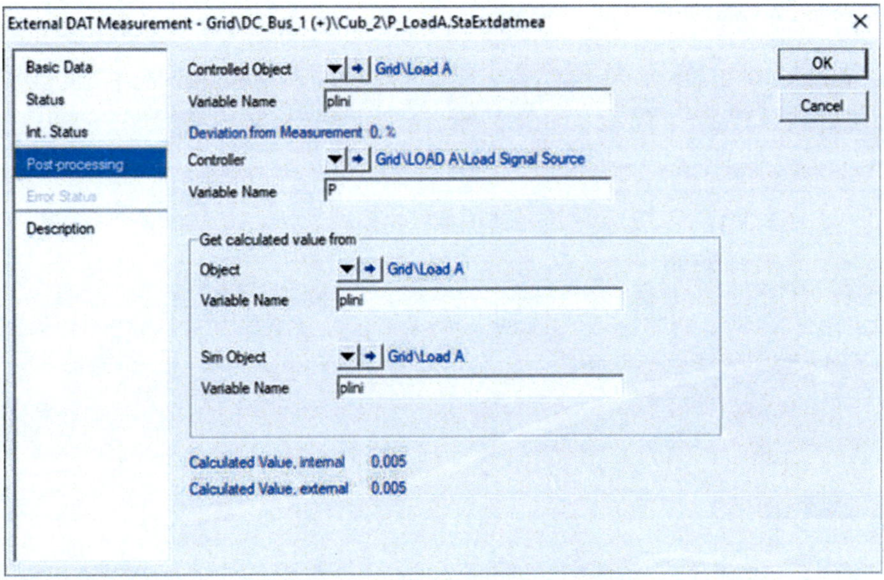

Fig. 4.23 Post-processing tab for "External DAT Measurement" for active power variables

managed in DIgSILENT PowerFactory under the tag "plini". Moreover, the variable "*P*" is a signal sent from a DSL command.

4.4.4 DSL Interface

The post-processing of the power signal for injection into the electric power system requires a DSL interface, which associates the signal in the net value of the external communication object with an element of the power system, which will be a load.

The elements of the DSL interface are detailed below:

- Composite frame

The composite frame allows defining the interactions between the different signals. In this case, it links the signal in the net value of the external communication object with the load, as shown in Fig. 4.24. There are two blocks in the composite frame, the first one refers to the *ElmDsl* type block (the load signal source) and the second one to the *ElmLod* type block (the load to be controlled).

- Common model

The load signal source, that characterises the external communication object, needs to be included by means of a common model (*ElmDsl*) whose initialisation is shown in Fig. 4.25.

- Composite model

The composite frame of Fig. 4.24 is used to implement the composite model that contains the common model *ElmDsl* (load signal source) and the network element *ElmLod* (Load A), as shown in Fig. 4.26. In this way, the signal in the net value of

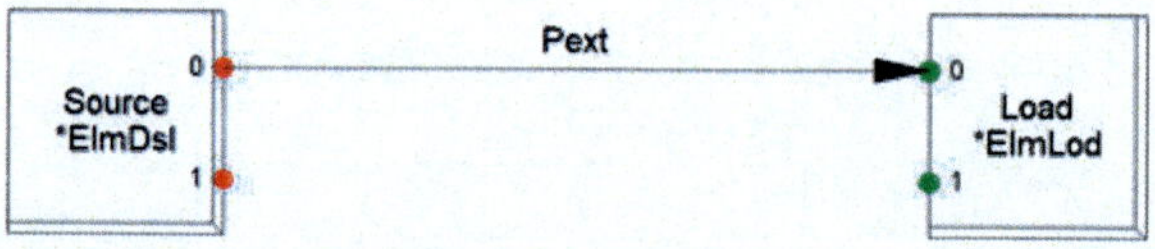

Fig. 4.24 Composite frame scheme

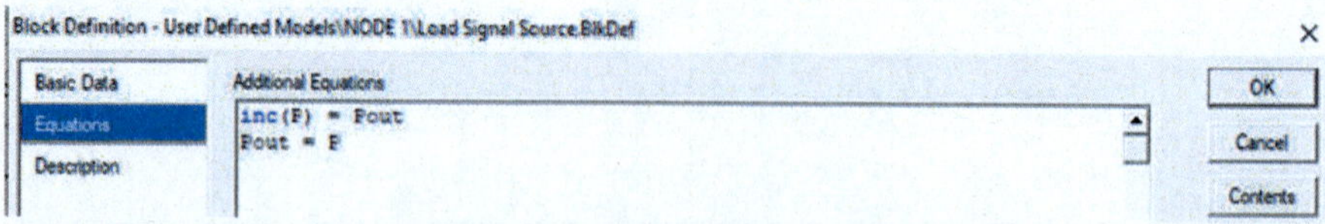

Fig. 4.25 Initialisation of the common model "load signal source"

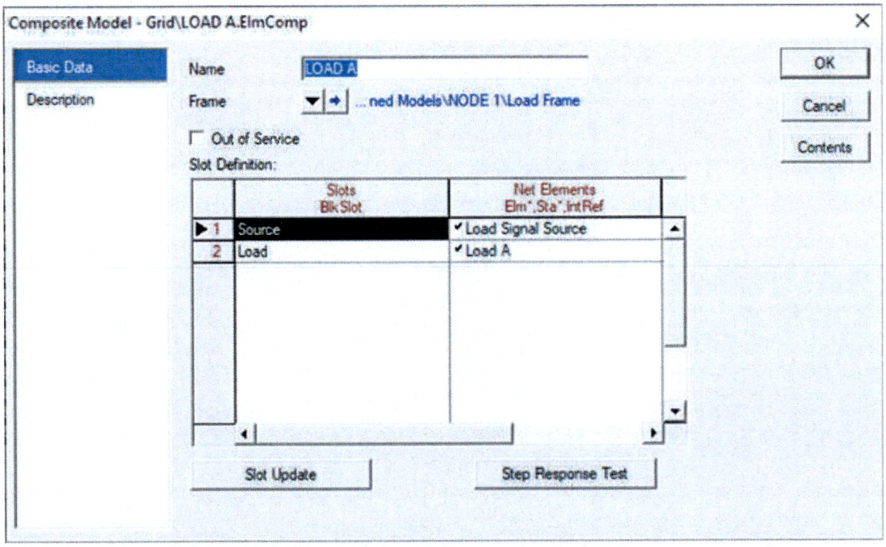

Fig. 4.26 Parameters of "composite model"

the external communication object is injected into the power system.

4.5 Real-Time Configuration in PowerFactory

The RMS/EMS simulation option is checked in the simulation mode toolbox, as shown in Fig. 4.27. This enables RMS simulation in PowerFactory.

The initial conditions calculation option is selected. Figure 4.28 shows the simulation method selection in the "Basic Options" tab: RMS values (electromechanical transients).

In the "Step Sizes" tab, as shown in Fig. 4.29, the simulation steps are established. It must be in accordance with the simulation steps in MATLAB/Simulink.

The Real-Time tab is set, as shown in Fig. 4.30:

(1) The type of synchronisation is set in "Synchronised by system clock" mode.
(2) Set a scale for the simulation time in the "Ratio between real time and calculation time" parameter. Its value does not have to be very high, as this affects the simulation speed. This parameter is varied according to the simulation required speed.
(3) The update time of data coming from the OPC server must be set. So, the interval of 0.001 s is established to prevent the loss of information.
(4) The default value is maintained, as there will be an asynchronous simulation between the OPC server and PowerFactory.

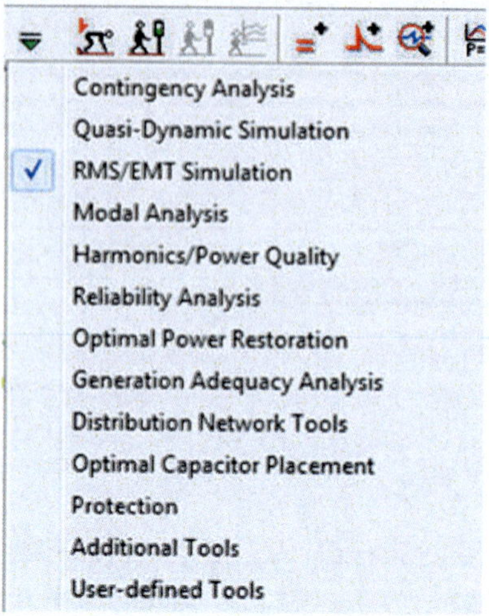

Fig. 4.27 PowerFactory toolbox showing RMS simulation mode

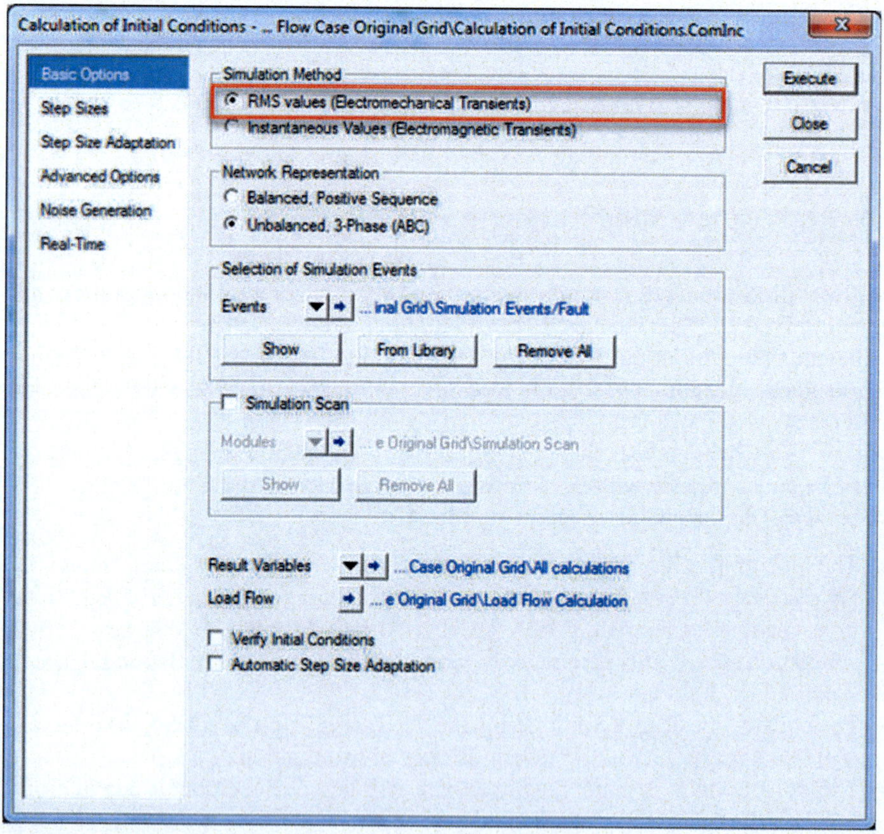

Fig. 4.28 Initialisation command editor. Simulation Method

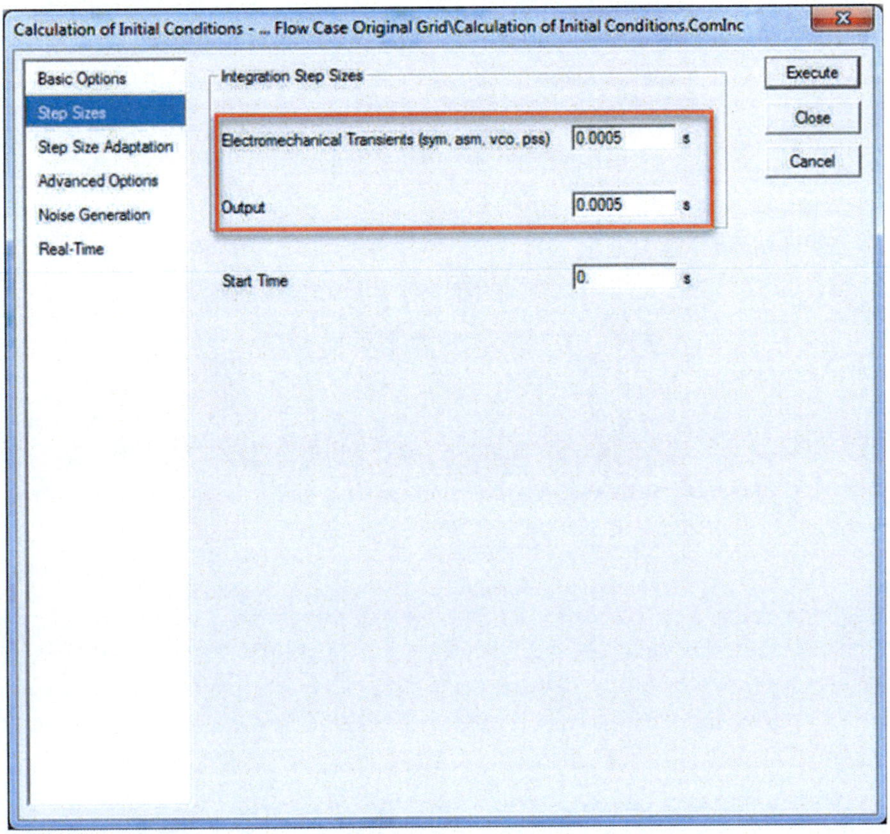

Fig. 4.29 Initialisation command editor. Step sizes

4.6 Co-simulation Execution

4.6.1 MATLAB/Simulink Function for Starting Co-simulation

In MATLAB/Simulink, a function embedded in a Simulink block is created in order to indicate the simulation start of the Simulink model via a two-state triggering variable that is written to a txt-type file named "RunTriggerPF". An initial state of value 0 indicates that the simulation has not yet started, so when the simulation starts, this state changes to value 1. This state change is written to the txt-type file. The MATLAB/Simulink function for starting co-simulation is presented as follows.

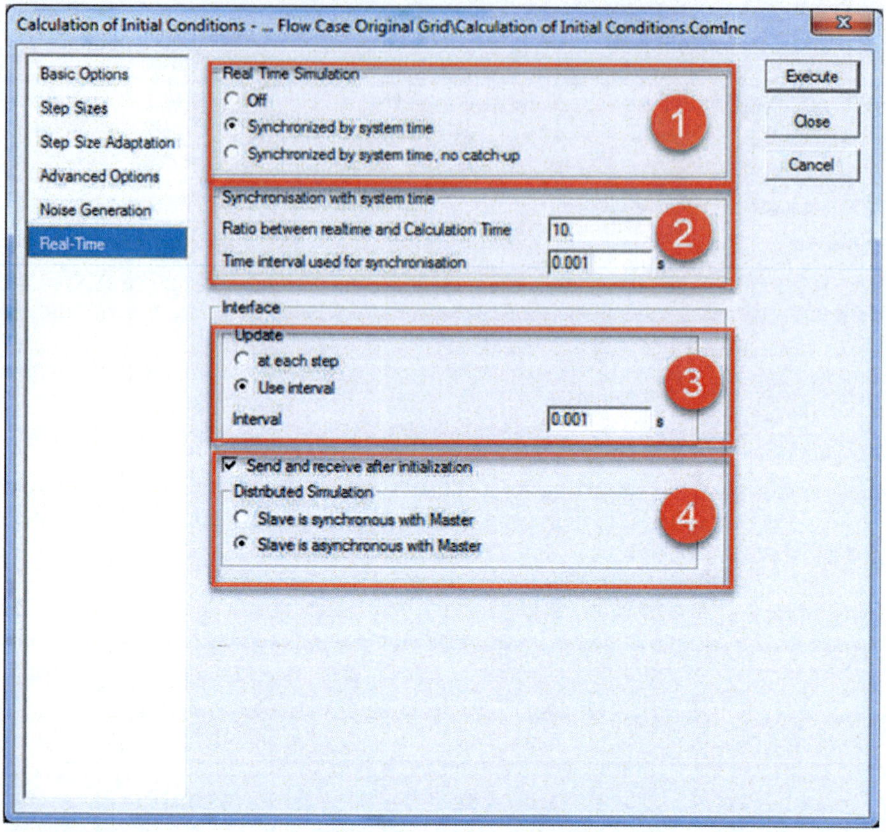

Fig. 4.30 Initialisation command editor. Real-time configuration

```
function fcn(u,w)
if w==1
    file = 'RunTriggerPF.txt';
    fileID = fopen(file,'w');
    fprintf(fileID, '%d', round(int8(u)));
    fclose(fileID);
end
```

4.6.2 *Python Script for Starting Co-simulation*

In PowerFactory, a Python Script named *Run_trigger.py*, linked to a PowerFactory Python Script Command (*Co_sim_exe.ComPython*), controls the simulation process. First, it initialises the required variables and commands in PowerFactory (even the simulation events reset, the "RunTriggerPF.txt" reset or the initial conditions calculation). Afterwards, it verifies if the simulation of the Simulink model has already began in order to allow the co-simulation to start by checking the state change of the triggering variable. As this regards, if the value is 0, the PowerFactory model remains in standby mode, whereas if the value changes to 1, PowerFactory immediately starts the simulation. In this way, a pairing between the simulation environments and their simulations starts. The proposed Python script is presented as follows.

```
import powerfactory

app = powerfactory.GetApplication()

ldf = app.GetFromStudyCase('ComLdf')
ini_cond = app.GetFromStudyCase('ComInc')
Run_sim = app.GetFromStudyCase('ComSim')
sim_events = app.GetFromStudyCase('IntEvt')
sim_events.Delete()

file = open('RunTriggerPF.txt','w')
file.write(str(0))
file.close()
trigger=0

ldf.Execute()
ini_cond.Execute()

while trigger <1:
    file = open('RunTriggerPF.txt','r')
    y = file.readline()

    if y.isdigit():
        trigger=int(y)

    file.close()

Run_sim.Execute()

app.PrintPlain('Done')
```

4.6.3 Running the Simulation

(1) Ensure that all simulation events in PowerFactory (*IntEvt*), as well as Results (*ElmRes*) in the active study case, are empty.
(2) The OPC link in PowerFactory (*External Data Link.ComLink*) must be initialised as detailed in [30].
(3) The simulation mode in the Simulink model is set to accelerator mode.
(4) Run the Python script in PowerFactory (execute the *Co_sim_exe.ComPython* object), so that the model will remain in standby mode until the simulation of the Simulink model starts.

(5) In the MATLAB editor, execute the script "*initial_conditions.m*" to establish the initial conditions and to start the simulation of the Simulink model "*train_unit_model_17b.mdl*".
(6) The simulation time in PowerFactory is according to the rotor speed profile and the Simulink simulation time.
(7) In case if there is a high deviation between PowerFactory and Simulink simulation time, stop the simulation and modify the "ratio between real time and calculation Time", increasing or diminishing as necessary (refer to point 2 of Fig. 4.30).
(8) The signals in Simulink are shown through the signals scopes. The most relevant signals are: rotor speed, power consumed by the train unit, voltage on the DC bus and electromagnetic torque.
(9) The plots in PowerFactory show the DC voltage signals on the bus, on the feeders on the high side and the power consumed by the load A.

4.7 Application Example

For instance, rotor speed during an operation stage between two passenger station appears in Fig. 4.31, which is used as a piece of input information in this model. The reference value is at the nominal speed value of 1,677 rpm.

Achieved results from the co-simulation environment let to analyse power consumption, electrical equipment loadability, voltage and current variations at the moment of starting or braking electric trains in specifics or multiple points of the test grid. Figure 4.32 presents the power consumption generated by each train as a result of the actuation of the proposed control model in MATLAB–Simulink (i.e. the power electronic converters).

Considering the stages of operation of the traction motors that are: start and acceleration, speed in steady-state and braking.

On the other hand, Fig. 4.33 shows the power behaviour on a DC bus that supplies electricity in DIgSILENT PowerFactory. Both figures allow showing the asynchronous interaction reached between MATLAB–Simulink and PowerFactory via the OPC communication.

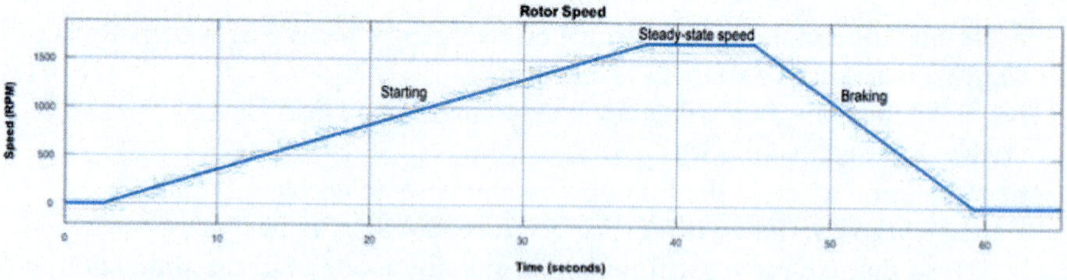

Fig. 4.31 Rotor speed curve for modelled traction motors in MATLAB–Simulink

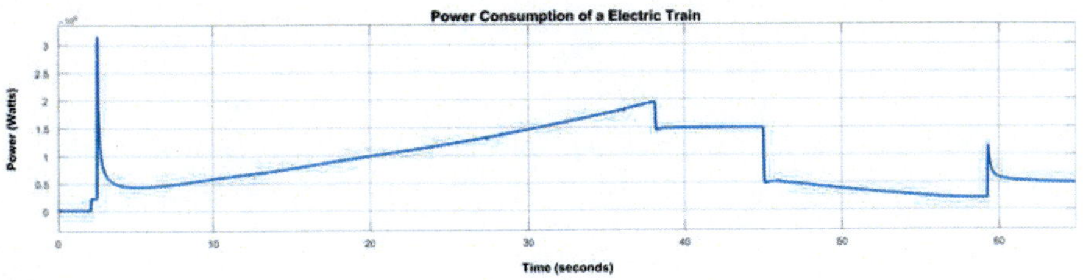

Fig. 4.32 Load profile for a train in MATLAB–Simulink

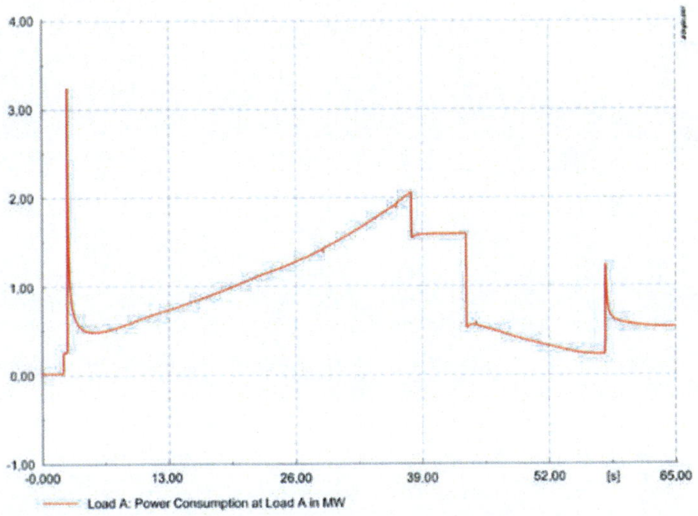

Fig. 4.33 Power behaviour at DC bus in DIgSILENT PowerFactory

An analysis of each of the stages of operation of the train unit, in relation to its power consumption, is performed. It can be observed that the greatest demand occurs during the start-up phase of the engines with an overshoot of approximately 87% over the steady-state value (1.6 MW approximately). Also, in each of the transient states, surges of approximately 25% of the steady-state value occur at each stage. At the end of the braking process, at zero motor speed, a power peak appears that follows the torque value in the load of the rotor. The torque value in the load is 85% of the nominal value.

In the same way, Fig. 4.34 illustrates the voltage behaviour at the DC bus of the electric train. As can be seen, a maximum voltage drop of around 3% is expected in DC buses for the simulated scenario when only a train is starting at the time. During the start-up stage of the motors, a voltage drop occurs up to a value of approximately 2% of the nominal value in steady state.

Another parameter to evaluate is the voltage in the feeders on the high-voltage side, about the traction transformer. Its value is shown, in per unit, in Fig. 4.35.

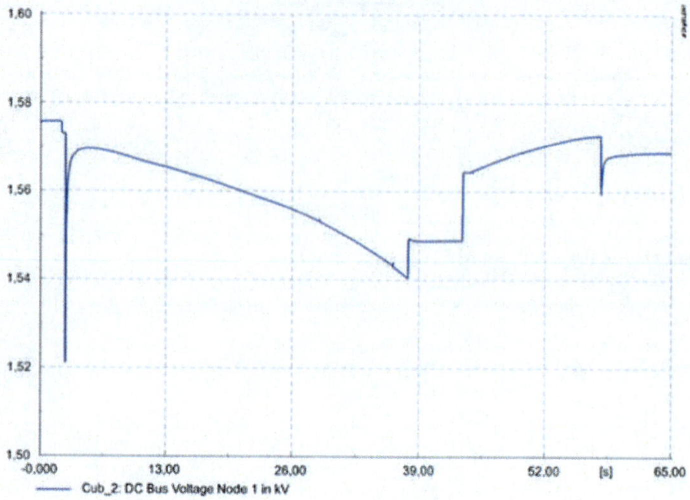

Fig. 4.34 Voltage behaviour at DC bus

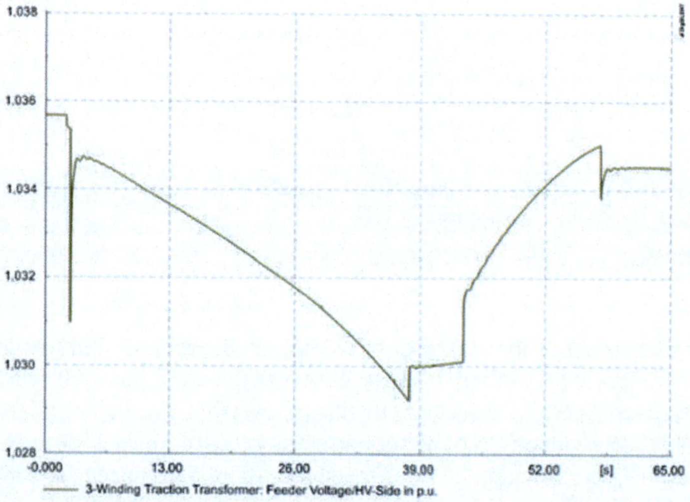

Fig. 4.35 Voltage behaviour at feeder in HV side in p.u.

The voltage behaviour shows a maximum voltage drop of around 0.5%. During the engine start stage, the voltage drop reaches a final value of up to a maximum value of 0.6% of the nominal value in a steady state.

Arriving at zero speed of the training unit when holding a load causes a voltage drop which depends on the percentage of load torque in the rotors. This effect is reflected both in the DC bus voltage and in the feeder voltage.

4.8 Conclusions

The present chapter provides a coordinated simulation tool to analyse the dynamic response of power electronic converters of electric massive mobility systems in which the operation of traction motors could generate disturbances on electric grids. This modelling and simulation proposal could serve as a tool for analysing the impact of electric railway systems in the power grid. For this purpose, a co-simulation environment between MATLAB–Simulink and DIgSILENT PowerFactory is proposed, which is based on the data exchange between these software tools by means of OLE for process control (OPC). PowerFactory is used for railway and utility power systems modelling and simulation. Meanwhile, MATLAB–Simulink simulates electrical drives as well as control and operation of asynchronous motors (i.e. the power electronic converters).

The combination of both computer programs through MatrikonOPC Simulation Server sets a powerful platform up to test complex control systems applied in traction systems of electric trains. The principal advantage of the proposed co-simulation environment is to a couple of simulation environments with external applications for designing and testing control systems applied in the modern power system. Moreover, the tool allows to assess the dynamic of power systems, which provide to system operators enough information for the following: (1) evaluate possible risks of operation limits that might be exceeded and power quality in the coupling points; (2) make decisions about network operation and control actions regarding risk minimisation; and (3) plan new operation strategies for underground transport systems. All the detailed aspects contribute to a substantial saving for consulting companies, utilities and research centres in specialised software acquisition.

References

1. G. De Filippo, V. Marano, R. Sioshansi, Simulation of an electric transportation system at The Ohio State University. Appl. Energy **113**, 1686–1691 (2014)
2. S. Habib, M. Kamran, U. Rashid, Impact analysis of vehicle-to-grid technology and charging strategies of electric vehicles on distribution networks—A review. J. Power Sources **277**, 205–214 (2015)
3. B. Barnard, Railway System Modelling—Not Just for Fun Keynote Address. IEEE Semin. Railw. Syst. Model. Just Fun, 2–6 (2004)
4. G. Celentano, R. Iervolino, Global modelling and simulation for analysis and design of a railway vehicle. SPEEDAM 2012—21st Int. Symp. Power Electron. Electr. Drives, Autom. Motion, 1490–1495 (2012)
5. A. Riofrio, M. Chamba, J. Cepeda, Y. Lecaro, F. Chimarro, M. Mora, Probabilistic assessment and risk management of grid loadability due to the Quito City subway commissioning considering Electric Trains Stochastic Movement. Rev. Técnica "energía**15**(II), 1–12 (2019)
6. R. Goodall, Control engineering challenges for railway trains of the future. IET Seminar Digest **4**, 1–10 (2010)
7. S. Bruni, R. Goodall, T.X. Mei, H. Tsunashima, Control and monitoring for railway vehicle dynamics. Vehicle Syste. Dyn. **45**(7–8), 743–779 (2007)

8. Y. Li, F. Ma, M. Kazerani, Research on the control strategy for the traction motor on the test bench of vehicular energy storage system. Proceeding of the CSEE **34**(21), 3481–3487 (2014)
9. R. J. Hill, J. Lamacq, Railway traction vehicle electromechanical simulation using SIMULINK. Trans Built Environ **18** (1996)
10. M. Quraan, J. Siam, Modelling and simulation of railway electric traction with vector control drive. 2016 IEEE Int. Conf. Intell. Rail Transp. ICIRT **2016**(1), 105–110 (2016)
11. M. Chen, T. Wen, W. Jiang, J. Luo, *Modelling and Simulation of New Traction Power Supply System in Electrified Railway*, IEEE Conf. (Intell. Transp. Syst. Proceedings, ITSC, 2015), pp. 1345–1350
12. A. Županan, A.T. Teklić, B. Filipović-Grčić, Modelling of 25 kV electric railway system for power quality studies. IEEE EuroCon **2013**, 844–849 (2013)
13. W. Mingli, C. Roberts, and S. Hillmansen, Modelling of AC feeding systems of electric railways based on a uniform multi-conductor chain circuit topology. IET Conf. Railw. Tract. Syst. (RTS 2010),**2010**(13342), 12–12 (2010)
14. R. Huang, R. Fan, J. Daily, A. Fisher, J. Fuller, Open-source framework for power system transmission and distribution dynamics co-simulation. Transmission Distribution IET Generation **11**(12), 3152–3162 (2017)
15. C. D. López, A. A. v. d. Meer, M. Cvetković, P. Palensky, A variable-rate co-simulation environment for the dynamic analysis of multi-area power systems, in*2017 IEEE Manchester PowerTech*, pp. 1–6 (2017)
16. M. Cvetković, H. Krishnappa, C. D. López, R. Bhandia, J. Rueda Torres, P. Palensky, Co-simulation and dynamic model exchange with consideration for wind projects, in*16th Wind Integration Workshop*, September 2017
17. F. Andrén, M. Stifter, T. Strasser, An environment for the coordinated simulation of power grids together with automation systems, in*2013 IEEE Grenoble Conference*, Grenoble, pp. 1–6 (2013)
18. J. Garcia-Villalobos, I. Zamora, M. Marinelli, P. Eguia, J. I. San Martin, Co-simulation with DIgSILENT PowerFactory and MATLAB: Optimal integration of plug-in electric vehicles in distribution networks,*Advanced Smart Grid Functionalities based on Power Factory (Green Energy and Technology)*, Springer (2017)
19. A. Latif, M. Shahzad, P. Palensky, Y. W. Gawlik,*An alternate PowerFactory MATLAB coupling approach*, Vienna: International Symposium on Smart Electric Distribution Systems and Technologies (EDST), IEEE (2015)
20. F. Gonzalez-Longatt, J. Rueda, Introduction to Smart Grid Functionalities, in*Advanced Smart Grid Functionalities Based on PowerFactory*, Green Energy and Technology, Springer (2018)
21. Stifter, F. Andrén, R. Schwalbe, W. Tremmel, Interfacing PowerFactory: Co-simulation, Real-Time Simulation and Controller Hardware-in-the-Loop Applications, in*PowerFactory Applications for Power System Analysis, Gewerbestrasse*, Cham, Switzerland: Springer International Publishing (2014)
22. MatrikonOPC, *MatrikonOPC Explorer. User's Manual*, Matrikon International, Edmonton, Canadá, pp. 1–79 (2018)
23. MatrikonOPC, "MatrikonOPC Simulation Server," 2019. [Online]. Available: https://www.matrikonopc.com/products/opc-drivers/opc-simulation-server.aspx. Accessed 03 Feb 2019
24. A. Riofrio, M. Chamba, J. Cepeda, Y. Lecaro, F. Chimarro, Y. M. Mora,*Probabilistic Assessment of Underground Railway Systems Impact Over Distribution Grids*, Innovative Smart Grid Technologies (ISGT), IEEE (2019)
25. Toshiba, *DC-AC Inverter Circuit*, TOSHIBA Electronic Devices & Storage Corporation, pp. 1–29 (2018)
26. B. K. Bose,*Modern Power Electronics and AC Drives*, Prentice Hall (2002)
27. T. Banerjee, J. Bera, S. Chowdhuri, and G. Sarkar, A comparative study between different modulations techniques used in Field Oriented Control Induction Motor Drive. 2nd Int. Conf. Control. Instrumentation Energy Commun. 358–362 (2016)
28. C. Liu and Y. Luo,*Overview of Advanced Control Strategies for Electric Machines*, vol. 3, no. 2 (2017)

29. Mathworks, “OPC Toolbox,” The Mathworks, Inc. [Online]. Available: https://la.mathworks.com/products/opc.html.
30. R. Srinivasan, PowerFactory as a Software Stand-in for Hardware in Hardware-In-Loop Testing, in *PowerFactory Applications for Power System Analysis*, eds. by F. M. Gonzalez-Longatt, J. L. Rueda, Springer International (2014)
31. N. Quang, *Vector Control of Three-Phase AC Machines* (Springer, Berlin Heidelberg, 2015)
32. W. Leonhard, *Control of Electrical Drives* (Springer, Berlin Heidelberg, 2001)
33. D. Grahame Holmes, Thomas A. Lipo,*Pulse Width Modulation for Power Converters: Principles and Practice*, Wiley-IEEE Press (2003)
34. V. K. Pavuluri, X. Wang, J. Long, G. Zhuo and W. Lian, Field Oriented Control of Induction Motors Using Symmetrical Optimum Method with Applications in Hybrid Electric Vehicles, in *2015 IEEE Vehicle Power and Propulsion Conference (VPPC)* (Montreal, QC, 2015) pp. 1–6

Chapter 5
Transient Stability Assessment of Power System Incorporating DFIM-Based Pumped Storage Hydropower and Wind Farm

Mohsen Alizadeh Bidgoli, Saman Atrian, Weijia Yang, and Francisco M. Gonzalez-Longatt

Abstract Variable-speed (VS) pumped storage technology has become a new trend for providing better support power system incorporating renewable energies. On the other hand, the transient stability of the power system can be affected by the application of such kind of storage unit. State-of-the-art pumped storage hydropower plants (PSHP) based on doubly fed induction machine (DFIM) known as variable-speed and conventional PSHP-based on synchronous machine (SM) known as fixed speed (FS) have a different effect on transient stability of a large-scale power system. This chapter intends to present the modelling and controller of DFIM-based PSHP of two types of PSHPs, i.e., VS and FS under generating operation mode in DIgSILENT software. Also, the IEEE 10-machine 39-bus system namely New England test system is adopted as a large power network. The results show that using DFIM-based VS-PSHP in the interconnected power grids, not only the oscillation modes of PSHP is eliminated, but also it can strongly improve rotor angle and voltage transient stability of the power system.

Keywords DIgSILENT dynamic simulation language (DSL) · Doubly fed induction machine (DFIM) · Transient stability · Variable-speed pumped storage hydropower · Windfarm

---

Electronic supplementary material The online version of this chapter (https://doi.org/10.1007/978-3-030-54124-8_5) contains supplementary material, which is available to authorized users.

M. Alizadeh Bidgoli (✉) · S. Atrian
Department of Electrical Engineering, Yadegar -E- Imam Khomeini (RAH), Share Rey Branch, Islamic Azad University, Tehran, Iran
e-mail: m.alizadeh.b@iausr.ac.ir

W. Yang
State Key Laboratory of Water Resources and Hydropower Engineering Science, Wuhan University, Wuhan 430072, China

F. M. Gonzalez-Longatt
1Institutt for elektro, IT og kybernetikk, Universitetet I Sørøst-Norge, Porsgrunn, Norway

F. M. Gonzalez-Longatt and J. L. Rueda Torres (eds.), *Modelling and Simulation of Power Electronic Converter Dominated Power Systems in PowerFactory*, Power Systems,
https://doi.org/10.1007/978-3-030-54124-8_5

5.1 Introduction

The main goal of renewable resource owners in the market-based system is to attain maximum profit by injecting the active power into the grid. Therefore, from point of view of power grids, using the flexibility of pumped storage hydropower plant (PSHP) is crucial to cope with the unexpected rise or drop in the production of renewable energies (REs). The first-worldwide DFIM-based PSHP, installed in Japan in the 1990s, has been employing the cycloconverter, while the first one in Europe has been commissioned in 2004 (Goldisthal in Germany). This practical strategy has been recently used in many projects worldwide using the state-of-the-art power electronic converter, i.e., back-to-back voltage source converter, such as Avce in Slovenia or Linthal in Switzerland.

Comparative study between fixed-speed and variable-speed PSHP connected to single machine infinite bus (SMIB) modelled in Simsen software has been performed [1], but this paper has not considered the performance of PSHP in a large power system. The dynamic performance of two 320 MW PSHP has been reported using simplified converter model in [2] using Simsen software; however, in this paper, challenges ahead for the power system such as reactive power support and damping of the oscillation modes has not been addressed. Alizadeh et al. [3] propose nonlinear controller in fault ride-through (FRT) condition for both generating and motor modes, and Azbe et al. [4] consider the transient stability of the large DFIM in a weak grid. However, both papers have been presented results of PSHP in a SMIB test system. Also, Yang et al. [5] investigate the advantage of variable-speed pumped storage units for mitigating wind power variations comparing with fixed-speed units. Moreover, simplified model of DFIM-based wind turbine for RMS-type simulation and risk assessment of oscillatory stability in power system, implemented in DIgSILENT PowerFactory, can be valuable for studying the large power systems [6, 7].

In summary, the transient stability of the power system with the presence of MW-size PSHP based on DFIM and wind power is an important issue. Moreover, DIgSILENT PowerFactory can be useful software to analyze the mentioned problem in the large power system. Therefore, in this chapter, transient stability of large power system incorporating DFIM-based pumped storage hydropower and wind farm is assessed.

This chapter is arranged as follows: Mathematical modelling is introduced in Sect. 5.2. The description and implementation of the models in DIgSILENT PowerFactory is presented in Sect. 5.3. Also, Sect. 5.4 studies and draws the reactive power capability curve of a DFIM. In Sect. 5.5, simulation results are presented and analyzed. Finally, Sect. 5.6 summarizes the main concluding remarks and outlines prospective future work.

5.2 Mathematical Model of PSHP

In this section, the variable-speed (VS) PSHP and fixed-speed (FS) PSHP are briefly introduced, and the corresponding modelling is presented. In this study, a hydro turbine (HT) coupled with a DFIM in comparison with the synchronous machine (SM) with the same capacity is considered. A DFIM, coupled to the HT, uses the back-to-back converter, i.e., machine-side converter (MSC) and grid-side converter (GSC), whose power rating is the partial scale of the DFIM-rated power. This converter is located between the rotor and stator terminal that can transfer the mechanical power through both the stator and rotor, as shown in Fig. 5.1a.

The controller of FS-PSHP consists of two parts, as shown in Fig. 5.1b. First, the static excitation system which regulates the voltage of stator winding and also an auxiliary input is known as power system stabilizer (PSS), is added to excitation input to damp local oscillations [8]. The active power of the unit could be controlled by the turbine governor in generating mode, while the active power of the motor cannot be adjusted, and the guide vane opening is optimized to increase the efficiency of the pump. It is because of that wicket gate throttling increases the vibrations and the losses in the pumping mode. Note that, the reference gate opening of the pump is extracted from the best efficiency point according to hill chart [3].

5.2.1 *Hydraulic System*

Considering a non-elastic water column with long length penstock without affecting the surge tank, the state space equations of the hydraulic subsystem in turbine mode are shown in the following equations [3]:

$$\dot{q}_t = (h_s' - h_t)/T_w \tag{5.1}$$

$$\dot{g} = -k_a g + k_a v_{\text{sm}} \tag{5.2}$$

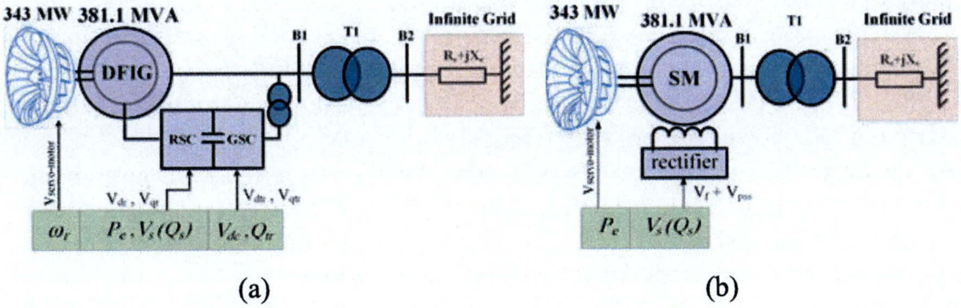

Fig. 5.1 PSHP connected to single machine infinite bus (SMIB) in generating mode,**a** DFIM-based,**b** SM-based

Also where q is the water flow, g is the gate opening, v_{sm} is the servomotor voltage, T_w is water starting time of the pipe, k_a is the constant gain for servomotor modelling, and f_{eq} is the equivalent frictional coefficient. Moreover, $h_s^{'}$, h_t are net static head and net head in the operating point of the turbine which are defined as follow.

$$\begin{aligned} h_s' &= h_s - f_{\text{eq}} q_t^2 \\ h_t &= (q_t/(g/(g_n - g_{\text{nl}})))^2 \\ T_w &= Lq_n/(g'\text{Ah}_n) \end{aligned} \tag{5.3}$$

where static head equal to 1 ($h_s = 1$),L is the length of the pipe, A is the area of the pipe, and g' is gravitational acceleration, which is equal to 9.81 m/s^2; note that, subscripts n and nl are nominal and no-load values, respectively. Additionally, the mechanical power of the turbine is defined as follows:

$$P_m = \eta p_r h_t (q_t - q_{\text{nl}}) \tag{5.4}$$

where η is efficiency of the turbine and $p_r = P_n(\text{MW})/S_n(\text{MVA})$ is a constant value for per unit value conversion from the turbine base to the electric machine (DFIM or SM) base; Also,P_n and S_n are nominal power of the turbine and nominal apparent power of the machine in generating mode, respectively.

5.2.2 *DFIM*

Depending on the goal of the analysis, it is possible to select the models of an appropriate detailing level, by choosing the type of simulation method. For stability analysis, RMS simulations are based on simplified electromechanical transient models. In the case of induction generators, the RMS simulation is using a third-order generator model, where the stator transients are neglected. The built-in model used in this study is the usual induction generator which extends by a MSC in series to the rotor impedance. In other words, in this model, the DC-link is hidden, and the user does not have any controls over the GSC which is suitable for the load flow calculations. Moreover, using this model, it is possible to define DFIM as a PV (to set active power and voltage) or PQ (to set active and reactive powers bus). While using GSC model in DIgSILENT, DFIM can only determine as PQ. Therefore, this study is generally interested in total power and not only either stator or power.

As shown in Fig. 5.1, the positive direction for the stator and rotor currents is assumed as motor conversion (into the machine). Also, all parameters and equations are given in per unit, except for time is sec. In the synchronous reference frame, *d–q* axis, dynamic equations of the DFIM are derived as follows:

$$\frac{1}{\omega_b}\frac{\mathrm{d}\varphi_{\text{dr}}}{\mathrm{d}t} = -R_r i_{\text{dr}} + s\varphi_{qr} + v_{\text{dr}} \tag{5.5}$$

$$\frac{1}{\omega_b}\frac{d\varphi_{\text{qr}}}{\text{d}t} = -R_r i_{\text{qr}} - s\varphi_{\text{d}r} + v_{\text{qr}} \tag{5.6}$$

$$\frac{d\omega_r}{\text{d}t} = \frac{1}{2H}(T_m - T_e) \tag{5.7}$$

where $i_{\text{d}r}$, i_{qr}, $v_{\text{d}r}$, v_{qr} are the$d$–q axis rotor currents and voltages; R_r is the rotor resistance; s is the slip, and $\omega_b = 377$ (rad/sec); T_m, T_e, and ω_r are the mechanical torque, electric torque, and rotor speed, respectively. Additionally, $\varphi_{\text{d}s}$, φ_{qs}, $\varphi_{\text{d}r}$, φ_{qr} are the d-q axis stator and rotor flux magnitude which can be written as follows:

$$\begin{cases} \varphi_{\text{d}s} = L_s i_{\text{d}s} + L_m i_{\text{d}r} \\ \varphi_{\text{qs}} = L_s i_{\text{qs}} + L_m i_{\text{qr}} \\ \varphi_{\text{d}r} = L_r i_{\text{d}r} + L_m i_{\text{d}s} \\ \varphi_{\text{qr}} = L_r i_{\text{qr}} + L_m i_{\text{qs}} \end{cases} \tag{5.8}$$

where $i_{\text{d}s}$, i_{qs} are d-q axis stator currents; L_s,L_r are the stator and the rotor inductances; L_m is the magnetizing inductance; Also, neglecting the stator transients, the following equations can be written:

$$v_{\text{ds}} = R_s i_{\text{ds}} - \varphi_{\text{qs}}$$

$$v_{\text{qs}} = R_s i_{\text{qs}} + \varphi_{\text{d}s} \tag{5.9}$$

$$T_e = (L_m/L_s)\left(-\varphi_{\text{d}s} i_{qr} + \varphi_{qs} i_{\text{d}r}\right)$$

$$Q_s = v_{\text{d}s} i_{\text{qs}} - v_{\text{qs}} i_{\text{d}s} \tag{5.10}$$

where T_e is electrical torque, Q_s is the stator reactive power, and P_e is the electrical active power. It should be mentioned that the modulation of the MSC is a sinusoidal pulse width modulation (PWM). For $|P_m| < 1$, the d-q components of rotor voltage are calculated as follows.

$$\begin{aligned} v_{\text{rd}} &= K_o p_{\text{md}} U_{\text{DC}} \\ v_{\text{rq}} &= K_o p_{\text{mq}} U_{\text{DC}} \end{aligned} \tag{5.11}$$

where K_o is a constant coefficient which depends on modulation method, and for sinusoidal modulation it is equal to $\frac{\sqrt{3}}{2\sqrt{2}}$; p_{md}, p_{mq} are d–q axis PWM coefficient which control parameter of the MSC in DIgSILENT.

5.2.3 Grid-Side Converter

For simplification, GSC can be represented either as a controlled current source or current-controlled voltage source. However, as mentioned earlier, in this studym the DC-link dynamics are neglected. A built-in model without DC link is available, PowerFactory. This model allows the user to define the DFIM as a PV bus.

5.2.4 Synchronous Machine

The model of a salient pole synchronous machine, namely fifth-order model, is derived in the rotor reference frame as Eqs. (5.12–5.16) [8, 9]. Note that, as shown in Fig. 5.1b, the positive direction for the stator and field currents are assumed as motor conversion (into the stator, field, and damper windings). Also, all parameters and equations are given in per unit except for time is sec.

$$\frac{1}{\omega_b}\frac{d\varphi_{ds}}{dt} = -R_s i_{ds} + \omega_r \varphi_{qs} + V_{ds} \tag{5.12}$$

$$\frac{1}{\omega_b}\frac{d\varphi_{qs}}{dt} = -R_s i_{qs} - \omega_r \varphi_{ds} + V_{qs} \tag{5.13}$$

$$\frac{1}{\omega_b}\frac{d\varphi_{fd}}{dt} = -R_f i_{fd} + V_{fd} \tag{5.14}$$

$$\frac{1}{\omega_b}\frac{d\varphi_{kd}}{dt} = -R_{kd} i_{kd} \tag{5.15}$$

$$\frac{1}{\omega_b}\frac{d\varphi_{kq}}{dt} = -R_{kq} i_{kq} \tag{5.16}$$

where $i_{ds} i_{qs}, i_{fd}, i_{kd}$, i_{kq} are *d–q* axis stator, field, and damper winding currents; V_{fd} is field winding voltage, R_s,R_{fd},R_{kd}, *R*kq are stator, field, and damper winding resistances, respectively. Also, ω_r is the rotor speed. Additionally,$\varphi_{/textds}$,φ_{qs},φ_{fd},φ_{kd}, φ_{kq} are the *d–q* axis stator, field, and damper windings flux magnitude, which can be written as follows:

$$\begin{aligned}
\varphi_{ds} &= L_d i_{ds} + L_{md} i_f + L_{md} i_{kd} \\
\varphi_{qs} &= L_q i_{qs} + L_{mq} i_{kq} \\
\varphi_f &= L_{fd} i_{fd} + L_{md} i_{ds} + L_{md} i_{kd} \\
\varphi_{kd} &= L_{kd} i_{kd} + L_{md} i_{fd} + L_{md} i_{ds} \\
\varphi_{kq} &= L_{kq} i_{kq} + L_{mq} i_{qs}
\end{aligned} \tag{5.17}$$

where L_d, $L_q$are d–q axis stator winding self-inductances, L_{kd}, L_{kq} are d–q axis damper windings self-inductances, L_{md}, L_{mq}, d–q axis magnetizing inductances, L_{fd} field self-inductance. Moreover, swing equations for the dynamic model of the generator rotor are written as follows:

$$\frac{\mathrm{d}\omega_r}{\mathrm{d}t} = \frac{1}{2H}(T_m - T_e) \tag{5.18}$$

$$\frac{\mathrm{d}\delta}{\mathrm{d}t} = \omega_r - \omega_0 \tag{5.19}$$

where δ is the rotor angle of the generator and ω_0 is base angular speed which is 1 p.u.

5.3 Creating a Model of PSHP Using DSL

5.3.1 Variable-Speed PSHP Model

In this section, an overview for creating of PSHP model in DIgSILENT is introduced. In this study, some slots (e.g., DFIG) are selected from the library, and other ones are created as common models and linked them to the relevant model definition. Figure 5.2 shows the model of the HT and DFIM/MSC, which was built as a composite model and contains the following components:

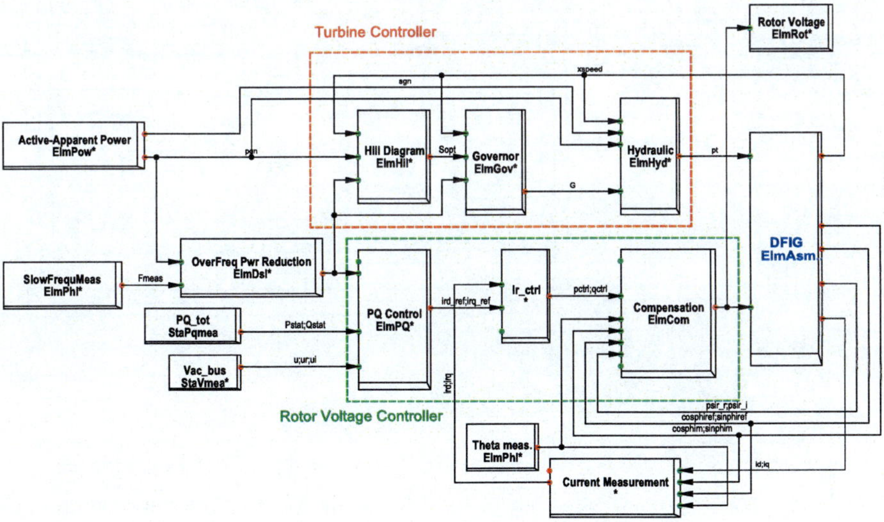

Fig. 5.2 Composite frame for VS-PSHP system

1—DFIG (*ElmAsm*): The standard PowerFactory model of DFIG and MSC; 2—current measurement: stator voltage-oriented reference transformation; 3—Ir_ctrl: current controller; 4—Vac_gen (*StaVmea*): stator voltage measurement; 5—Vac_bus (*StaVmea*): bus voltage measurement; 6—PQ_tot (*StaPqmea*): total active and reactive power measurement of machine; 7—PQ control (*ElmPQ*): active and reactive controller for MSC; 8—Theat_meas (*ElmPhi*): angle measurement; 9—hill diagram (*ElmHil*); reference speed computation 10—Over Freq Pwr Reduction: preventing of over frequency; 11—Slow Freq Meaa (*ElmPhi*): frequency measurement; 12—Rotor Voltage (*ElmRot*); 13—Hydroulic (*ElmHyd*); 14—Compensation (*ElmCom*); 15—protection(*ElmPro*): crow bar protection, and 16—SlowFrequMeas (*ElmPhi*).

Variables of Fig. 5.2 are introduced as follows; *Pt* is HT power in per unit (p.u.), *usr, usi* are real and imaginary parts of rotor voltage in p.u., *xspeed* is rated rotor speed in p.u., *pgn* is nominal power of HT in kW, *sgn* is apparent power of DFIM in kVA, *sinphim, cosphim* are sin and cos of rotor angle, *sinphi, cosphi* are sin and cos of power angle, *psir_r, psir_i* are real and imaginary parts of rotor flux in p.u., i_{rd}, i_{rq} are *d–q* axis rotor current in p.u.

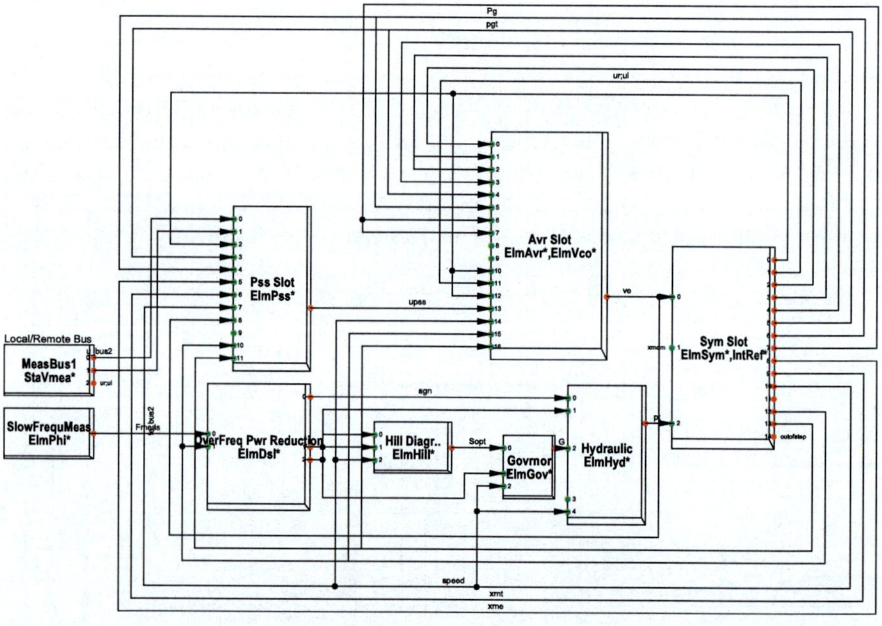

Fig. 5.3 Composite frame for FS-PSHP system

5.3.2 Fixed-Speed PSHP Model

Figure 5.3 shows the composite model of FS-PSHP. Note that, there is already a built-in model for the synchronous generator, hydro governor, and other elements except hill diagram and therefore does not require an additional definition.

5.3.3 Hydro Turbine Model

The speed variation range of a VS-PSHP depends on two key factors. The first is the efficiency improvement in the turbine mode and the second is the participation percent of VS-PSHP in the frequency control for pumping mode. On the other hand, the size of converters is determined according to speed variation range of the machine. Efficiency improvement for the turbine mode is introduced on the hill diagram of the HT which is explained in following.

The hill chart curves of a Francis HT are furnished by manufacture or designer of HT for a specific speed number which presents its efficiency in different operating point. For a typical Francis turbine, the efficiency hill chart is given in unit values N_{11}–Q_{11}. For this study, however, the hill curves are used in unit values N_{11}–P_{11}. The unit values, unit power (P_{11}), unit speed (N_{11}), and unit discharge (Q_{11}), are defined as follows.

$$P_{11} = \frac{P}{D_1^2\sqrt{H^3}}, N_{11} = \frac{D_1 N}{\sqrt{H}}, Q_{11} = \frac{Q}{D_1^2\sqrt{H}} \tag{5.20}$$

where D_1, Q, N, P, and H are the diameter of the turbine runner in meter, the discharge in m^3/sec, the rotational speed in rpm, the power input of the turbine in kW, and the net head in meter, respectively.

For deduction from the above explanations, the maximum of efficiency improvement for a VS-PSHP can be obtained by decreasing the speed when both the head and output power of the unit are low in the turbine mode. In other words, the minimum speed of a VS-PSHP is determined by the cost analysis of the efficiency improvement in the turbine mode at partial load. It should be noted that the power variation range in pumping mode is another parameter which impact the required speed variation range. More detailed explanations is given in [3].

Figure 5.4 shows the composite model of reference speed computation named as hill diagram slot. As stated before, the hill chart curves (P_{11}- N_{11}) of HT prepared by the designer of HT presents its efficiency in different operating point. At first, rotor speed and active power are converted to unit power (P_{11}) and unit speed (N_{11}), respectively. Parameters defined in this model are reference power (*pref*) and rated rotor speed (*xspeed*), inner diameter (*D*), and static head (*H*). Using maximum points on the hill diagram, optimal unit speed ($N\text{-}_{11o}$) is computed, and then, $N\text{-}_{11o}$ is converted to reference rotor speed (*Sopt*).

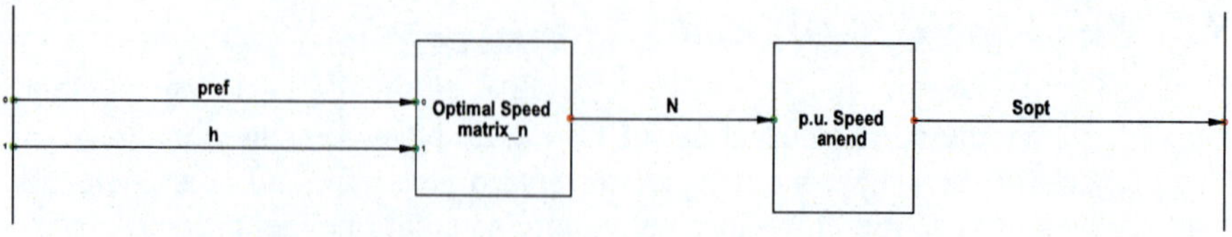

Fig. 5.4 Composite frame for PSHP system

5.4 Model Initialization

In DIgSILENT, all dynamic models and variables such as network voltages, active and reactive power flows, loadings, etc., are initialized according to the steady-state condition which is computed in load flow calculation. Therefore, it is recommended that the steady-state load flow is configured correctly before running a time-domain simulation. The initialization equations for each dynamic block of VS-PSHP shown in Fig. 5.2 are defined as follows:

Governor Block

```
inc(x)=0
inc(x1)=x4
inc(x2)=x1*R_t
inc(x4)=G
inc(xspeed)=Sopt
inc(Sopt)=1
inc(G)=pref
```

Hydraulic Block

```
inc(pturb) = (x-Q_nl)*h
inc(x)     = sqrt(H_s/(f_p1+f_p2+1/(G*G)))
inc(x1)    = x
inc(x2)    = H_s-x*x*f_p2
inc(h)     = pow((x/G),2)
inc0(G)    = 1
```

Active Power Reduction Block

```
inc(x1)=yi1
inc(x)=Fmeas
inc(pref_out)=pref
inc(pgn)=P
vardef(P)='KW';'Rated Active Power'
```

Hill Diagram Block

```
inc(pturb) = (x-Q_nl)*h
inc(x)     = sqrt(H_s/(f_p1+f_p2+1/(G*G)))
inc(x1)    = x
inc(x2)    = H_s-x*x*f_p2
inc(h)     = pow((x/G),2)
inc0(G)    = 1
```

5.5 Reactive Power Support

P-Q capability curve of a machine is needed not only for designing the controller of the plant but also for the power flow of the grid. Therefore, it is necessary to consider the P-Q capability curves of both SM and DFIM-based PSHPs in this section. A FS unit, equipped with the salient pole SM, can only control the reactive power of the plant through the excitation system, which adjusts the output voltage of the machine. Unlike the FS unit, the VS one using the DFIM can control the reactive power via both the stator winding, through the MSC and the GSC. Although using the DFIM stator and the GSC concurrently can be a cost-effective strategy to control reactive power, uncoordinated control of the reactive power can be caused the reactive current circulation between the GSC and the machine stator. Since the GSC is not considered in the model of DFIM, the power factor of the GSC is supposed to be unity for drawing the PQ capability curve. In other words, in this study, DFIM can only produce reactive power from the stator side. This work allows the user of DIgSILENT to define DFIM as PV bus in simulations. The active and reactive currents of understudy DFIM, in the per-unit system, can be listed in Table 5.1.

Figure 5.5 shows the active–reactive current capability diagram of the DFIM comparing to the SM. For VS units, the over-excited operation is restricted to the rated current of the MSC, i.e., the rotor winding current, while the rated current of the field winding limits over-excited operation of the FS unit. It should be noted that DFIM-based PSHP can control the reactive power via both the stator winding and the GSC. On the other hand, the total apparent current of the GSC is designed based on

Table 5.1 Nominal current of DFIM in p.u.

Power factor of the stator and the GSC are 0.9 and 1.0

Current	Active	Reactive	Apparent
Unit (p.u)	1.000	0.484	1.111
Stator (p.u)	0.935	0.484	1.053
MSC/rotor (p.u)	0.935	0.734	1.189
GSC (p.u)	0.0785	0.000	0.0785

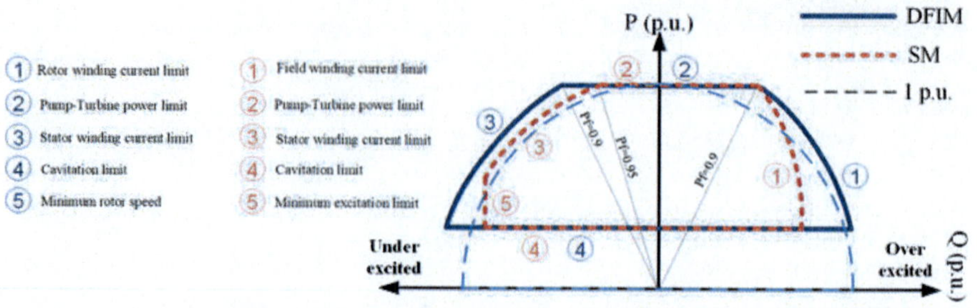

Fig. 5.5 Active–reactive current capability diagram of the DFIM and SM in generating mode

vector summation of the real and reactive current. It means that the capability of the exchanging reactive power of the GSC depends on the rotor speed operating point, i.e., the output real power of the PSHP. Note that, if the reactive power is needed at full real power production, the GSC must be over-sized.

5.6 Simulation Results

5.6.1 *Case Study*

New England 39-bus transmission network consists of 10 generators, 12 transformer, and 46 transmission lines as shown in Fig. 5.6. The technical parameters for the generation system and the transmission network are obtained from [10]. It should be mentioned that G_1, located at bus 39, is the interconnection of New England network to New York one. As reported in Table 5.2, all generators except G_{10} and G_1, i.e.,

Table 5.2 Parameters of generators for understudy 39-bus test system

Name	Bus#	P (MW)	Bus type	S (MVA)	V (p.u.)	Rated voltage (kV)	Plant type
G_1	39	1000	PV	10,000	1.03	345	–
G_2	31	–	Slack	700	0.982	16.5	Steam
G_3	32	650	PV	800	0.9831	16.5	Steam
G_4	33	632	PV	800	0.9972	16.5	Steam
G_5	34	254	PV	300	1.0123	16.5	Steam
G_6	35	650	PV	800	1.0493	16.5	Steam
G_7	36	560	PV	700	1.0635	16.5	Steam
G_8	37	540	PV	700	1.0278	16.5	Steam
G_9	38	830	PV	1000	1.0265	16.5	Steam
G_{10}	30	343	PV	381	–	18	Hydro

G_2-G_9, are steam plants. In other words, G_2-G_9 are represented by the synchronous generator with a round rotor which is modelled using six order, while salient pole rotor SM is modelled by fifth-order model [11]. The last generator (G_{10}) which is located at bus 30 is a hydro plant. In this study, G_{10} is substituted by either SM-based (salient pole rotor) or DFIM-based PSHP. DC exciters of all generators are IEEE type 1 available in DIgSILENT PowerFactory, which are equipped with a power system stabilizer (PSS). Also, all generators are equipped with generic governors, representing steam and hydro turbines. Moreover, DFIM is represented by a typical third-order model neglecting the stator transients which was described in Sect. 5.2.2.

5.6.2 Transient Stability of VS-PSHP in Comparison with FS-PSHP

According to grid codes of a different country, the PSHPs must be connected to the grid during the severe faults [13]. For example, a PSHP must not be disconnected from the grid when a 100% voltage drop occurs shorter than 150 ms. When the stator voltage drops to zero instantly, the DFIM-based PSHP may suffer from two important problems. First, over-currents in rotor and stator windings are due to containing the DC component of stator flux during the fault. Second, DC-link overvoltage is due to unbalancing between injected real power from the MSC and the GSC. In order to cope with the problems, deactivating the MSC during the fault with activating the crowbar is proposed as reported in Avce PSHP in Sloveni [4]. However, ABB in Grimsel 2 PSHP in Switzerland [14] uses the DC-link brake chopper, to activate during the fault in parallel to DC-link capacitors. It should be noted that the transient stability issues of DFIM-based PSHP during the severe fault is more than DFIM-based wind turbine when the crowbar protection method is used [3]. Also, the oscillations of the electrical torque on the rotor shaft are inevitable in both protection methods. While in the FS-PSHP, the power system stabilizer (PSS) is commonly used to attenuate the power swing in the large disturbances [1].

As aforementioned, hydro plant of 39-bus system placed at bus 30 is replaced by DFIM-based and SM-based PSHP units. In order to calculating the power flow, selecting PV type for bus 30, and available oscillations analysis in a large power system, model of GSC is ignored as mentioned in Sect. 5.2.2. In this study, a 150 ms three-phase short-circuit fault is applied at bus 30 in the network. Note that, the type of fault protection is crowbar in this study which DFIM converted to induction machine during of crowbar activation.

5.6.3 *Interconnected Power System Oscillations*

Several different modes of oscillation can be found in the interconnected system. One of them occurred when an energy exchange through the transmission lines arising from relative rotor oscillations between the network's various power plants. Their most famous are known as local and inter-area modes; (i) Local mode: Local modes are often associated with the oscillation angle of a the generator unit rotor to the rest of the system (infinite bus) or oscillations of the generator unit against another generator unit within the same area of the system, with oscillation frequencies varying from 0.7–2.0 Hz. On the other hand, aforementioned electromechanical oscillations are seen in SM-based generators, while those oscillations are not seen in DFIM-based PSHP. Also, DFIM does not introduce any new oscillation modes to the system due to its allowable speed variation range of known as flywheel impact [1]. Therefore, a PSS with different input signal such as machine speed, grid frequency, terminal voltage, and oscillating power is designed for SM-based generators to damp low-frequency oscillations in the range of 0.1–2 Hz. (ii) Inter-area mode: Such oscillations involves a group of generators belonging to a system area that is found when a network area oscillates in the range of 0.1–0.8 Hz against another area. Local and inter-area oscillations are analyzed for this chapter with a three-phase temporary disturbance near to PSHP bus bar, i.e., 150 ms short circuit at bus 30. Moreover, a 39-bus test system (New England network) is selected to assess effect of both DFIM-based and SM-based PSHP on the electromechanical oscillations (local and interarea modes) considering wind power plant.

5.6.3.1 Simulation Results with PSHP and Without Wind Farm

For the New England test system, shown in Fig. 5.6, it can be said that there are two areas in this system; Area 1 referring to the New York system is compactly represented by generator 10 and Area 2 represents the New England system and consists of nine generators. Local electromechanical oscillations are seen in FS-PSHP and another SM-based generator in steam plants, i.e., G2-G9. Figure 5.7a–e compare the simulation results in terms of FS and VS-PSHP for rotor speed, output voltage, reactive power, active power, and stator current, respectively. If the PSHP is the FS, frequency of the post-fault oscillations, associated with local mode, is 1.23 Hz in generating mode for this study. However, as obvious in Fig. 5.7, the VS-PSHP does not experience power swing (Fig. 5.7d) and the rotor speed oscillation (Fig. 5.7a) for generating mode. It should be noted that the rotor speed of the DFIM in the generating mode increases about 0.083 p.u. when the fault occurs and stabilizes after 41 s. Moreover, the stator voltage oscillations of the SM is more than the DFIM because of the reactive power absorption from the network (Fig. 5.7c). In both cases

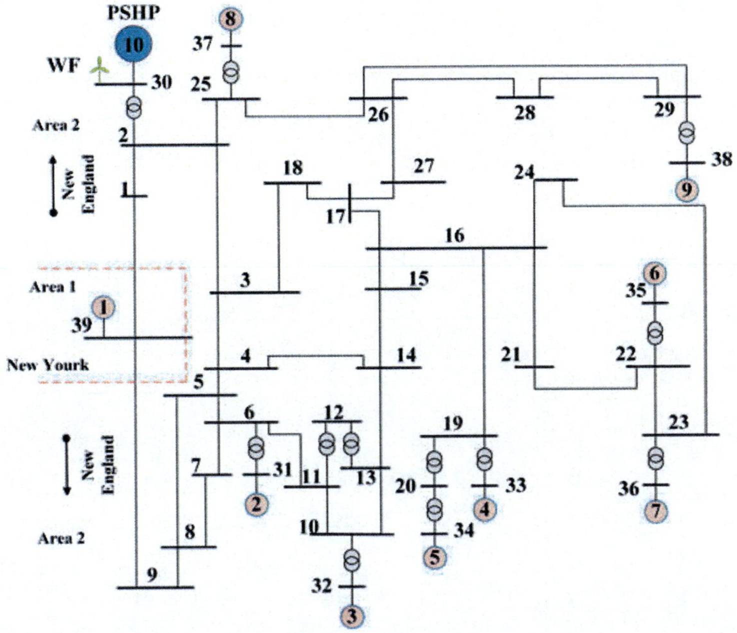

Fig. 5.6 Single line diagram of New England 39-bus test system [12]

(PSHP be either DFIM-based or SM-based), the output voltage of PSHP, i.e., voltage of bus 30 (Fig. 5.7b), reaches to zero during fault ($t = 10$ to 10.15 s) while the voltage recovers quicker to nominal value (1 p.u.) when the PSHP is DFIM comparing to SM. Another note should be mentioned that is about reactive power oscillations. The SM operates in both cases under-excitation and over-excitation after that the fault is cleared, which, according to Fig. 5.5, the SM limits from the under-excitation operation more than DFIM.

In the following, effect of DFIM-based and SM-based PSHP on inter-area oscillations between two areas available in 39-bus test system is considered. These oscillations are found when an area of the network swinging against another area. According to Fig. 5.8b, if the PSHP is a FS unit in the generating mode, inter-area mode due to swinging Area 1 (δ_1) against Area 2 (δ_2 is reference) has a frequency of 0.45 Hz, while the frequency of inter-area oscillation is 0.5 Hz, if the G_{10} is VS-PSHP. On the other side, maximum relative rotor angle separation can be used to evaluate the transient stability of a power system using a defined index as follows.

$$\eta = \max|\delta_i - \delta_{i0}| \quad (5.21)$$

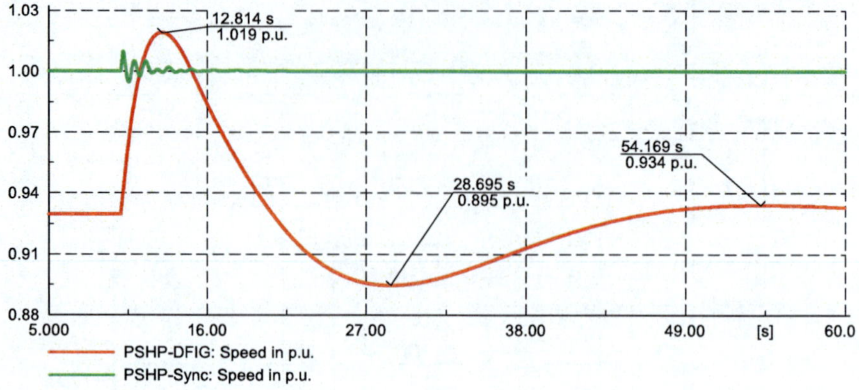

(a)

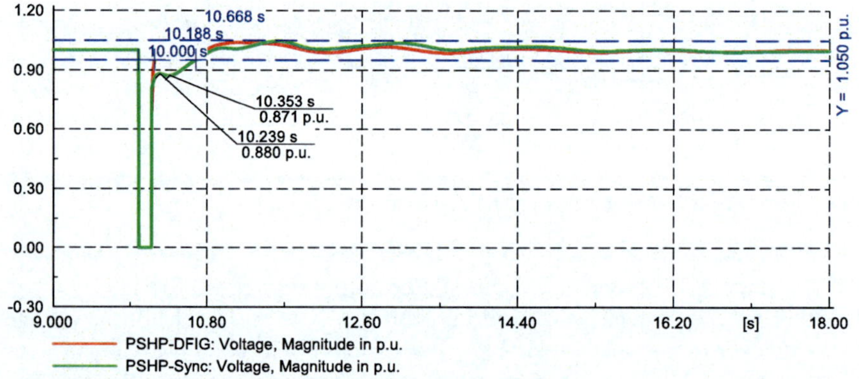

(b)

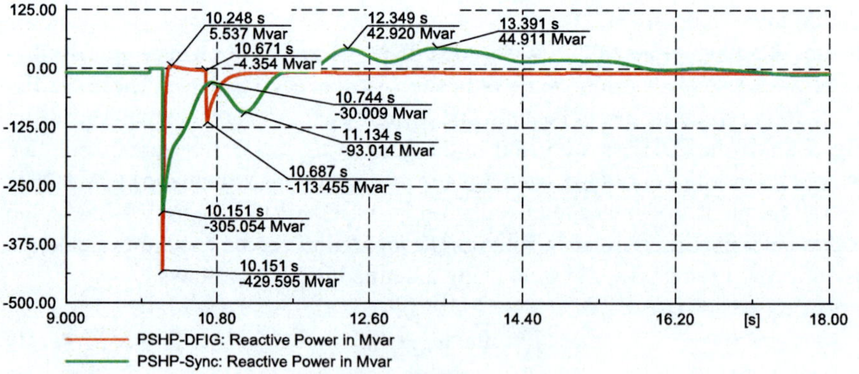

(c)

Fig. 5.7 Simulations for both PSHPs in generating mode without wind power,**a** rotor speed,**b** output voltage,**c** reactive power,**d** active power, and**e** stator current

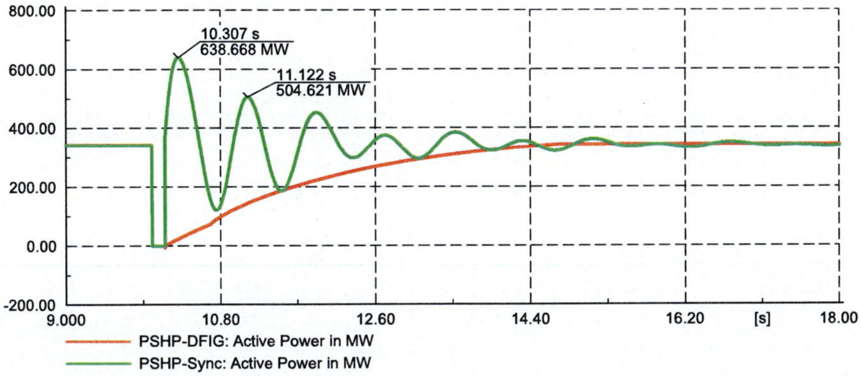

(d)

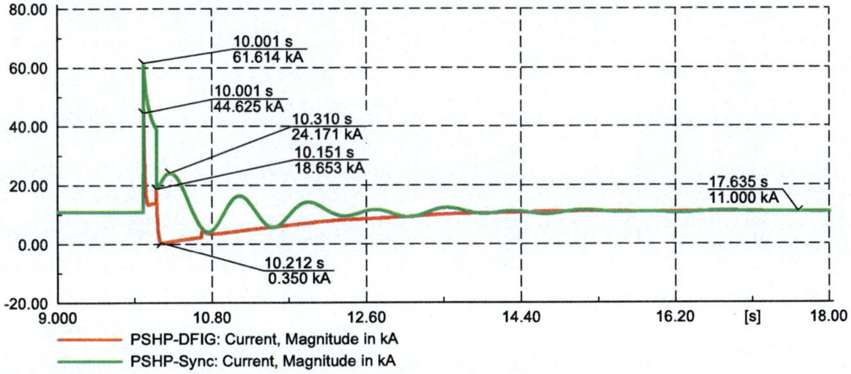

(e)

Fig. 5.7 (continued)

where η is transient stability index (TSI), δ_i is maximum relative rotor angle separation of*i*th SM (in degree), and δ_{io} is initial rotor angle of *i*th SM (in degree).

According to Fig. 5.8 a–b, TSI in term of δ_{12}, which is the inter-area mode, is 19, 27 for DFIM- and SM-based PSHP. Therefore, in summary, the existence of VS-PSHP can improve TSI for inter-area mode and voltage recovery time of power system comparing to FS-PSHP.

5.6.3.2 New England 39-Bus Test System with the Presence of Wind Farm

In this section, simulation results are presented when both DFIM-based PSHP and wind power simultaneously connected to bus 30 (100 MW wind power and 243 MW PSHP). Figure 5.9a–e compare the simulation results in term of FS and VS-PSHP

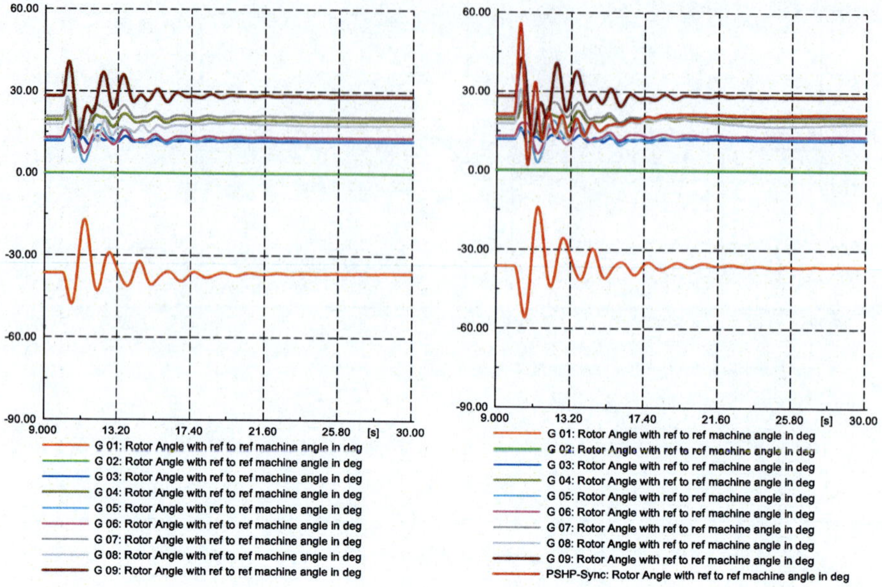

(a) (b)

Fig. 5.8 Relative rotor angle of G_1-G_{10} against G_2 without wind farm and PSHP**a** with DFIM-based,**b** with SM-based

for rotor speed, output voltage, reactive power, active power, and stator current, respectively. Same results can be found in those figures when wind power is added in bus 30. However, TSI for relative rotor angle of FS-PSHP is much smaller than the latter case (when the DFIM-based wind power is not). According to Fig. 5.10a–b, TSI in term of δ_{12}, which is the interarea mode, is 18, 27 for DFIM and SM-based PSHP in bus 30. Also, the voltage recovery time is significantly improved by utilizing of VS-PSHP instead of FS-PSHP.

5.7 Conclusions

This chapter presents modelling and controller of both DFIM-based and SM-based PSHP under generating operation mode using RMS-type simulation in DIgSILENT *PowerFactory*. Also, 10-machine 39-bus test system is used to assess effect of both types of PSHP on transient stability of large power system. Moreover, reactive power

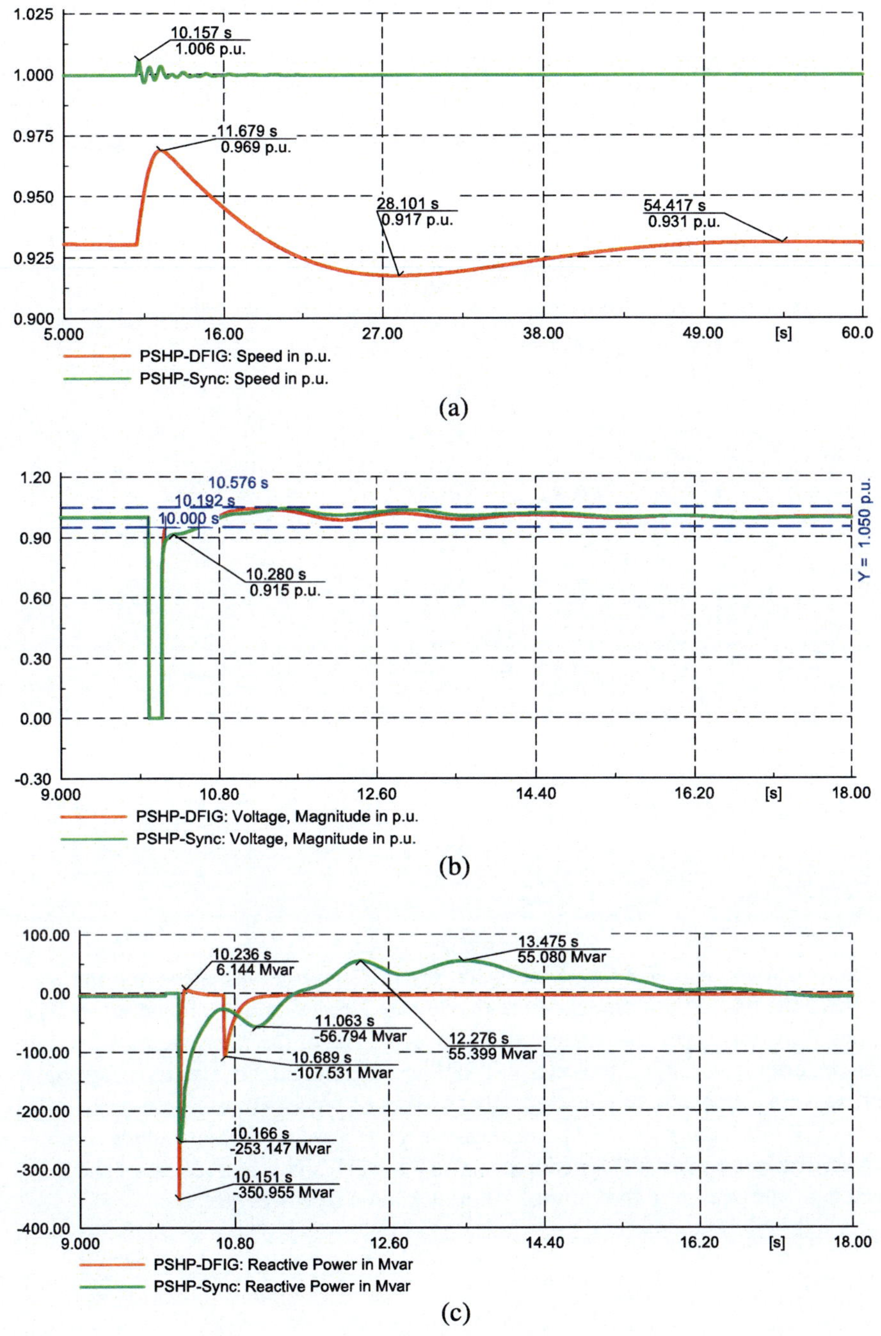

Fig. 5.9 Simulations for both PSHPs in generating mode with wind power **a** rotor speed **b** output voltage **c** reactive power **d** active power, ande stator current

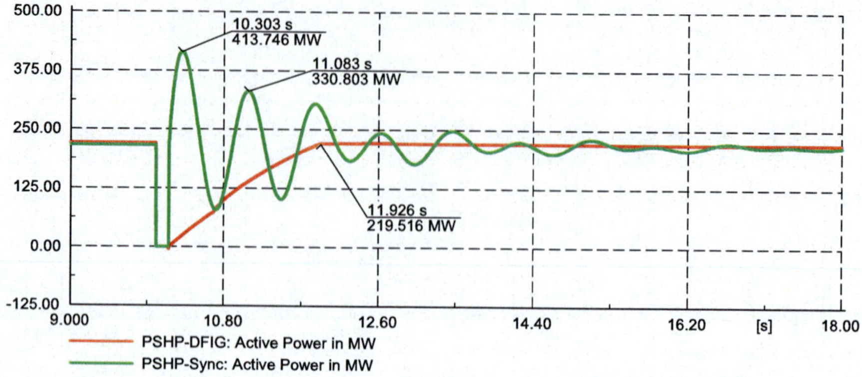

(d)

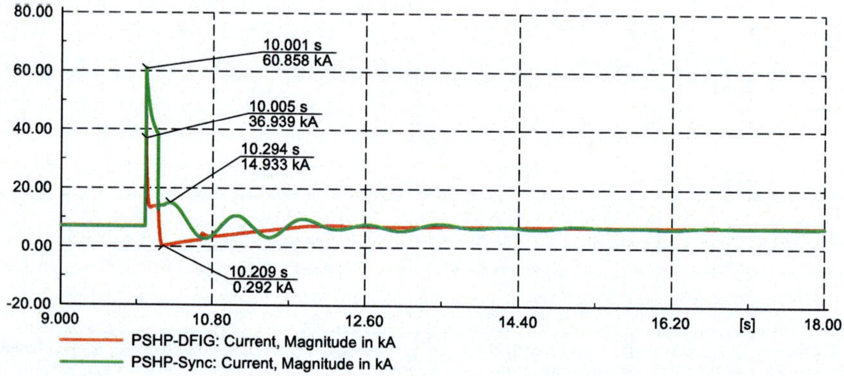

(e)

Fig. 5.9 (continued)

capability curves of both FS and VS were drawn. These curves show that the VS-PSHP has the high important role in both over- and under-excitation operation. The simulation results show that using DFIM-based PSHP in the interconnected power grids, not only the oscillation modes of PSHP are eliminated, but also it can strongly improve rotor angle and voltage transient stability of power system. In other words, using VS-PSHP can improve TSI in terms of inter-area mode and reduce voltage recovery time of power system comparing to FS-PSHP. This is because of the fast-injected power storing in the rotor of VS unit-known flywheel effect.

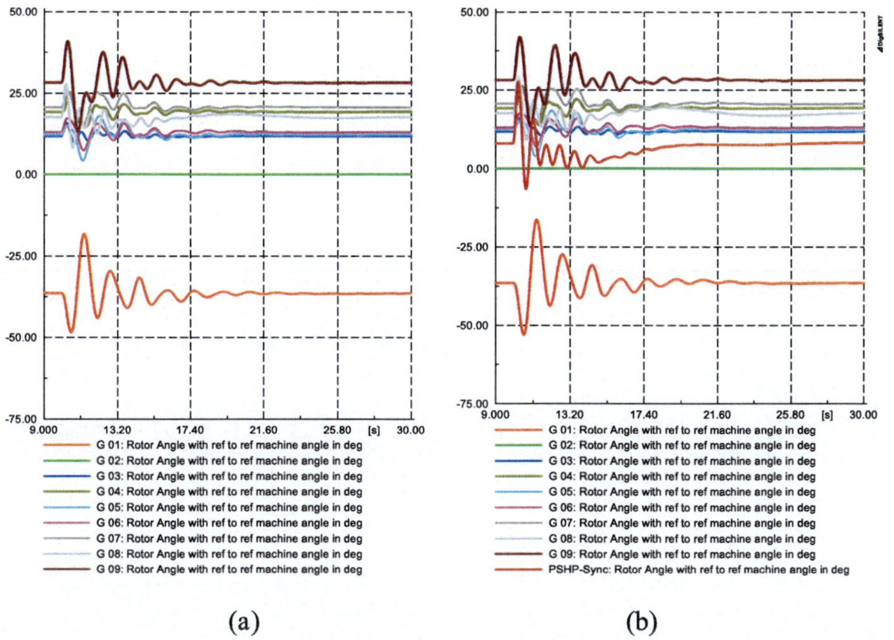

Fig. 5.10 Relative rotor angle of G_1-G_{10} against G_2 with wind farm and PSHP **a** with DFIM-based **b** with SM-based

Acknowledgements This chapter is the output of a research project supported by Yadegar -e-Imam Khomeini (RAH) Shahre Rey Branch, Islamic Azad University Iran. Also, the authors are thankful for the support from the National Natural Science Foundation of China (No.51809197, No.51879200).

References

1. A.C. Padoan et al., Dynamical behavior comparison between variable speed and synchronous machines with PSS. IEEE Trans. Power Syst. **25**(3), 1555–1565 (2010)
2. Y. Pannatier, B. Kawkabani, C. Nicolet, J.-J. Simond, A. Schwery, P. Allenbach, Investigation of control strategies for variable-speed pump-turbine units by using a simplified model of the converters. IEEE Trans. Ind. Electron.**57**(9), 3039–3049 (2010)
3. M.A. Bidgoli, H.A. Mohammadpour, S.M.T. Bathaee, Advanced vector control design for DFIM-based hydropower storage for fault ride-through enhancement. IEEE Trans. Energy Convers. **30**(4), 1449–1459 (2015)
4. V. Azbe, R. Mihalic, Transient stability of a large doubly-fed induction machine in a pumped-storage plant. Electr. Power Syst. Res. **142**, 29–35 (2017)
5. W. Yang, J. Yang, Advantage of variable-speed pumped storage plants for mitigating wind power variations: integrated modelling and performance assessment. Appl. Energy **237**, 720–732 (2019)

6. J. L. Rueda, A. W. Korai, J. C. Cepeda, I. Erlich, and F. M. Gonzalez-Longatt, "Implementation of Simplified Models of DFIG-Based Wind Turbines for RMS-Type Simulation in DIgSILENT PowerFactory," in *PowerFactory Applications for Power System Analysis*, Springer, 2014, pp. 197–220
7. J. L. Rueda, J. C. Cepeda, I. Erlich, A. W. Korai, and F. M. Gonzalez-Longatt, "Probabilistic Approach for Risk Evaluation of Oscillatory Stability in Power Systems," in *PowerFactory Applications for Power System Analysis*, Springer, 2014, pp. 249–266
8. M. Alizadeh Bidgoli and S. M. T. Bathaee, Full-state variables control of a grid-connected pumped storage power plant using non-linear controllers, *Electr. Power Components Syst.*, vol. 43, no. 3, pp. 260–270 (2015)
9. H. K. Pamsari, M. A. Bidgoli, M. Rajabzadeh, S. M. T. Bathaee, and S. Ozgoli, Application of a new multivariable sliding mode controller for the single machine infinite bus systems, in *2011 2nd Power Electronics, Drive Systems and Technologies Conference*, 2011, pp. 211–216
10. M. A. Pai, *Energy function analysis for power system stability*. Springer Science & Business Media (2012)
11. T. W. Stegink, C. De Persis, A. J. Van Der Schaft, An energy-based analysis of reduced-order models of (networked) synchronous machines, Math. Comput. Model. Dyn. Syst.**25**(1), pp. 1–39 (2019)
12. T. Surinkaew, I. Ngamroo, Hierarchical co-ordinated wide area and local controls of DFIG wind turbine and PSS for robust power oscillation damping. IEEE Trans. Sustain. Energy **7**(3), 943–955 (2016)
13. V. D. N. V. E.ON Netz GmbH,*TransmissionCode 2007. Network and System Rules of the German Transmission System Operators*, August, no. August 2007
14. S. Auber, Power on tap from variable speed pumped water storage scheme, *Energize*, no. July, 2012

Chapter 6
Implementation of a Generic Type 3 Wind Turbine Model in DIgSILENT PowerFactory

R. Villena-Ruiz, A. Honrubia-Escribano, and E. Gómez-Lázaro

Abstract In order to ensure grid stability and due to the growth of newly installed wind power capacity worldwide, conduction of transient stability analyses is one of the current challenges for Transmission System Operators (TSOs) and Distribution System Operators (DSOs). TSOs and DSOs need to provide a safe, reliable and sustainable service to consumers, safeguarding the security of electricity supply. Due to the variable nature of wind, the integration of wind power into power systems has an impact on power grid planning and operation that needs assessment. The implementation and subsequent dynamic simulation of wind turbines connected to the grid are, therefore, necessary to achieve all the above requirements. This is where the International Electrotechnical Commission (IEC) comes in. The IEC 61400-27-1 defined four generic simulation models of wind turbines—Types 1, 2, 3 and 4—for transient stability analysis, covering the four main topologies of real wind turbines available in the market. Given that the Type 3 wind turbine, which represents a doubly-fed induction generator, is the most common topology installed and the most technologically advanced wind turbine model, it is implemented in this chapter. Using the DIgSILENT Simulation Language (DSL) functionality available in DIgSILENT PowerFactory, this chapter presents an extensive description of the generic Type 3 wind turbine model, and its implementation and simulation, following the IEC

Electronic supplementary material The online version of this chapter (https://doi.org/10.1007/978-3-030-54124-8_6) contains supplementary material, which is available to authorized users.

R. Villena-Ruiz (✉)
Researcher at the Wind Power and Power Systems Section, Renewable Energy Research Institute, Universidad de Castilla-La Mancha, Calle de La Investigación, 1, Albacete, Spain
e-mail: Raquel.Villena@uclm.es

A. Honrubia-Escribano · E. Gómez-Lázaro
Researcher at the Wind Power and Power Systems Section, Renewable Energy Research Institute/Escuela Técnica Superior de Ingenieros Industriales, Universidad de Castilla-La Mancha, Calle de La Investigación 1, Albacete, Spain/Avda. de España, S/N, Albacete, Spain
e-mail: Andres.Honrubia@uclm.es

E. Gómez-Lázaro
e-mail: Emilio.Gomez@uclm.es

F. M. Gonzalez-Longatt and J. L. Rueda Torres (eds.), *Modelling and Simulation of Power Electronic Converter Dominated Power Systems in PowerFactory*, Power Systems, https://doi.org/10.1007/978-3-030-54124-8_6

61400-27-1 guidelines. In so doing, it details the step-by-step process followed to build the model and the adaptations required to simulate it in this powerful software tool.

Keywords Dynamic simulation · IEC 61400-27-1 · Transient stability · Type 3 (DFIG) wind turbine · Wind power

6.1 Introduction

Renewable energy sources are now essential to reduce the current heavy dependence on fossil fuels worldwide. In particular, wind power is positioned as the most important renewable energy source, with a global installed capacity of nearly 591 GW at the end of 2018 [1], which is expected to increase significantly in the coming years. Furthermore, large-size Wind Power Plants (WPPs) are being installed, which involves conducting in-depth transient stability analyses to ensure grid stability, also partly due to the variable nature of wind. The overall performance of a WPP, as well as the behaviour of each individual wind turbine (WT), must, therefore, be analysed.

Studies on detailed WT models, i.e. vendor-specific models, can already be found in the scientific literature [2, 3]. Contributions such as [4] analyse the operation procedure for the fault ride-through capability of a WPP composed of actual WTs. Further studies, such as [5 and 6], perform the validation of an actual mechanical model and a commercial DFIG WT model, respectively, while [7 and 8] address simulation analyses of very specific and complex WT simulation models. However, the generic WT models defined by Standard IEC 61400-27-1, which was published in 2015, are intended to faithfully represent the behaviour of WTs produced by any manufacturer, so that Transmission System Operators (TSOs) and Distribution System Operators (DSOs) do not depend on the complex, private WT models analysed in the works previously mentioned. In this regard, it should be noted that in October 2009 the International Electrotechnical Commission (IEC) formed a working group to begin developing the generic WT models. As some of the IEC Committee members had already been working with other entities related to the wind power industry, such as the Western Electricity Coordinating Council (WECC), the results achieved were highly positive [9]. The IEC working group developed two parts of the Standard: Part 1, focused on the development of the WT models and their validation procedures, and Part 2, focused on the WPP models and their validation procedures. The generic models were developed during 2011 [10], with the IEC 61400-27-1 Final Draft International Standard (FDIS) being issued three years later, in 2014, with its final publication being in February 2015 [11]. As IEC 61400-27-1 was already published and Part 2 duplicated information from Part 1, the initial structure of IEC 61400-27 was modified, thus bringing together both the WT and the WPP models in Standard IEC 61400-27-1 and their validation procedures in IEC 61400-27-2. The 2nd edition of IEC 61400-27-1, based on the new structure and including both parts, is currently under development [12].

In accordance with the considerations mentioned above, the four main topologies of actual WTs included in IEC 61400-27-1 are represented by the following types of generic WT models: (i) Type 1, which consists of an asynchronous generator directly connected to the grid, with no power converter; (ii) Type 2, consisting of an asynchronous generator equipped with a variable rotor resistance; (iii) Type 3, which uses a Doubly-Fed Induction Generator (DFIG), where the stator is directly connected to the grid and the rotor is connected through a back-to-back power converter; (iv) and Type 4, which is connected to the grid through a full-scale power converter.

Time-domain simulations of generic WTs constitute a useful tool to assess the dynamic performance of power systems, as well as to plan and ensure network operation. Furthermore, IEC-based generic WT models are intended to be used in fundamental frequency electromechanical stability studies, so that Root Mean Square (RMS) simulations, based on simplified models of electromechanical transients, are used. Working at this simulation level, the electromagnetic transient dynamics are not considered in the pertinent calculations, reducing the computational costs required. Moreover, IEC generic WT models are intended to represent the active and reactive power behaviour at the point of interconnection of the WT with the grid, i.e. at the WT terminals. Nevertheless, due to the recent publication of IEC 61400-27-1, the number of generic model implementations is currently quite low. Based on [13], and considering the development stages of Standard IEC 61400-27-1, it may be noted there are few studies related to the simulation of generic WT models, and even fewer when it comes to generic models simulated using DIgSILENT PowerFactory [14]. For example, studies such as [15 , 16] perform simulation or validation tasks on generic WT models using multidisciplinary simulation software tools such as MATLAB Simulink. The complete implementation process of the generic Type 3 WT using the DIgSILENT Simulation Language (DSL) available in DIgSILENT PowerFactory will be described in detail in this chapter [17], showing the most important control models as well as explaining how they work. Also, we address the modification process of some control blocks provided by IEC 61400-27-1 [11], so that they can be adapted to the features of this specific power system simulation software [18]. Finally, a transient stability analysis is carried out, analysing the performance of the WT by conducting a voltage dip of a certain magnitude and duration. The work developed is thus of great interest to TSOs, DSOs, research institutes, private companies and all stakeholders involved in the wind power industry.

The chapter is organised as follows: Sect. 6.2 addresses the description of the dynamic sub-models of the generic Type 3 WT defined by IEC 61400-27-1. Section 6.3 explains the implementation process of the dynamic sub-models in DSL, describing the modifications required and showing the models' structure once implemented. Section 6.4 describes the initialisation process and how dynamic simulations are conducted in DIgSILENT PowerFactory. Finally, Sect. 6.5 presents the results obtained when conducting the pertinent transient stability analyses, while Sect. 6.6 summarises the conclusions obtained.

6.2 Generic Type 3 WT Model Defined by IEC 61400-20-1

As mentioned in Sect. 6.1, Type 3 WT corresponds to the DFIG WT topology. Depending on the strategy used by the WT manufacturer to tackle voltage disturbances [13], the power converter is usually equipped with a chopper circuit, i.e. an insulated gate bipolar transistor connected in parallel with the power converter direct current (DC) circuit, which avoids WT disconnection due to voltage dips. Moreover, some Type 3 WTs incorporate a crowbar system [19, 20], which short-circuits the rotor against electromagnetic transients when a voltage dip occurs. During this time, the generator behaves like a squirrel-cage induction generator. In fact, the implementation of these protection systems is necessary to comply with the requirements of the existing legislation [21] on Type 3 WT behaviour against voltage dips. The emergence of a voltage dip in a DFIG WT will result in high induced currents in the rotor, as well as a large injection of power in the DC circuit that connects both sides of the power converter. Therefore, the WT must be protected against over-voltages and over-currents. Hence, the solutions adopted by WT manufacturers to evacuate the surplus power during a fault may be either a DC chopper system and/or an alternating current (AC) crowbar device [22].

Consequently, the WT design will vary depending on the manufacturer. The recently developed Standard IEC 61400-27-1 thus aims to bring together the capabilities of Type 3 WTs [23] to design a generic Type 3 WT model able to represent the behaviour of real ones faithfully. Furthermore, IEC 61400-27-1 distinguishes between two types of generic Type 3 WTs: Type 3A [24], and Type 3B, depending on the generator model used. The output signal of both generator models is a current injected through a current source with a parallel impedance, and the losses in the generator are neglected since the generator air gap power is equal to the measured power at the wind turbine terminals (WTT). The Type 3B generator model, as stated in [11], "*is the state-of-the-art simplification of the 3A generator set model with the addition of a crowbar model*". Therefore, as the generator system modelled in this chapter includes a crowbar system, the generic Type 3B WT model is the one implemented (henceforth referred to as the generic Type 3 WT model). The general structure of the generic Type 3 WT implemented is therefore explained in detail in the present section.

Five main dynamic sub-models are found within the modular structure of the generic Type 3 WT, Fig. 6.1: (i) *Pitch Control Model*; (ii) *Aerodynamic Model*; (iii) *Mechanical Model*; (iv) *Generator System Model*; and (v) *Generator Control Model*. *Generator Control Model* is, in turn, composed of four dynamic sub-models: (i) *P Control Model* (active power Control Model); (ii) *Q Control Model* (reactive power Control Model); (iii) *Q Limitation Model* (reactive power limitation model); (iv) *Current Limitation Model*. The operation of all these control models is explained in Sect. 6.3.3.

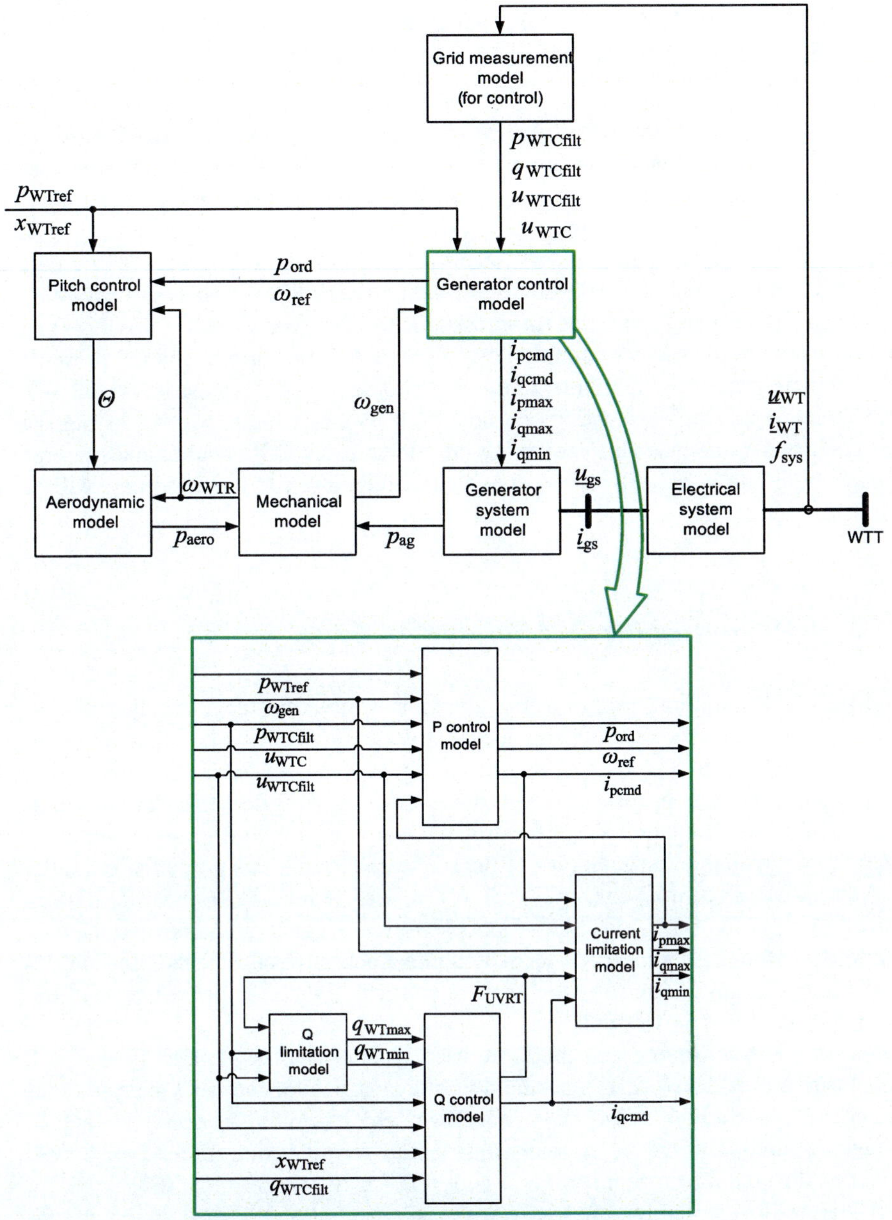

Fig. 6.1 Modular structure of the generic Type 3 WT (adapted from [11])

6.3 Implementation of Generic Type 3 WT Using the Dynamic Simulation Language (DSL)

DIgSILENT PowerFactory (PF) is regarded as one of the most advanced software tools in the field of power system analysis [25, 26], establishing itself as one of the world's leaders in wind energy systems because of the priority given to these types of renewable power plants, meeting the needs of various groups of interested parties, from research institutes to WT manufacturers. PF also has several WT models (known as general templates) available in the software library, such as fully-rated converted WTs (Type 4), variable rotor resistance WTs (Type 2) or DFIG WTs (Type 3). However, for the reasons already given in Sect. 6.1, the current chapter presents the complete process of implementation and simulation of the generic Type 3 WT [27]. It also provides explanations on how each dynamic model works, listing the particularities found when implementing this model using DSL. Names and symbols of each model's signals will also be specified, and their initialisation process will be detailed.

6.3.1 *Definition of the Test Network*

DIgSILENT Simulation Language is a dynamic simulation language that permits interaction between the electrical devices that are part of a network and their corresponding control models. One of the main characteristics of PF is the need to link the control models to at least one electrical device in the grid [28]. This means it is impossible to simulate a set of control models not linked to a predefined electrical device available in the library of the software [18]. In this particular case, the modular structure of the generic Type 3 WT model defined by IEC 61400-27-1 is entirely composed of dynamic control models—even the generator, known as *Generator System Model*, is defined by a set of dynamic blocks instead of a predefined DFIG available in the simulation software used, in contrast to the modular structure of the generic Types 1 and 2 WTs [11].

An AC Current Source was therefore used to interface the generic Type 3 WT model with the test network. Hence, the simplified test network consists of an external grid connected to a high-voltage terminal (HV), which is in turn connected to a low-voltage terminal (LV) through a two-winding transformer. The generic Type 3 WT, modelled through the current source, is connected to the LV terminal (Fig. 6.2).

A hierarchical structure is followed when working with DSL [29], in which two basic elements are considered: the grid objects or elements, equivalent to physical devices in power systems, including control devices, and their corresponding types stored in the library. Therefore, as explained in [18, 30], there may exist several identical *Elements* (*.Elm*) in the grid (*Network Data* side), and all of these may be related to the same *Type* (*.Typ*) within the library (*Library* side). This basic idea applies to all the components that are part of the dynamic model implemented: each *Element* in the

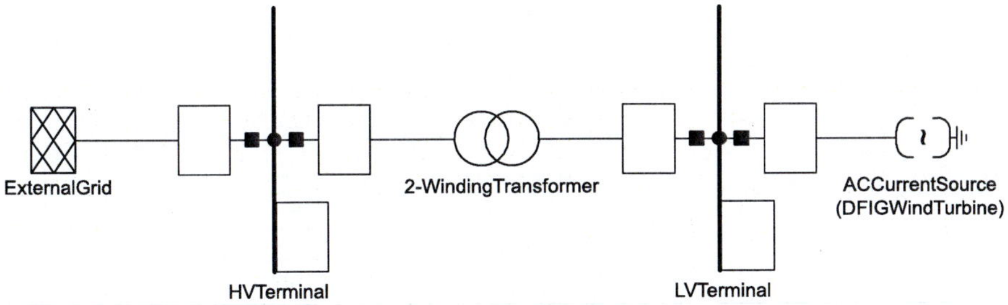

Fig. 6.2 Test Network defined in DIgSILENT PowerFactory

network has its corresponding *Type* stored in the library. Hence, following the steps detailed in [30], it can be said that the *Composite Frame* (*.BlkDef*) and the *Model Definition* (*.BlkDef*) are the first *Types* to be created, corresponding to the modular structures and control models of the WT in the *Library* side, respectively. Then, the equivalent models in the *Network Data* side are created, i.e. the *Composite Model* (*.ElmComp*) and the *Common Models* (*.ElmDsl*) are defined (*Elements*). Finally, the *Network Elements*, required to interface the control models with the external grid, are introduced.

6.3.2 Composite Frame of the Generic Type 3 WT

As a first step, following the guidelines established in [30], the *Slots* containing the WT dynamic control models, as well as the electrical device which connects them to the grid, must be defined in a new *Block/Frame Diagram* (Insert New Graphic → Block/Frame Diagrams). It is important to note that these *Slots* do not contain mathematical information such as equations or standard macros (gains, limiters or logic functions), but just the names of their input and output signals. Input signals are represented by green circles and output signals by red circles. Moreover, the signals establishing the relationships among all the models are also defined in this step. Figure 6.3 shows the *Composite Frame* (*.BlkDef*) of the generic Type 3 WT modelled in the present work, based on Fig. 6.1.

As Fig. 6.3 shows, a total of 13 *Slots* appear in the *Composite Frame* (*.BlkDef*), 8 of which (also*.BlkDef* in the *Library* side) are the main control models that simulate the behaviour of a generic Type 3 WT: *Pitch Control Model*, *Aerodynamic Model*, *Mechanical Model*, *Q Limitation Model*, *Q Control Model*, *Current Limitation Model*, *P Control Model* and *Generator System Model*. These are all user-defined models implemented by using the DSL language and are therefore referred to as*.ElmDsl* elements in the *Network Data* side. The *AC Current Source* is the built-in electrical device used to interconnect the WT model with the electrical network, i.e. an *ElmIac* element (the Type of element used can be consulted when selecting, creating or

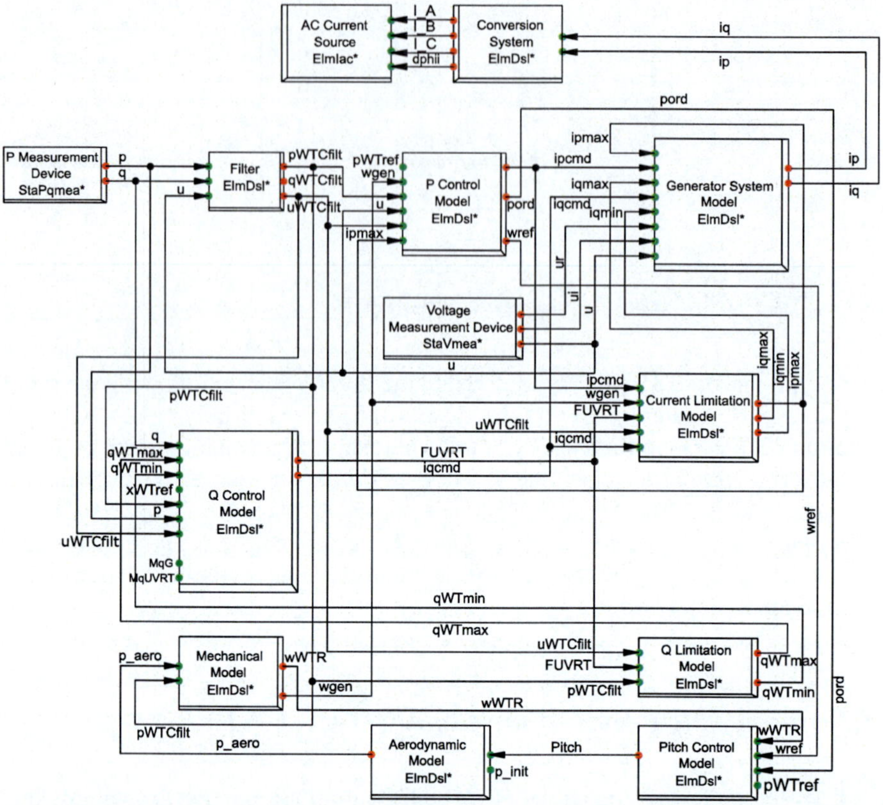

Fig. 6.3 Composite Frame of the Generic Type 3 WT Model

editing a device from the library). *Conversion System* and *Filter* models are also user-defined models necessary for adapting the output signals from the IEC WT model, so they are also identified as *ElmDsl* elements. Voltage and power measurement devices (*Voltage Measurement Device* and *P Measurement Device*) are used to provide the voltage, active power and reactive power signals to the corresponding models, as these types of signals cannot be used directly from the calculation parameters of the *AC Current Source. Voltage Measurement Device* and *P Measurement Device* are also predefined models available in PowerFactory and are referred to as *StaVmea* and *StaPqmea,* respectively.

6.3.3 Model Definitions

The second step is to create the *Model Definitions* (*BlkDef*) corresponding to each of the control models of the WT on the *Library* side. This is done by defining new

Block/Frame Diagrams (Insert New Graphic → Block/Frame Diagrams). However, instead of containing *Slots*, the *Model Definitions* will contain different mathematical blocks and multipliers, divisors, switches or summation points. These mathematical blocks may be directly selected from the PF global library or may be defined by the user, and will contain the equations characterising the behaviour of the block, all of which are defined according to the DSL language syntax [30]. Gains, delays, limiters, logic functions or first and second-order functions are some of the predefined or standard blocks available in the library. The generic Type 3 WT *Model Definitions* are shown and explained below. For a better understanding of the models, please see Fig. 6.3. Furthermore, a table summarising the symbol, base unit and description of all parameters, which are first described in the following sub-sections, is included in the Appendix.

6.3.3.1 Aerodynamic Model

The *Aerodynamic Model* (*BlkDef*) [31] implemented is the one-dimensional model, which neglects the effects of the rotor speed, but considers the pitch angle [11]. The output aerodynamic power (*p_aero*) depends on the wind speed (or initial wind power, *p_init*, the value of which must be set by the user), the pitch angle (*Pitch*) of the WT blades provided by the *Pitch Control Model,* the initial value of the pitch angle (*Pitch_wo*) and the aerodynamic gain (*K_a*), and can be simplified as shown in (1) and (2). Thus, the higher the value of the *Pitch*, the lower is the value of the aerodynamic power *p_aero*. The *Aerodynamic Model* provides the *Mechanical Model* with the aerodynamic power (*p_aero*), that will ultimately generate the injection of active power equal to that set by the user in both the *P Control Model* and the *Pitch Control Model*, i.e. equal to the active power reference (*pWTref*, see Fig. 6.3).

$$p\_aero = p\_init - \Delta P \tag{6.1}$$

$$p\_aero = p\_init - K\_a \cdot Pitch \cdot (Pitch - Pitch\_wo) \tag{6.2}$$

6.3.3.2 Pitch Control Model

The *Pitch Control Model* (*BlkDef*) [32], [13], shown in Fig. 6.4, calculates the value of the *Pitch* angle required to adjust the difference between the power order (*pord*) deriving from the *P Control Model* and the active power reference (*pWTref*, which is the user-defined active power setpoint at the wind turbine terminals, WTT), as well as between the wind turbine rotor rotational speed (*wWTR*, calculated by the *Mechanical Model*), and the reference rotational speed (*wref*) coming from the *P Control Model* (see Fig. 6.3). This means that the position angle of the WT blades is adjusted with the ultimate aim of reaching the active power setpoint value, avoiding

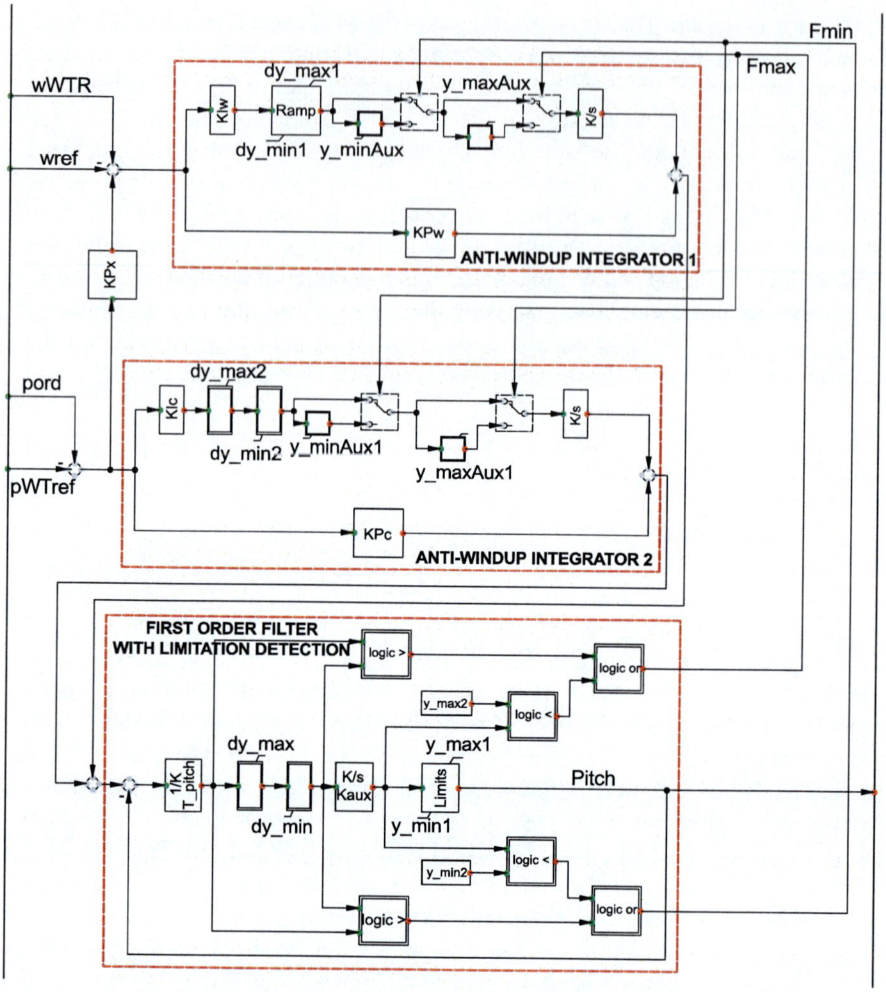

Fig. 6.4 Pitch Control Model implemented in DIgSILENT PowerFactory

a sudden increase in rotor speed. This is done through two PI controllers called *Anti-Windup Integrators*, which avoid the output signal becoming uncontrolled when very low or very high values of *Pitch* are reached. In brief, as stated in [11], the integrators will stop winding up if signal $Fmax = 1$ and stop winding down if $Fmin = 1$.

6.3.3.3 Mechanical Model

The *Mechanical Model* (*BlkDef*) is implemented as a two-mass model [33], in which the low-speed and the high-speed sides are represented [10, 11]. The WT shaft is modelled as a spring and a damper, represented by a drive train stiffness coefficient

(*K_drt*) and a drive train damping coefficient (*C_drt*). The WT rotor and the electrical generator are represented by their inertia constants (*H_WTR* and *H_gen*, respectively). Equations (6.3) and (6.4) define the relationships between these parameters. In these equations, T_{WTR} and T_{gen} are the aerodynamic and electrical torque values, respectively. The angles of the WT rotor and the generator are defined by parameters Θ_{WTR} and Θ_{gen}. Hence, under transient periods, when the WT is submitted to a torque imbalance, these equations represent the acceleration of the WT rotor and the decrease of the generator rotational speed, attempting to balance both sides of the shaft.

$$2H\_WTR \cdot \frac{dwWTR}{dt} = T_{\text{WTR}} - K\_drt \cdot (\theta_{\text{gen}} - \theta_{\text{WTR}}) - C\_drt \cdot (wWTR - wgen) \tag{6.3}$$

$$2H\_gen \cdot \frac{dwgen}{dt} = -T_{\text{gen}} + K\_drt \cdot (\theta_{\text{gen}} - \theta_{\text{WTR}}) + C\_drt \cdot (wWTR - wgen) \tag{6.4}$$

The generator rotational speed (*wgen*) and the *wWTR* are the output signals from this mechanical model, while the filtered active power which is measured at the WTTs (*pWTCfilt*), and *p_aero* are the required input signals. Figure 6.5 shows the two-mass mechanical model.

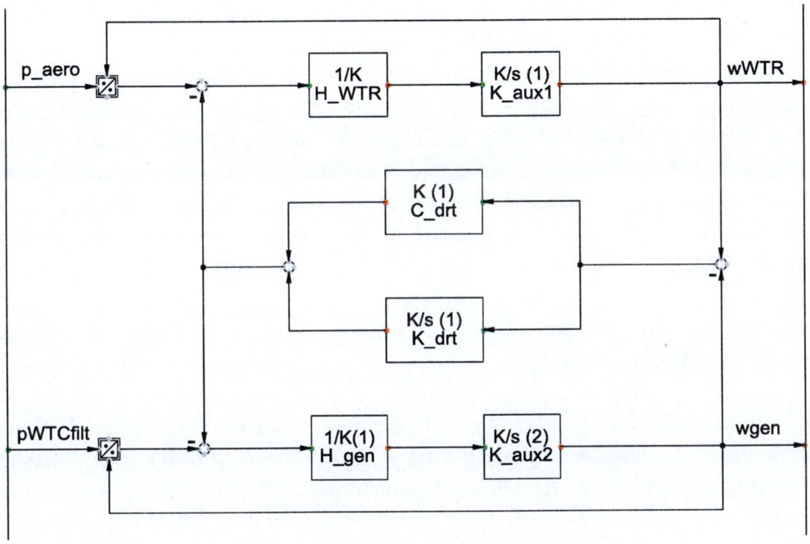

Fig. 6.5 Mechanical Model implemented in DIgSILENT PowerFactory

6.3.3.4 P Control Model

The *Torque PI*, the most important sub-model incorporated within the *P Control Model* (*BlkDef*) [11, 15], is intended to calculate the torque value required to rectify the error between *wgen* and *wref* (*wref* is the output of a lookup table which provides a rotational speed value depending on the measured active power at the WTTs once it is filtered, *pWTCfilt*). The output of the *Torque PI* sub-model, together with the measured voltage (*u*), the maximum active current (*ipmax*, coming from the *Current Limitation Model*) and *wgen*, form the power order from the WT controller, *pord*, which, as mentioned earlier, is the input to the *Pitch Control Model*. *pWTref* is another input signal used in the calculation of the torque value, while the active current command (*ipcmd*) is another output signal resulting from the *P Control Model,* which is in turn the input to the *Generator System Model*. In addition to its influence on the calculation of the torque value, *wref* also serves as input to the *Pitch Control Model.*

6.3.3.5 Q Control Model

The *Q Control Model*, (*BlkDef*) is able to operate under five general WT control modes, depending on the value of the general reactive power control mode *(MqG)*: voltage control, reactive power control, open-loop reactive power control, power factor control and open-loop power factor control [11, 15]. When operating in a reactive power control mode, the user can define the desired WTT reactive power reference (*xWTref*). The *Q Control Model*, through the reactive current command (*iqcmd*, which serves as input to the *Generator System* and *Current Limitation Models*, see Fig. 6.3), will thus be able to control the injection of reactive power at the WTTs following the *xWTref* signal. *MqUVRT*, the under-voltage ride-through control mode, sets the reactive current injection when a voltage dip occurs. *iqcmd* also depends on the under-voltage ride-through flag signal (*FUVRT*), which modifies its value according to the operation stage of the WT: normal operation, during a fault or a post-fault period.

6.3.3.6 Q Limitation Model

The *Q Limitation Model* (*BlkDef*) [11] or reactive power limitation model, shown in Fig. 6.6, is responsible for providing the *Q Control Model* with the maximum and minimum reactive power at the WTTs (*qWTmax* and *qWTmin*, see Fig. 6.3), which are dynamic values, i.e. values that may change during the simulation period: Its variation depends on *FUVRT*, *pWTCfilt*, and the measured voltage once it is filtered (*uWTCfilt*).

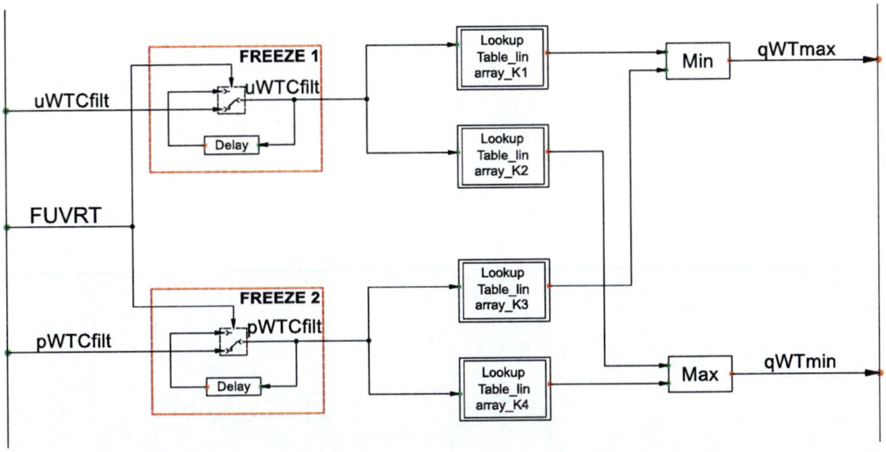

Fig. 6.6 Q Limitation Model implemented in DIgSILENT PowerFactory

6.3.3.7 Current Limitation Model

The *Current Limitation Model* (*BlkDef*) model sets the maximum and minimum limit values of reactive current allowed, as well as the maximum value of active current [11] (*iqmax*, *iqmin* and *ipmax*, respectively), as shown in Fig. 6.3. The variables resulting from the model depend on *uWTCfilt*, *ipcmd*, *iqcmd*, *wgen* and *FUVRT*.

6.3.3.8 Generator System

The *Generator System* (*BlkDef*) is divided into two different parts: the *crowbar* system [21] and the *Reference Frame Rotation* (Fig. 6.7). The *crowbar* system multiplies by zero the active and reactive power commands, *ipcmd* and *iqcmd*, when the measured voltage *u* is below a specific threshold, i.e. when a voltage dip of a specified depth occurs [34, 20]. In such situations, the generator behaves as a squirrel-cage asynchronous generator [19], as explained in Sect. 6.2. The *Reference Frame Rotation* coordinates the current phase according to the phase of the measured voltage, providing the *AC Current Source* with the final value of active and reactive current that will be injected into the grid, *ip* and *iq*. Nevertheless, these output signals from the IEC *Generator System* are expressed in rectangular coordinates, while the input current signals required by the predefined *AC Current Source* are the magnitude and phase of the current (phases A, B and C). For this reason, a user-defined *Conversion System* model is required, the behaviour of which is explained in Sect. 6.3.3.9.

There is a further difference between the *Generator System* implemented and that specified in Standard IEC 61400-27-1. As shown in the original generator model presented in [11], the value of the air gap power (output signal from the *Generator System,* which serves as input to the *Mechanical* model) is the real part of the

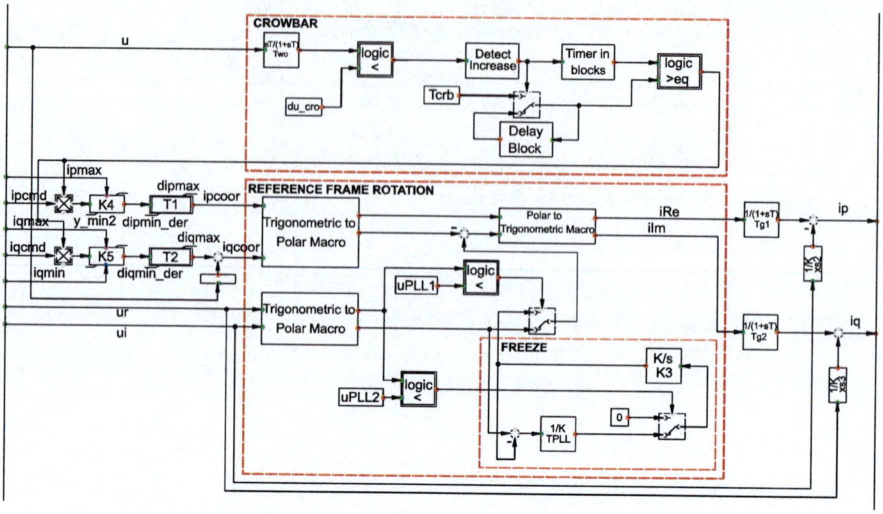

Fig. 6.7 Generator System Model implemented in DIgSILENT PowerFactory

result obtained from the multiplication of the voltage by the current. However, it is unnecessary to perform that calculation in PF: the *Power Measurement Device* (*StaPqmea*), connected at the WTTs, directly provides the *Mechanical* model with the power value.

6.3.3.9 Conversion System

The *Conversion System* (*BlkDef*), shown in Fig. 6.3, is based on a set of equations that convert the active and reactive current values, *ip* and *iq*, into the polar coordinates of the current, the magnitude of which is measured in kA. In the predefined *AC Current Source* model, since it is possible to independently control the magnitude and phase of the three components of the current (phases A, B and C), the input signals which must be defined are: I_A, I_B, I_C, φ_A, φ_B, and φ_C. However, as generic WT models are only focused on the analysis of positive-sequence balanced power systems, the magnitude of the three components of the current is equal (Eq. (6.5)); and the angles of phases A, B and C are always shifted 120 degrees from each other. In this regard, based on the technical documentation of the *AC Current Source* [35], which states that the angle of phase A is the sum of a user-defined angle named *dphii* and *phii* (the latter has a value of zero as the network frequency conditions do not change during the simulation, see Eq. (6.6)), the only two signals that must be calculated are *I_A* and *dphii*.

Therefore, Eqs. (6.7), (6.8) and (6.9) define the rest of the input signals. It should also be noted that φ_A, φ_B, and φ_C are automatically calculated by the software based on the *dphii* input signal.

$$I\_A = I\_B = I\_C \tag{6.5}$$

$$phii = 0 \tag{6.6}$$

$$\varphi_A = phii + dphii = dphii \tag{6.7}$$

$$\varphi_B = \varphi_A - 120^{\circ} = dphii - 120^{\circ} \tag{6.8}$$

$$\varphi_C = \varphi_A + 120^{\circ} = dphii + 120^{\circ} \tag{6.9}$$

6.3.3.10 Filter Model

The *Filter Model* (*BlkDef*) uses first-order delay blocks to filter the values of the active power, reactive power and voltage, *p*, *q* and *u* (see Fig. 6.3). The output resulting signals are *pWTCfilt*, *qWTCfilt* and *uWTCfilt*.

Once the *Composite Frame* and the *Model Definitions* of the control models in the *Library* side have been defined, the *Composite Model* (*ElmComp*) and the *Common Models* (*ElmDsl*), equivalent to the *Composite Frame* (*BlkDef*) and the *Model Definitions* (*BlkDef*) in the *Network Data* side, respectively, must be created.

6.3.4 Composite Model

Within the *Data Manager* window, the *Composite Model* (*ElmDsl*), which is equivalent to the *Composite Frame (BlkDef)* in the *Network Data* side, must be defined by the user. It must be created within the grid of the active project by clicking on the *New Object* button (Network Data → Grid → New Object → Composite Model). The previously defined *Composite Frame* must be linked to the *Composite Model* to automatically make all the *Slots* appear in the pop-up window (Frame → Select Project Type → Composite Frame).

6.3.5 Common Models

The fourth step is to create the *Common Models* (*ElmDsl*) in the *Network Data* side, which are the equivalent items to each of the *Model Definitions* (*BlkDef*) in the *Library* side. Within the pull-down menu that appears in the *Data Manager* window, the *Common Models* must be defined one by one inside the *Composite Model* created in the previous step (New Object → Common Model). After that, each *Common Model* created must be linked to its corresponding *Model Definition* to automatically make all the parameters appear (Model Definition → Select Project Type → corresponding Model Definition). The values of all these parameters can also be edited and set in this step (gains, time constants…) [10].

6.3.6 Common Models to Composite Model Size Cross-Reference

The final step is to interconnect all the objects in both the *Network Data* and the *Library* sides. In the present case, the network element used to interface the generic Type 3 WT model with the test grid, the *AC Current Source* (*ElmIac*), as well as all the *Common Models (ElmDsl)* corresponding to the dynamic control models that are part of the WT, must be placed within their corresponding *Slot* in the *Composite Model* (*ElmComp*). Figure 6.8 presents the window that pops up when relating the *Common Models* and the *AC Current Source* to the *Slots* that appear in the *Composite Model.*

6.4 Model Initialisation and Dynamic Simulation of Generic Type 3 WT

Since IEC dynamic generic WT models are designed to provide accurate responses for transient stability analyses, simulations are conducted by using RMS models,

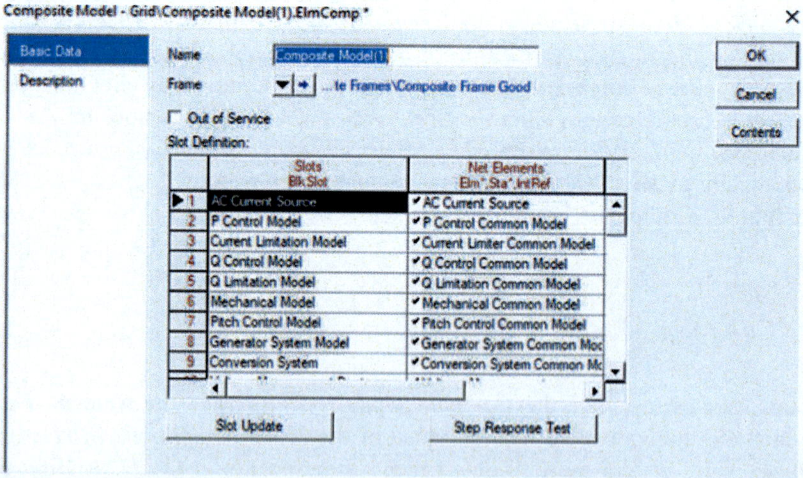

Fig. 6.8 Common Models to Composite Model Size Cross-Reference

which are based on simplified models of electromechanical transients. Before starting the simulation, the initial condition of the power system must be known to avoid fictitious transients during the simulation. The first step is to perform a load flow calculation ('*Calculate Load Flow*') to know the steady-state condition of the power system. In fact, electrical devices are initialised in this way when using this simulation software. To do this, the user must provide these devices with their required input signals, such as, for example, the rated current in the case of the AC Current Source. Hence, once the load flow calculation is executed and all the electrical components of the power system are initialised, the initialisation process of the dynamic models implemented using DSL start. User-defined control models must be initialised in a sequential manner, in such a way that the known signals deriving from a model are used as the input signals to initialise another model. This will be done by setting to zero the derivatives of all state variables and defining, within each model, the unknown signals in accordance with the known ones.

The following sub-sections are devoted to explaining and defining the initial condition equations set in some of the dynamic control models. They also aim to provide an overall understanding of the way in which dynamic models are initialised in PF.

6.4.1 Setting of Initial Conditions

The *Mechanical Model* is taken as an example of the initialisation process (Fig. 6.5). When working under steady-state conditions, the mechanical torque and the electromagnetic torque must be equal to avoid the WT being accelerated or decelerated (i.e. under steady-state conditions the WT works at constant speed). Hence, it can be said that the aerodynamic power (*p_aero*) is equal to the filtered active power (*pWTCfilt*), and the WT rotor rotational speed (*wWTR*) equal to the generator speed (*wgen*), both in turn equal to the reference rotational speed (*wref*). *wref* depends on a lookup table (within the *P Control* Model) that provides a value depending on *pWTCfilt*. Thus, since the WT is operating at full load conditions (*pWTCfilt* $\approx$ 1.0 pu), 1 pu is the value provided by the lookup table in this case, and 1.0 pu is hence the value set by the user as the initial condition for *wref*. *p_aero* and *pWTCfilt* are equal to the active power reference (*pWTref*), which is the user-defined active power setpoint at the WTTs (set also as 1.0 pu in steady-state).

Therefore, once we know the unknown input and output signals, and which are provided by other dynamic models, the initialisation process may begin [30]. In this case, the input signals *p_aero* and *pWTCfilt* are provided by the *Aerodynamic* and *Filter* models, respectively. *wgen* is provided by the *P Control* Model, in which the initial condition "*wgen* = *wref*" is set. Hence, the initial condition of *wWTR* must be set in the '*Equations*' window of the *Model Definition* of the two-mass *Mechanical* model. Furthermore, as previously mentioned, the derivatives of the state variables within each model must be set to zero (in this case *x*, *x1* and *x2*), which means that working under a steady-state condition, their output values must be constant. Hence, as there exists no error between the signals, they can be easily defined according to the input and output signals of the model.

The following paragraphs define the initial conditions set within the '*Equations*' window of the two-mass *Mechanical Model* (double click on the corresponding diagram, which is the *Model Definition* of the Mechanical model). The definition of the parameters and the initial conditions are set in this window.

```
! ### Definition of Parameters and Initial Conditions of the Two-Mass Mechanical Model
###
vardef(H_WTR) = 's';'Inertia constant of WT rotor' !Definition of Parameters
vardef(H_gen) = 's';'Inertia constant of electrical generator'
vardef(K_drt) = 'p.u.';'Drive train stiffness coefficient'
vardef(C_drt) = 'p.u.';'Drive train damping coefficient'

inc(wWTR) = wgen !Initial Condition of Unknown Output Signal

inc(x) = wWTR !Initial Conditions of State Variables
inc(x1) = wgen
inc(x2) = (pWTCfilt/wgen)
```

Based on Fig. 6.3, Eq. (6.1) of Sect. 6.3.3.1 and the explanations given in the previous paragraphs, the following lines present the definition of the parameters and the initial conditions set within the *Model Definition* of the *Aerodynamic* model.

```
! ### Definition of Parameters and Initial Conditions of the Aerodynamic Model ###

vardef(Pitch_wo) = 'deg';'Initial pitch angle' !Definition of Parameters
vardef(K_a) = '-';'Aerodynamic gain'

inc(p_aero) = 1 !Initial Conditions of Unknown Signals
inc(p_init) = 1
inc(Pitch) = sqrt((p_init-p_aero)/K_a)
```

6.4.2 *Running a Simulation*

Once the initial conditions are set, the buttons '*Calculate Load Flow*' and '*Calculate Initial Conditions*' must be pressed. In the window that pops up when clicking the second button, in the '*Network Representation*' level, the option '*Balanced, Positive Sequence*' must be chosen. Then, the '*Start Simulation*' button, after setting the absolute simulation time, must be executed.

6.5 Results

For this study, the behaviour of the generic Type 3 WT model is studied under fault conditions. Hence, a voltage dip is defined through the conduction of a balanced three-phase short-circuit of certain values of resistance and reactance at the LV terminal represented in Fig. 6.2. In order to do so, the '*Simulation Events/Fault*' option within the active '*Study case*' pull-down menu must be selected. Once there, click on the button '*New Object*'. In the window that appears, select the option '*Short-Circuit Event (EvtShc)*'. Now, in the new window, choose the LV terminal as the '*Object*' in which the short-circuit is defined, also defining the '*Fault Type*' (3-Phase Short-Circuit), the '*Execution Time*' desired and the fault's resistance and reactance values (these values will determine the residual voltage, i.e. the minimum voltage value reached during the voltage dip). Then, another '*Short-Circuit Event (EvtShc)*' must be defined in the same terminal ('*Object*' option) to clear the short-circuit and hence, the '*Clear Short-Circuit*' option will be chosen as the '*Fault Type*'. In this second case, the '*Execution Time*' will determine the timeframe in which the short-circuit occurs, as it indicates the precise simulation time in which the fault is cleared.

Working at full load operation, a voltage dip with a duration of 2000 ms and a residual voltage of 0.76 pu is defined at the LV terminal of the test network to which the generic Type 3 WT is connected (Fig. 6.2). The active and reactive power responses of the WT are shown in Figs. 6.9 and 6.10, respectively.

When the voltage drop in the WT terminals is above a specific value, the crowbar system, which short-circuits the rotor against electromagnetic transients and makes the generator behave like a squirrel-cage induction generator, is activated, as shown in Fig. 6.7 and explained in Sects. 6.2 and 6.3.3. The crowbar system will, therefore, be activated both when the voltage dip occurs and when it is later cleared, and its effects can be observed both in the active and reactive power responses of the WT (Figs. 6.9 and 6.10). At those points, a reduction of the active power injection is observed, particularly at the beginning and the end of the voltage dip, in which a value near

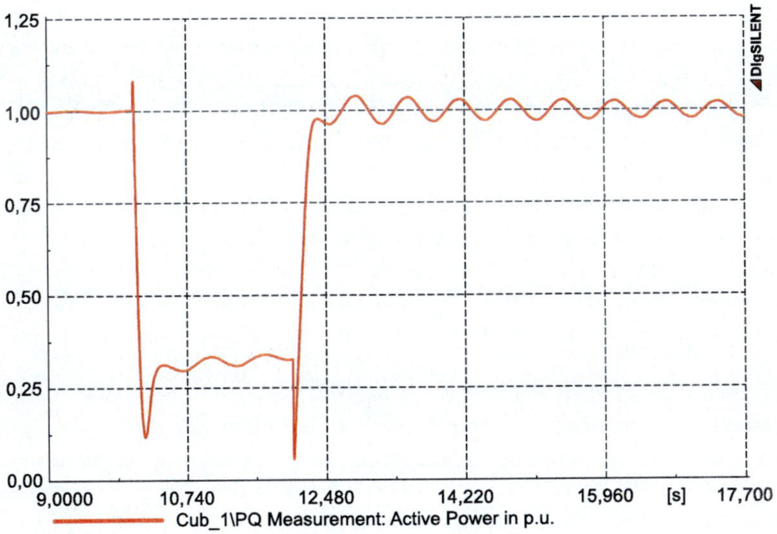

Fig. 6.9 Active power response. Full load conditions, voltage dip with a duration of 2000 ms and a residual voltage of 0.76 pu

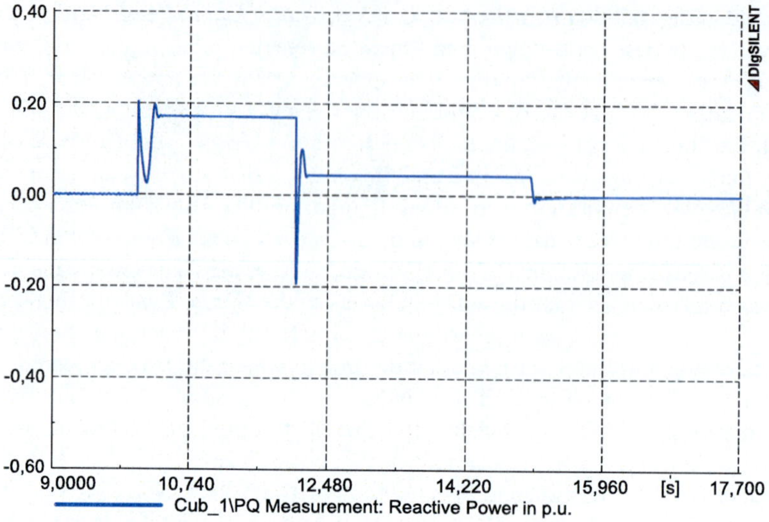

Fig. 6.10 Reactive power response. Full load conditions, voltage dip with a duration of 2000 ms and a residual voltage of 0.76 pu

to zero is reached. Nevertheless, the development of a generic crowbar model is challenging since each crowbar system differs depending on the WT manufacturer. This is also usually associated with a low level of accuracy when attempting to represent the sub-synchronous behaviour of actual WTs [14, 36]. Regarding the simulation period after the voltage dip, it can be said that the amplitude and the phase shift of the WT's active power response depend greatly on the setting of the two-mass mechanical model parameters. Concerning the reactive power response of the WT, a post-fault reactive current is injected into the grid for a few seconds, which causes the injection of reactive power (see Fig. 6.10). In this sense, the injection of reactive power during the voltage dip aims to stabilize the voltage. Finally, it is worth noting that the simulation results obtained in the present study are highly consistent with the simulation responses of the generic Type 3 WT model and the field tests conducted in an actual DFIG WT, shown in [14 and 36].

6.6 Conclusions

The conduction of time-domain simulations focused on the dynamic performance of power systems is undoubtedly important for TSOs and DSOs. Furthermore, the analysis of the behaviour of WTs connected to the grid is also of specific importance, as wind energy is currently a rapidly expanding technology. Proof of this is the reports provided by entities such as *Wind Europe* (formerly known as the *European Wind Energy Association*, EWEA), which state that, among other aspects, Europe added 11.7 GW of wind energy in 2018. However, as most of the existing dynamic simulation models of WTs are developed by private companies and their behaviour is typically very specific, generic and publicly available dynamic models of WTs are also required to represent the behaviour of actual WTs for transient stability analysis. Hence, the International Electrotechnical Commission (IEC), in its Standard IEC 61400-27-1, published in February 2015, developed four generic-also known as simplified or standard- dynamic WT models, representing the four main types of technologies.

As the Type 3 model is the most technologically advanced WT and currently the most widely installed, this chapter has addressed, starting from scratch, the implementation and simulation of the generic Type 3 WT model, using DIgSILENT PowerFactory, a highly advanced software in the field of power system analysis, which has not been greatly used in conducting dynamic analyses of IEC-developed WT models. DSL, a simulation language originally developed for this software, was the main tool used to model the WT. The chapter has specifically detailed the steps that must be followed by a DSL user to completely implement a generic Type 3 WT simulation model, with these steps also being useful to learn how to build other types of dynamic power systems. It has also addressed the specific adaptation tasks carried out to be able to connect the generic Type 3 WT dynamic model to the required external grid and has provided an overview of the initialisation process of the dynamic models (including the setting of initial conditions within them). Finally, with the aim of testing the transient performance of the WT under fault conditions, a balanced three-phase short-circuit was conducted at the LV WTT. The results obtained in the present study are in line with the findings of other studies published, in which field tests of actual DFIG WTs in operation and simulation responses of IEC Type 3 WT models are analysed and compared.

In light of the above, it can be concluded that DIgSILENT PowerFactory is a versatile and multifunctional software that allows the user to implement and simulate complex dynamic power systems from the beginning, and to test their performance. It also offers a wide range of possibilities regarding the types of analyses that can be conducted, being at the forefront of renewable energy technologies.

Acknowledgements This work was partially supported by the Spanish Ministry of Economy and Competitiveness and the European Union-FEDER Funds, ENE2016-78214C2-1-R-.

References

1. GWEC, *Global Wind Report 2018*, Global Wind Energy Council, Tech. Rep., 2019
2. S. Seman, J. Niiranen, R. Virtanen, J. P. Matsinen, Low voltage ride-through analysis of 2 MW DFIG wind turbine—Grid code compliance validations. IEEE Power. Energy. Soc. Gen. Meet. pp. 1–6 (2008)
3. A. Timbus, P. Korba, A. Vilhunen, G. Pepe, S. Seman, J. Niiranen, in *Simplified Model Of Wind Turbines with Doubly-Fed Induction Generator*, Proceedings of the 10th Wind Integration Workshop, 2011, p. 6

4. F. Jimenez, E. Gómez-Lázaro, J.A. Fuentes, A. Molina-García, A. Vigueras-Rodríguez, Validation of a DFIG wind turbine model and wind farm verification following the Spanish grid code. Wind. Energy. **15**(4), 645–659 (2012)
5. F. Jiménez, A. Vigueras-Rodríguez, E. Gómez-Lázaro, J.A. Fuentes, A. Molina-García, Validation of a mechanical model for fault ride–through: Application to a Gamesa G52 commercial wind turbine. IEEE Trans. Energy Convers. **28**(3), 707–715 (2013)
6. F. Jiménez, E. Gómez-Lázaro, J.A. Fuentes, A. Molina-García, A. Vigueras-Rodríguez, Validation of a DFIG wind turbine model submitted to two-phase voltage dips following the Spanish grid code. Renew. Energy **57**, 27–34 (2013)
7. V. Akhmatov, B. Andresen, J. N. Nielsen, K. H. Jensen, N. M. Goldenbaum, J. Thisted et al. Unbalanced short-circuit faults: siemens wind power full scale converter interfaced wind turbine model and certified fault-ride-through validation. Eur Wind. Energy. Conf. Exhib. p. 9 (2010)
8. L. Trilla, O. Gomis-Bellmunt, A. Junyent-Ferre, M. Mata, J.S. Navarro, A. Sudria-Andreu, Modeling and validation of DFIG 3-MW wind turbine using field test data of balanced and unbalanced voltage sags. IEEE Trans. Sustain. Energy. **2**(4), 509–519 (2011)
9. P. Pourbeik, *Proposed Modifications to the WT3 and WT4 Generic Models*. Tech. rep. Electric Power Research Institute, (2011)
10. A. Honrubia Escribano, E. Gómez-Lázaro, A Vigueras-Rodríguez, A. Molina-García, J.A. Fuentes, E. Muljadi, *Assessment of DFIG Simplified Model Parameters Using Field Test Data.* In Proceedings of the IEEE symposium on power electronics and machines for wind application, pp. 1–7 (2012)
11. IEC 61400-27-1. Electrical simulation models—wind turbines, Feb 2015
12. J. Fortmann, N. Miller, Y. Kazachkov, J. Bech, B. Andresen, P. Pourbeik et al., *Wind plant models in IEC 61400-27-2 and WECC—latest developments in international standards on wind turbine and wind plant modeling*. In Proceedings of the 14th wind integration workshop, p. 5 (2015)
13. A. Honrubia-Escribano, E. Gomez-Lazaro, J. Fortmann, P. Sørensen, and S. Martin-Martinez, generic dynamic wind turbine models for power system stability analysis: a comprehensive review. Renew. Sustain. Energy. Rev. **81**, Part 2 (2018)
14. R. Villena-Ruiz, A. Honrubia-Escribano, J. Fortmann, E. Gómez-Lázaro, Field validation of a standard type 3 wind turbine model implemented in DIgSILENT-PowerFactory following IEC 61400-27-1 guidelines. Int. J. Electr. Power. Energy. Syst. **116** (2020)
15. A. Lorenzo-Bonache, R. Villena-Ruiz, A. Honrubia-Escribano, and E. Gómez-Lázaro, *Operation of active and reactive control systems of a generic type 3 WT model*, in Compatibility, Power Electronics and Power Engineering (CPE-POWERENG), 2017 11th IEEE International Conference on. IEEE, 2017, pp. 606–610
16. R. Villena-Ruiz, F. Jimenez-Buendia, A. Honrubia-Escribano, A. Molina-García, E. Gomez-Lazaro, Compliance of a generic type 3 wt model with the spanish grid code. Energies **12**, 1631 (2019)
17. A.D. Hansen, C. Jauch, P. Sørensen, F. Iov, F. Blaabjerg, *Dynamic wind turbine models in power system simulation tool DIgSILENT*, Riso-DTU National Laboratory, 2007
18. D. GmbH, DIgSILENT PowerFactory v14—user manual (2012). Available from: http://www.digsilent.de/
19. F. Jiménez Buendía, B. Barrasa Gordo, *Generic Simplified Simulation Model For Dfig With Active Crowbar*. In Proceedings of the 11th wind integration workshop, p. 6 (2012)
20. M.B.C. Salles, K. Hameyer, J.R. Cardoso, A.P. Grilo, C. Rahmann, Crowbar system in doubly fed induction wind generators. Energies **3**, 738–753 (2010)
21. M. Seyedi, *Evaluation of the DFIG wind turbine built-in model in PSS/E, [Master's thesis]* (Chalmers University of Technology, Gothenburg, Sweden, 2009)
22. J.J. Justo, F. Mwasilu, J.-W. Jung, Doubly-fed induction generator-based wind turbines: a comprehensive review of fault ride-through strategies. Renew. Sustain. Energy Rev. **45**, 447–467 (2015)
23. I. Hiskens, Dynamics of Type-3 wind turbine generator models. IEEE Trans. Power Syst. **27**(1), 465–474 (2012)

24. C. Subramanian, D. Casadei, A. Tani, P. Sorensen, F. Blaabjerg, P. McKeever, *Implementation of electrical simulation model for IEC standard Type-3A generator*. In Proceedings of the european modelling symposium, pp. 426–431 (2013)
25. F.M. Gonzalez-Longatt, J.L. Rueda (eds.), *PowerFactory Applications for Power System Analysis* (Springer, 2014)
26. H. Zhao, *Coordinated control of wind power and energy storage*, CEE-PhD-2014 (Kongens Lyngby, 2014)
27. A.D. Hansen, G. Michalke, Fault ride-through capability of DFIG wind turbines. Renew. Energy **32**(9), 1594–1610 (2007)
28. A. Lorenzo-Bonache, R. Villena-Ruiz, A. Honrubia-Escribano, A. Molina-García, E. Gómez-Lázaro, *Comparison of a Standard Type 3b Wt Model With A Commercial Build-In Model*. In Electric machines and drives conference (IEMDC), IEEE International, pp. 1–6 (2017)
29. R. Villena-Ruiz, A. Lorenzo-Bonache, A. Honrubia-Escribano, E. Gómez-Lázaro, *Implementation of a Generic Type 1 Wind Turbine Generator for Power System Stability Studies*. In Proceedings of the international conference on renewable energies and power quality (Malaga, Spain, 2017), pp. 4–6
30. DIgSILENT PowerFactory, *Application Guide – Dynamic Modelling Tutorial*, DIgSILENT Technical Documentation, July 2014
31. W.W.Price, J.J. Sánchez-Gasca, *Simplified Wind Turbine Generator Aerodynamic Models for Transient Stability Studies*. In IEEE PES power systems conference and exposition, pp. 986–992 (2006)
32. J. Zhang, M. Cheng, Z. Chen, X. Fu, *Pitch Angle Control for Variable Speed Wind Turbines*. In Proceedings of the third international conference on electric utility deregulation and restructuring and power technologies, pp. 2691–696 (2008)
33. S. Muyeen, M. Ali, R. Takahashi, T. Murata, J. Tamura, Y. Tomaki et al., Comparative study on transient stability analysis of wind turbine generator system using different drive train models. IET Renew. Power Gener. **1**(2), 131–141 (2007)
34. I. Erlich, H. Wrede, C. Feltes, *Dynamic Behavior of DFIG-Based Wind Turbines During Grid Faults*. In Proceedings of the power conversion conference, pp. 1195–200 (2007)
35. DIgSILENT PowerFactory, *AC Current Source*, DIgSILENT technical reference documentation, 2011
36. A. Honrubia-Escribano, F. Jimenez-Buendia, E. Gomez-Lazaro, J. Fortmann, Field validation of a standard type 3 wind turbine model for power system stability, according to the requirements imposed by IEC 61400-27-1. IEEE Trans. Energy Convers. **33**(1), 137–145 (2018)

Chapter 7
Battery Energy Storage System Modelling in DIgSILENT PowerFactory

Mirza Nuhic and Guangya Yang

Abstract The current trend of increased penetration of renewable energy and reduction in the number of large synchronous generators in existing power systems will inevitably lead to general system weakening. The inherent characteristics of traditional synchronous machines will have to be replaced by converter-interfaced sources. The intermittent nature of renewable sources points to a need for high capacity energy storage. Battery energy storage systems (BESS) are of a primary interest in terms of energy storage capabilities, but the potential of such systems can be expanded on the provision of ancillary services. In this chapter, we focus on developing a battery pack model in DIgSILENT PowerFactory simulation software and implementing several control strategies that can address some of the issues mentioned previously. The last section of the paper contains a demonstration of the capabilities of the battery system and evaluation of the implemented functions in various operating scenarios.

Keywords Battery modelling · DSL · Fault-Ride-Through

7.1 Introduction

The increasing presence of renewable sources in power systems is introducing a serious challenge for system operators due to a reduction in total inertia of the system, limitation in voltage support, reduced frequency support capabilities, general system weakening, etc. As more renewable sources are being installed, the number of large synchronous generators will steadily decrease, thus creating a significant strain on

Electronic supplementary material The online version of this chapter (https://doi.org/10.1007/978-3-030-54124-8_7) contains supplementary material, which is available to authorized users.

M. Nuhic (✉) · G. Yang
Center for Electrical Power and Energy, Department of Electrical Engineering, DTU Technical University of Denmark, Elektrovej Building 325, 2800 Kgs. Lyngby, Denmark
e-mail: mirnuh@elektro.dtu.dk

F. M. Gonzalez-Longatt and J. L. Rueda Torres (eds.), *Modelling and Simulation of Power Electronic Converter Dominated Power Systems in PowerFactory*, Power Systems,
https://doi.org/10.1007/978-3-030-54124-8_7

the power system operation, reliability, and security [1]. In that regard, the battery energy storage systems (BESS) are attracting major interest as a technology that can provide ancillary services required for stable system operation [2]. The fast response combined with various functions and capabilities of a battery system makes it a very viable solution that can address some of the issues that the future power systems face [3].

One of the best readily available battery chemistries are the lithium-based batteries. They are characterized by high energy density, long lifetime, excellent efficiency (> 90%), wide temperature range, etc. The main drawback of this battery technology is that it is still relatively expensive [4]. There are three general types of battery models:

1. Chemical Models; the chemical models are used when the accuracy of the model is essential for the investigation of internal chemical processes of the battery. The models are very detailed, complex, and computationally expensive [5].
2. Mathematical models; these models are used when only the simple and approximate battery models are required. They are usually easy to build and do not require significant computational power [6].
3. Electrical models; they are based on equivalent electrical circuits. The chemical processes within the battery are represented by electrical components. The models provide a decent trade-off between complexity/accuracy and computational burden [7].

For power system simulations, the models based on equivalent electrical circuits are widely used, as they provide reasonable accuracy (1–5%) and are easy to model. The model that can be found in use in most of the publications that is related to battery modelling in power systems is based on the so-called "Double Polarization Model". The model captures the two distinct phenomena within the lithium-based battery cells, namely the separating concentration polarization (short-term dynamics), and the electrochemical polarization (long-term dynamics). Two RC circuits represent the two processes [8]. The electrical performance of the battery is characterized by the capacity, internal impedance, open-circuit voltage, and self-discharge. The parameters of the equivalent circuit are dependent on state of charge (SOC), temperature, charging/discharging current (C-rate), and state of health of the battery cell [1].

A more precise method for impedance-based battery modelling is by using a so-called ZARC element to represent the two distinct processes. The ZARC element is a parallel connection of a constant phase element (CPE) and resistance, which is an arc-shaped impedance [9]. The ZARC element can be expressed as:

$$Z_{\mathrm{ZARC}} = \frac{1}{1/R + (j\omega)^N Q}. \tag{7.1}$$

N is the depression factor, and it can be between 0 and 1 (0-pure resistance, 1-pure-capacitance). The CPE element is used to represent a non-ideal capacitor. The method that is used to extract the battery parameter (dependent on SOC, temperature and C-rate) is called the "Electrochemical Impedance Spectroscopy". The impedance is measured for a frequency spectrum and usually represented by a Nyquist plot. Five RC circuits can approximate one ZARC element with very low approximation error. The model that uses CPE elements provides high accuracy, but the measurements can be time-consuming [9].

The control structure of the battery model can provide flexibility in terms of ancillary services and the speed of the response. If we consider the frequency support capabilities, the battery systems can provide a significant contribution with fast response and readily available stored energy. The three key parameters that characterize a frequency disturbance are: (1) rate of change of frequency (RoCoF), (2) frequency nadir, and (3) steady-state frequency. The reduced inertia in a system will result in an increased RoCoF and lower frequency nadir, while the steady-state frequency relies solely on the primary frequency response (PFR). The reactive power capability of a converter can provide support under severe voltage dips and stay connected to the grid in extreme conditions. The fault-ride-through capability of a power converter is becoming a requirement in most grid codes. One of the biggest challenges in renewable energy dominated power systems is the intermittent nature of the renewable sources (i.e. wind and solar power), which can lead to significant forecast errors. Battery systems can, to a large extent, address this issue, and possibly provide additional services such as oscillation damping, harmonic filtering, etc. [1].

In this chapter, we try to provide detailed instruction on how to build a model of a battery energy storage system, including several functions, which can serve as ancillary services and provide support to the grid during disturbance and transient operating conditions.

7.2 Battery Model

The model that is widely used in the literature is the "Double Polarization Model". The equivalent electrical circuit is shown in Fig. 7.1. The model captures the two distinct chemical processes within the battery, namely separation polarization and electrochemical polarization (the short-term and the long-term dynamics, respectively). The element R_O represents the internal resistance of the battery cell. The

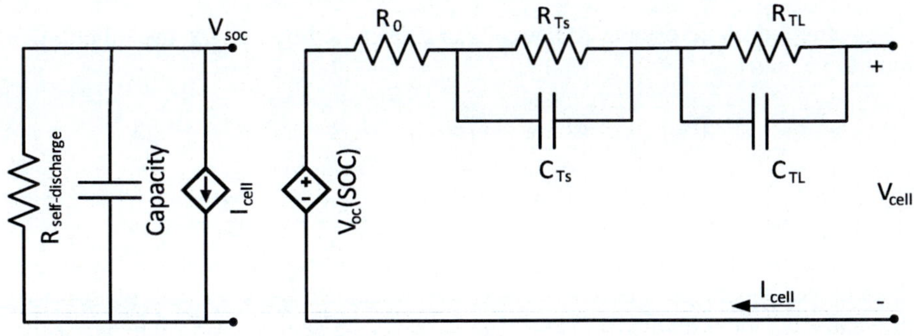

Fig. 7.1 Equivalent electrical circuit of a battery cell

R_{Ts} and C_{Ts} represent the separation polarization process, while the R_{TL} and C_{TL} represent the electrochemical polarization process within the battery cell. V_{OC} is the open-circuit voltage, also referred to as the internal voltage of the battery cell. The $R_{\text{self-discharge}}$ characterizes the loss of charge over a longer period. This parameter is usually not relevant for short-term studies because the loss of charge is negligible (up to 5% per month) [8].

The parameters of the equivalent circuit are dependent on the state of charge (SOC), temperature, and the operating state of the battery cell (charging or discharging). The most influential factor is the state of charge, and for this reason, estimating the state of charge is of the primary concern with respect to battery modelling. There are several methods for SOC estimation, and the two most widely used methods are Coulomb counting and OCV-based estimation [10]. The Coulomb counting method is a very accurate method and simple to implement. The main drawback is that it requires an accurate estimation of the initial state of charge. The OCV-based SOC estimation is a straightforward method, but the main disadvantages are that it requires sufficient relaxation time to enable accurate OCV measurement and it can be challenging to estimate the SOC for flat areas of the OCV curve. In practice, these two methods are combined in order to provide a simple and efficient approach for SOC estimation. Some of the other methods used in practice are an impedance-based estimation, model-based estimation, estimation using fuzzy logic and machine learning, etc. [10]. The method that we implemented is the Coulomb counting method. This method represents the charge transfer measurement to and from the battery, and it requires a current measurement. The SOC estimation is based on the following equation:

$$\mathrm{SOC} = \mathrm{SOC}_0 - \frac{1}{C_{\text{cell}}} \int_0^t I_{\text{cell}} \mathrm{d}t, \tag{7.2}$$

where SOC_0 represents the initial state of charge, C_{cell} is the capacity of the battery cell, and I_{cell} is the charging/discharging current. Once we know the state of charge, the other parameters can be estimated based on a lookup table or a parameterized function. The terminal voltage of the battery cell as per the circuit in Fig. 7.1 is:

$$U_{\text{cell}} = U_{\text{OC}} - I_{\text{cell}} R_0 - U_{\text{Ts}} - U_{\text{TL}}. \tag{7.3}$$

The voltage drop across the RC circuits is determined by solving the following differential equations:

$$\begin{aligned} I_{\text{cell}} &= \frac{U_{\text{Ts}}}{R_{\text{Ts}}} + C_{\text{Ts}} \frac{dU_{\text{Ts}}}{dt} \rightarrow \frac{dU_{\text{Ts}}}{dt} = \frac{I_{\text{cell}}}{C_{\text{Ts}}} - \frac{U_{\text{Ts}}}{R_{\text{Ts}} C_{\text{Ts}}}, \\ I_{\text{cell}} &= \frac{U_{\text{TL}}}{R_{\text{TL}}} + C_{\text{TL}} \frac{dU_{\text{TL}}}{dt} \rightarrow \frac{dU_{\text{TL}}}{dt} = \frac{I_{\text{cell}}}{C_{\text{TL}}} - \frac{U_{\text{TL}}}{R_{\text{TL}} C_{\text{TL}}}. \end{aligned} \tag{7.4}$$

A complete battery pack is comprised of several battery cells stacked together to form a battery pack with a specific voltage, power, and energy rating. The number of cells in a single string determines the battery terminal voltage. Since the nominal voltage of a battery cell is known, we can calculate the number of cells in a string n_S based on the required DC voltage level:

$$n_S = \frac{U_{\text{batt}}}{U_{\text{Cell}}}. \tag{7.5}$$

If the power requirement for the battery is known, then we can calculate the rated current as:

$$I_{\text{Batt}} = \frac{P_{\text{Batt}}}{U_{\text{Batt}}}. \tag{7.6}$$

The number of strings that need to be connected in parallel in order to achieve the rated current can be calculated as:

$$n_p = \frac{I_{\text{Batt}}}{I_{\text{Cell}}}. \tag{7.7}$$

I_{Cell} is the rated current of a single battery cell. The capacity of a single battery cell is usually expressed in Ah. C-rate represents the charging or discharging rate of a battery expressed in terms of the total battery capacity in Ah. 1C rate represents the current for which the battery is fully charged or discharged in one hour. The capacity of the battery pack is determined by the capacity of single-cell multiplied by the number of parallel-connected strings. Based on this, the total energy stored in a battery pack can be determined from:

$$E_{\text{Batt}} = U_{\text{batt}} C_{\text{Batt}}. \tag{7.8}$$

The battery pack model can be extrapolated from a single battery cell model. The model in PowerFactory is represented as in Fig. 7.2.

The parameters of the equivalent circuit are updated for every new value of SOC by using a special DSL function in PowerFactory "*lapprox*" [12], which uses a linear

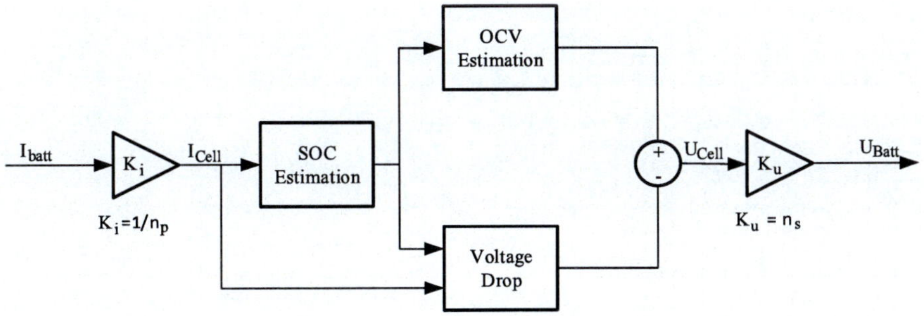

Fig. 7.2 Block diagram of a single cell in PowerFactory

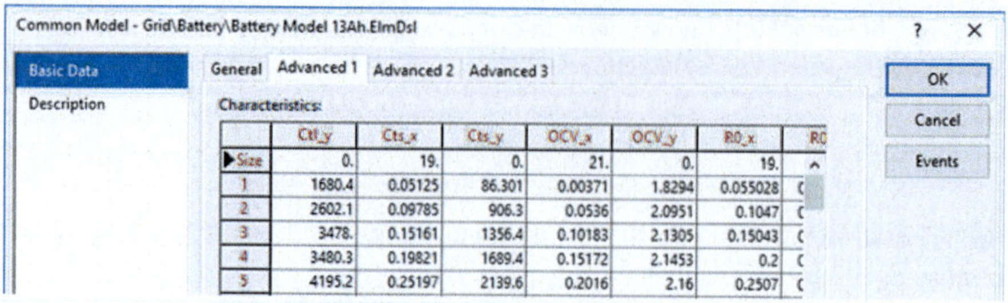

Fig. 7.3 PowerFactory common model of a battery—Equivalent circuit parameters

approximation method to evaluate the parameter based on the state of charge, as shown in the following DSL code of the model.

```
!Estimating the open circuit voltage based on SOC
OCV=lapprox(SOC,array_OCV)

!Estimating the parameters based on SOC
R_0=lapprox(SOC,array_R0)
R_Ts=lapprox(SOC,array_Rts)
R_TL=lapprox(SOC,array_Rtl)
C_Ts=lapprox(SOC,array_Cts)
C_TL=lapprox(SOC,array_Ctl)

!Calculating the voltage drop
U_0=Icell*R_0
U_Ts.=(Icell/C_Ts)-U_Ts/(R_Ts*C_Ts)
Uts=U_Ts
U_TL.=(Icell/C_TL)-U_TL/(R_TL*C_TL)
Utl=U_TL
U_drop=U_0+Uts+Utl
```

The values of the parameters are stored in an array under the tab "*Advanced 1*" of the DSL common model, as shown in Fig. 7.3. This model of the battery is based on the data provided in [11].

The parameters of the equivalent circuit can be given in the form of a function of SOC. The parameters of this model are based on the model that is given in [8]. The DSL code for this model is given below.

```
!Calculating the open-circuit voltage of the battery cell
OCV=-1.031*exp(-35*SOC)+3.685+0.2156*SOC-0.1178*sqr(SOC)+0.3201*pow(SOC,3)

!Calculating the parameters of the equivalent circuit
R_0=0.1562*exp(-24.37*SOC)+0.07446
R_Ts=0.3208*exp(-29.14*SOC)+0.04669
R_TL=6.603*exp(-155.20*SOC)+0.04984
C_Ts=-752.90*exp(-13.51*SOC)+703.60
C_TL=-6056*exp(-27.12*SOC)+4475

!Calculating the voltage drop
U_0=Icell*R_0
U_Ts.=(Icell/C_Ts)-U_Ts/(R_Ts*C_Ts)
U_TL.=(Icell/C_TL)-U_TL/(R_TL*C_TL)
U_drop=U_0+U_Ts+U_TL
```

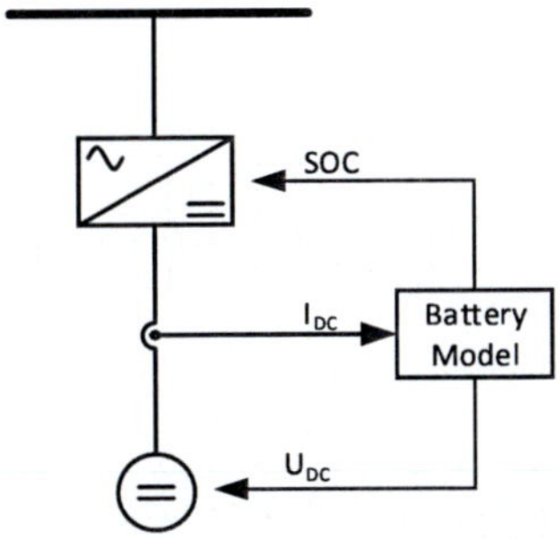

Fig. 7.4 Representation of BESS in PowerFactory

The battery in DigSilent PowerFactory is represented by a voltage-controlled DC voltage source. The battery model controls the voltage level of the DC voltage source. The voltage at the battery terminal is determined by following the procedure as described below:

1. The DC current is controlled by the power converter through which the battery is connected to the grid,
2. The DC current is measured by the DC Current Measurement block in PowerFactory,
3. Based on the current state of charge, the parameters of the equivalent circuit are evaluated, and by knowing the parameters of the circuit and the current, the actual voltage at the terminal of the battery is evaluated and used as a reference point of the DC voltage source.
4. The estimated state of charge is constantly updated, and information is provided to the power converter controller.

The complete system is placed within a composite frame in PowerFactory (*Battery.ElmComp*) with three slots included; slot for the battery (based on an equivalent circuit), DC Current Measurement slot, and the slot for the DC Voltage Source. The diagram of the model is shown in Fig. 7.4.

7.3 Power Converter Model

A standard two-level converter is responsible for controlling the battery pack. The block diagram of the control structure is given in Fig. 7.5. The block diagram shows the components of the complete battery system (converter, battery, and measurement components), as well as the main control blocks (frequency droop, active and reactive power control, and charging and discharging control). This structure is implemented in PowerFactory as a composite frame (*BESS.ElmComp*).

There are two main loops in the control structure, namely the active and reactive power loop. PLL is used to synchronize the power converter voltage angle with

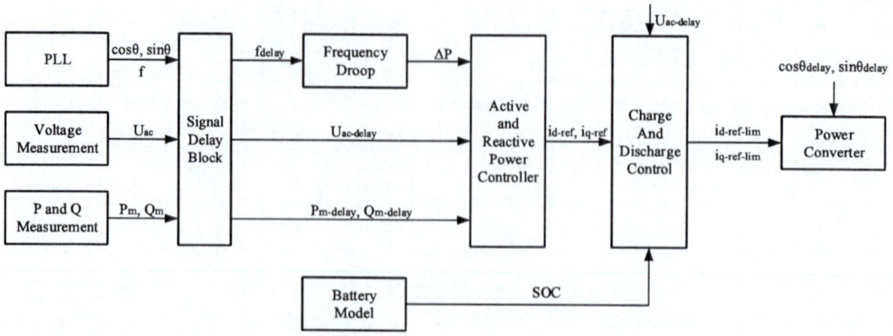

Fig. 7.5 The block diagram of BESS in as modelled in PowerFactory

the grid and measure the system frequency. The delay block is used to represent the communication and response latency of the battery system. The "Charge and Discharge Control" block has three main functions:

1. To control the charging and discharging mode of the battery (stopping the discharge at minimum SOC, or stopping the charge at maximum SOC),
2. To limit the charging current when the battery is at high SOC in order to prevent dangerous overvoltage at the battery terminal,
3. To limit the current of the power converter,
4. To switch to fault-ride-through operating mode under severe voltage dips (below 0.9 pu).

7.3.1 Active Power Control Loop

In the active power control loop, there are three basic functions:

1. Frequency droop control,
2. ROCOF based control,
3. Active power set-point.

Combining the droop and RoCoF functions can be considered a fourth function of the active power control loop. The frequency control based on the rate of change of frequency (RoCoF) and the frequency droop is a part of the frequency control loop of the battery system, as shown in Fig. 7.6. The frequency droop coefficient can be calculated as:

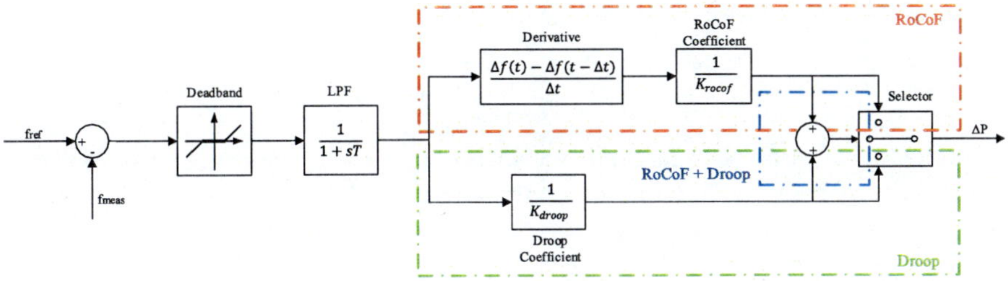

Fig. 7.6 Frequency Droop/RoCoF Controller

$$R_{\text{droop}} = \frac{1}{K_{\text{droop}}} = \frac{f_{\text{ref}} - f_{\text{meas}}}{P - P_{\text{ref}}} = \frac{\Delta f}{\Delta P}. \tag{7.9}$$

The parameter R_{droop} determines the active power change of the battery system, ΔP, for a change in the frequency of Δf. In most cases, the grid code defines an insensitive region of the frequency change known as deadband, which is a tolerated deviation in system frequency (usually between 0.02 and 0.05 Hz). The low pass filter prevents the controller reaction to high-frequency transients. The output of this control loop is an active power change signal to the active power controller. For under-frequency disturbances, the output will be positive, and the battery system will switch to discharging mode, while for the over-frequency disturbances, the output will be negative, and the battery will start charging.

The RoCoF function generates an active power reference, which is proportional to the rate of change of frequency. The rate of change of frequency or the slope of the frequency change is a derivative of frequency over time, and the relationship between RoCoF and the active power can be expressed as:

$$\Delta P = K_{\text{rocof}} \frac{\mathrm{d}f}{\mathrm{d}t}. \tag{7.10}$$

The time window of the RoCoF calculation will have a significant impact on the result, the smaller the time interval between the frequency measurements, the higher the response will be. If we want a larger time interval than the sampling time of the simulation, we can use the following expression:

$$\frac{\mathrm{d}f}{\mathrm{d}t} = \frac{f(t) - f(t - \Delta t)}{\Delta t}. \tag{7.11}$$

In PowerFactory, the RoCoF is calculated by using the "*delay*" function [12], which stores the value of a variable $x(t)$ and returns the value $x(t-t_{\text{delay}})$, as shown in the following DSL code.

```
! Derivative with adjustable time step
limits(T_rocof)=(0,)
f_dt=select(time()>T_rocof,delay(d_f,T_rocof),d_f)
df_r=(d_f-f_dt)/T_rocof
```

The RoCoF function becomes active when the simulation time exceeds the RoCoF calculation time interval. The initial conditions of the 'Frequency Droop' common model are defined as the following:

```
!Initial Conditions
inc(p0)=dpref
inc0(f0)=frq
inc(xf)=0
inc(xrc)=0
inc(df_dt)=0

vardef(db)='pu';'Deadband for frequency control'
vardef(droop)=;'Droop coefficient'
vardef(p_min)='pu';'Lower limit for droop control'
vardef(p_max)='pu';'Upper limit for droop control'
vardef(Tf)='s';'LPF time constant'
vardef(T_rocof)='s';'Derivation time step for RoCoF'
```

The active power control block (Fig. 7.7) generates the d-axis reference current. The inputs provided to the control block are the set-point active power control (P_{ref}), the frequency support reference (ΔP), as well as the Δi_d. The Δi_d signal comes from the converter current limiter, and it prevents the PI controller integrator windup in cases where the reference current exceeds the limit. The low pass filter eliminates the measurement noise and limits the rate of change of active power. The d-axis current $i_{d\text{-ref}}$ is the reference current generated by the active power controller.

The next step in the active power control loop is to limit the active power current according to the rating and determine the mode of the battery (charging, discharging or standby).

The block diagram of the PowerFactory common model (*Charge-Discharge.ElmDsl*) is shown in Fig. 7.8. Four functions characterize this control block:

1. Battery mode selector—determines the operation mode of the battery (charging, discharging, and standby).

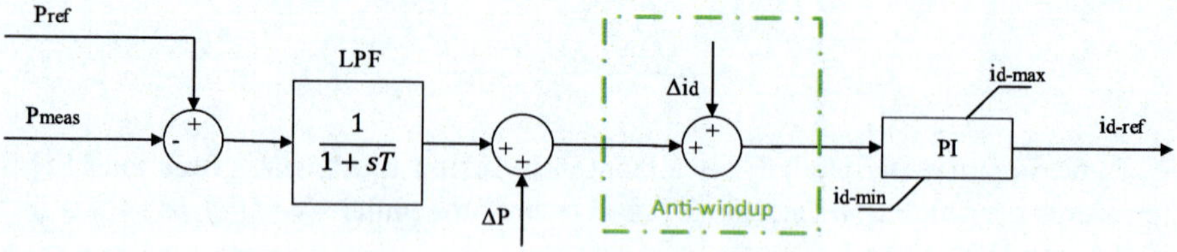

Fig. 7.7 Active power control loop

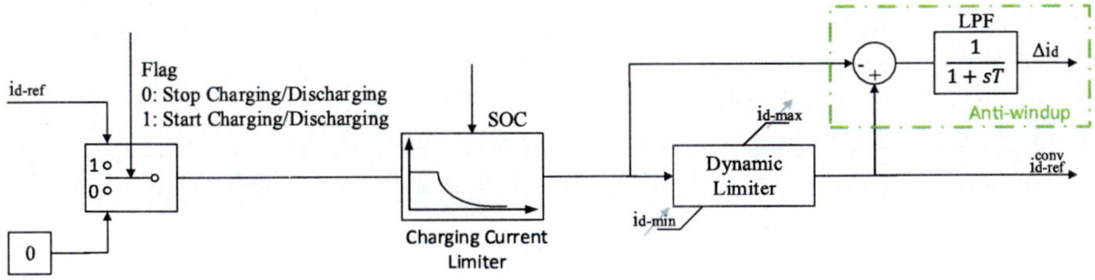

Fig. 7.8 Charging/Discharging control and current limiter block diagram—d-axis current

2. Charging current limiter—keeping the battery voltage constant at high SOC (> 80%) by reducing the charging current.
3. Dynamic limiter—the d-axis current limits are recalculated when the system is operating in fault-ride-through mode.
4. Anti-windup loop—generates the Δi_d signal to prevent the integrator windup of the active power block PI controller.

The mode selector switches between the charging and discharging modes based on the sign of the input current of the block. The algorithm of the decision procedure is shown in Fig. 7.9. Two conditions need to be satisfied in order to initialize the charging or discharging mode. If the input current is positive, and the state of charge

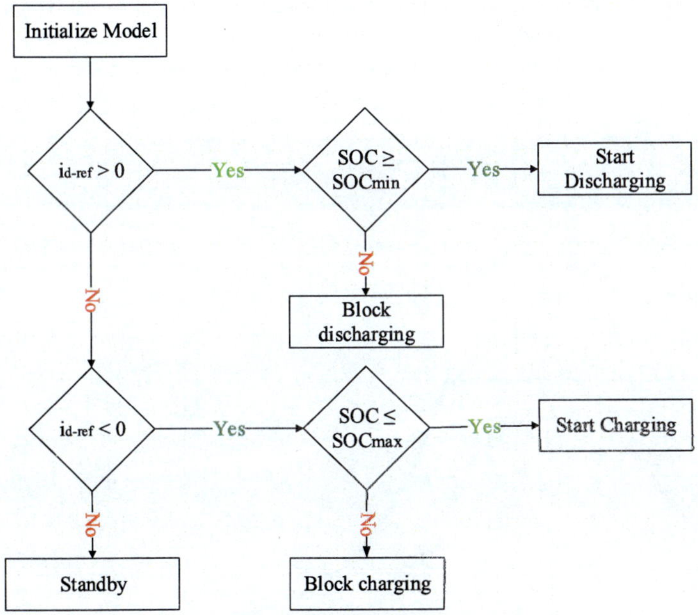

Fig. 7.9 Flow chart of the charging/discharging algorithm

of the battery is larger than the minimum state of charge (minimum SOC represents the discharged state), the discharging mode will be activated. For a negative current and the SOC lower than the maximum, the battery will be in the charging mode. The standby mode (no charging nor discharging) is initiated if the reference current is equal to zero. The charging and discharging of the battery will be blocked as long as the battery is fully charged ($SOC \geq SOC_{max}$) or fully discharged ($SOC \leq SOC_{min}$), respectively.

As previously said, the battery voltage needs to be kept below the maximum permissible level when the battery is charging because the voltage level tends to rise exponentially beyond the physical limits when the SOC is high (>80%), and it can cause a thermal runaway in the cell. Therefore, the charging current needs to be controlled in such a way that the battery voltage stays constant once it reaches the maximum level for continuous operation. This means that the charging current needs to be reduced at higher levels of SOC. The charging current can be expressed as:

$$I_{batt} = \frac{U_{OC} - U_{batt}}{R_0 + R_{Ts}\left(1 - e^{-\frac{t}{\tau_{Ts}}}\right) + R_{TL}\left(1 - e^{-\frac{t}{\tau_{TL}}}\right)},$$
$$\tau_{Ts} = R_{Ts}C_{Ts}, \quad \tau_{TL} = R_{TL}C_{TL}. \tag{7.11}$$

U_{batt} is the voltage at the battery terminal, and it is kept constant during the whole process of charging. The other parameters in the equation will change depending on the SOC. Once we calculate the charging current for each value of SOC, we can implement a limiter in PowerFactory as a lookup table, where the current limit is a function of SOC. The process of battery cell charging is exemplified in Fig. 7.10.

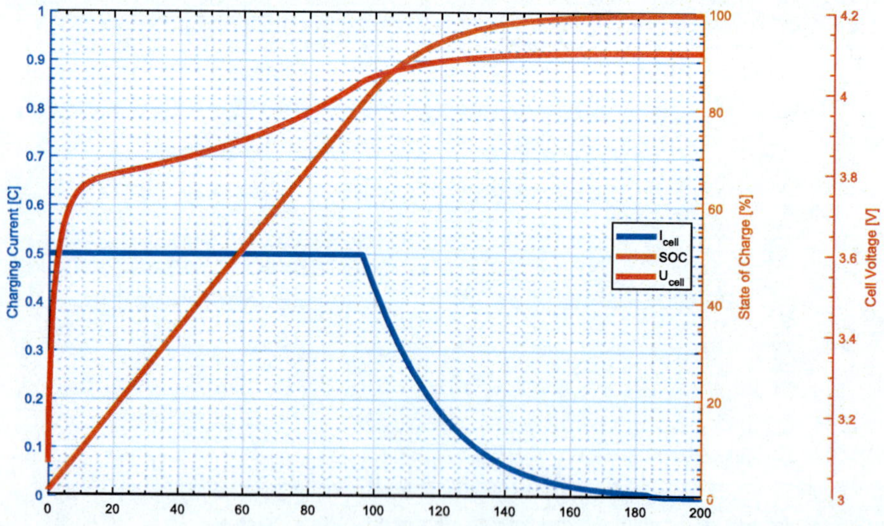

Fig. 7.10 CCCV battery charging method

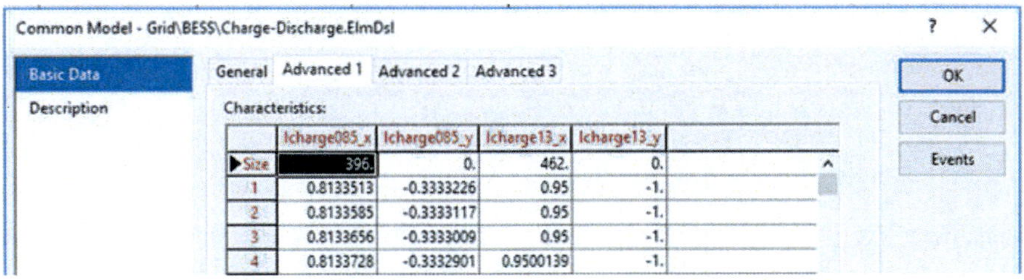

	Icharge085_x	Icharge085_y	Icharge13_x	Icharge13_y
Size	396.	0.	462.	0.
1	0.8133513	-0.3333226	0.95	-1.
2	0.8133585	-0.3333117	0.95	-1.
3	0.8133656	-0.3333009	0.95	-1.
4	0.8133728	-0.3332901	0.9500139	-1.

Fig. 7.11 Battery charging current limiter characteristic

The battery cell was charged with a constant current of 0.5C, until the terminal voltage reached the maximum permissible value. After this point, the battery was charged in constant voltage mode until it was fully charged. In the literature, this process of battery charging is known as "*Constant Current—Constant Voltage*" (CCCV) charging method.

The current limit lookup table is stored under the "*Advanced 1*" tab of the '*Charge-Discharge.ElmDsl*' common model within the "*BESS.ElmComp*" composite frame, as shown in Fig. 7.11.

The dynamic limiter can operate in two modes. The first mode is the *normal operation mode*, where the priority is given to active power injection. The current limits are based on the battery ratings. The other mode of operation is the *fault-ride-through* mode (voltage drops below 0.9pu), where the priority is given to reactive power injection, and the d-axis current limits are given as:

$$i_{\text{d-ref}}^{\text{conv}} = \sqrt{i_{\max}^{2} - i_{\text{q-ref}}^{\text{conv}\,2}}. \tag{7.12}$$

The anti-windup effect is accomplished by subtracting the input of the limiter block from the output. The sum is passed through the low pass filter in order to avoid forming of an algebraic loop.

In PowerFactory, the charging algorithm is represented by a DSL block reference "*Charging.BlkRef*". The flag variables for charging and discharging states (logic variables with values 0 or 1) are obtained by using a special DSL function, "*flipflop*" [12]. The variable state will change from 0 to 1 when the logical variable "*startc*" is 1 and the logical variable "*stopc*" is 0. The "*startc*" and "*stopc*" are also logical variables that indicate if the charging process has started or if it was stopped. The variables are obtained by using a special DSL function "*picdro*" [12].

```
!Initial conditions, charging variables
limits(Te)=[0.001,]
inc(charging)=flipflop(startc,stopc)
inc(startc)=picdro({SOC<lim(maxSOC,0.2,1) .and. idr<-0.0001},0,0)
inc(stopc)=picdro({SOC>lim(maxSOC,0.2,1)-0.001 .or. idr>-0.0001},0,0)
inc(xcharge)=max(idr,Ir)

!Initial conditions, discharging variables
inc(startdc)=picdro({SOC>lim(minSOC,0.1,0.9) .and. idr>0.0001},0,0)
inc(stopdc)=picdro({SOC<lim(minSOC,0.1,0.9) .or. idr<0.0001},0,0)
inc(discharging)=flipflop(startdc,stopdc)
inc(xdischarge)=lim(idr,0,1.5)

!Charging current limiter
Ir=select(BattSel<1.5,lapprox(SOC,array_Icharge085),lapprox(SOC,array_Icharge13))
!Charging current, low pass filter
xcharge.=(select(charging=1,max(idr,Ir),0)-xcharge)/0.005
icharge=xcharge

!Limiting the rate of change of the battery current upon sudden interruption
inc(Tc)=time()
inc(aux1)=0
inc(aux2)=0
aux1=aflipflop(time(),stopc,startc)
aux2=aflipflop(time(),stopdc,startdc)
Tc=select(charging<0.5 .and. stopc>0.5 .and. idr<0, aux1, select(discharging<0.5 .and.
stopdc>0.5,aux2,delay(Tc,0.01)))
idc=max(idr,Ir)*exp(-(1/Te)*(time()-Tc))

!Conditions to begin charging
startc=picdro({SOC<lim(maxSOC,0.2,1) .and. idr<-0.0001},0.01,0)
!Conditions to stop charging
stopc=picdro({SOC>lim(maxSOC,0.2,1)-0.001 .or. idr>-0.0001},0.01,0)
!Charging flag variable, 0 or 1
charging=flipflop(startc,stopc)
!Conditions to begin discharging
startdc=picdro({SOC>lim(minSOC,0.1,0.9) .and. idr>0.0001},0.01,0)
!Conditions to stop discharging
stopdc=picdro({SOC<lim(minSOC,0.1,0.9) .or. idr<0.0001},0.01,0)
!Discharging flag variable, 0 or 1
discharging=flipflop(startdc,stopdc)

!Discharging current, low pass filter
xdischarge.=(select(discharging=1,lim(idr,0,1.5),0)-xdischarge)/0.005
idischarge=xdischarge

!The reference id current
idref=select(charging,icharge,select(discharging,idischarge,idc))
```

This function evaluates the logical expression(s) (in our case those are the conditions for starting or stopping the charging state), and it pics up a logical state of 1 if the logical expression is *true* for the predefined pick-up time. A low pass filter with low time constant is implemented to prevent fast transients and eliminate noise.

7.3.2 Reactive Power Control Loop

The reactive power control loop allows for three basic functions:

1. Reactive power set-point,
2. Automatic voltage control,
3. Fault-ride-through mode.

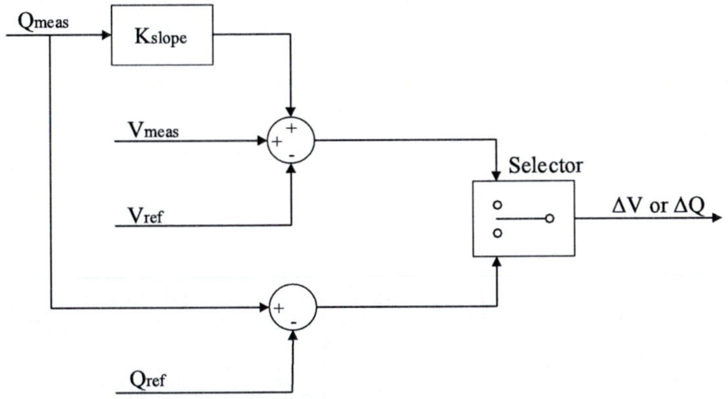

Fig. 7.12 Reactive power controller

Under normal operating conditions ($V_{meas} > 0.9$ pu), the reactive power control loop can provide voltage support with two basic functions. The block diagram is shown in Fig. 7.12. The reactive power set-point control adjusts the reactive power output as per user input. The automatic voltage control compares the reference voltage and the measured voltage and generates reactive power accordingly in order to reduce the voltage deviation. As the voltage is measured at the HV side of the connecting transformer, we must compensate for the transformer impedance with an additional signal $K_{slope}Q_{meas}$. The slope coefficient determines the allowed error between the measured voltage and the reference voltage.

The compensation block serves an additional function of ensuring an equal reactive power-sharing between the BESS system and other voltage support components connected in parallel (synchronous machine, wind turbine, etc.). The resulting voltage error can be written as:

$$\Delta V = V_{ref} - V_{meas} - K_{slope} Q_{meas}. \tag{7.13}$$

The user can choose which strategy should be employed for a simulation scenario. The resulting signal error is further forwarded to a PI controller, as shown in Fig. 7.13. The low pass filter is used to prevent fast transients and eliminate noise. The Δi_q signal serves the same purpose as in the active power control loop, to prevent the PI controller windup. The control block generates the q-axis reference current $i_{q\text{-ref}}$ (*PQ Controller.ElmDsl*).

The q-axis current reference $i_{q\text{-ref}}$ is the input to the current limitation and fault-ride-through mode selection block (*Charge-Discharge.ElmDsl*). This block includes three functions for the reactive power control loop:

1. Activating fault-ride-through mode for extreme voltage dips ($V_{meas} < 0.9$ pu).
2. Dynamic limiter—the q-axis current limit is recalculated if the system is operating under fault-ride-through mode.

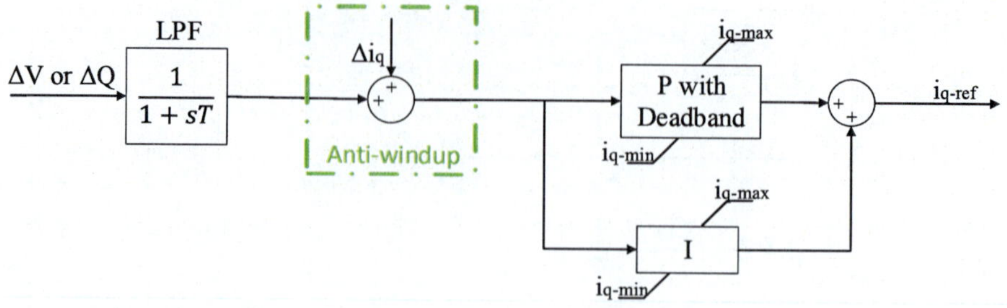

Fig. 7.13 Reactive power control loop

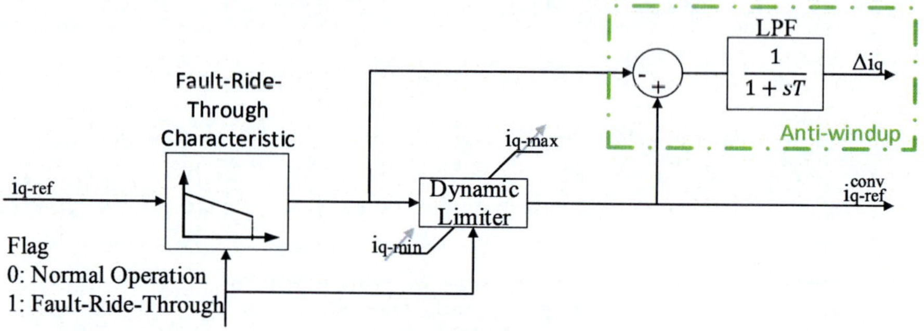

Fig. 7.14 Fault-ride-through mode and current limiter

3. Anti-windup loop—generates the Δi_q signal to prevent the integrator windup of the PI controller of the reactive power control block.

The block diagram is shown in Fig. 7.14.

The fault-ride-through mode is activated if the system voltage drops below 0.9 pu (the parameter is user adjustable). The flowchart of the fault-ride-through algorithm is given in Fig. 7.15. If the measured voltage dips below 0.9 pu, the controller will enter the fault-ride-through mode and give priority to the reactive power injection. As per the UK grid code [13], the total reactive current injected by the converter is a sum of the pre-fault value of the reactive current and the current estimated based on an I-V fault characteristic modelled as per the grid code. In most European grid codes, the requirement is to inject 1.00 pu of reactive power once the voltage ($V_{\min}$) drops to 0.5 pu. The current injected during the fault-ride-through operation mode can be expressed as:

$$i_{\mathrm{q-ref}}^{\mathrm{conv}} = i_{\mathrm{q-ref}}^{\mathrm{pre-fault}} + K_{\mathrm{frt}} \Delta V. \tag{7.14}$$

The slope of the characteristic is K_{frt}, which is determined by the following equation:

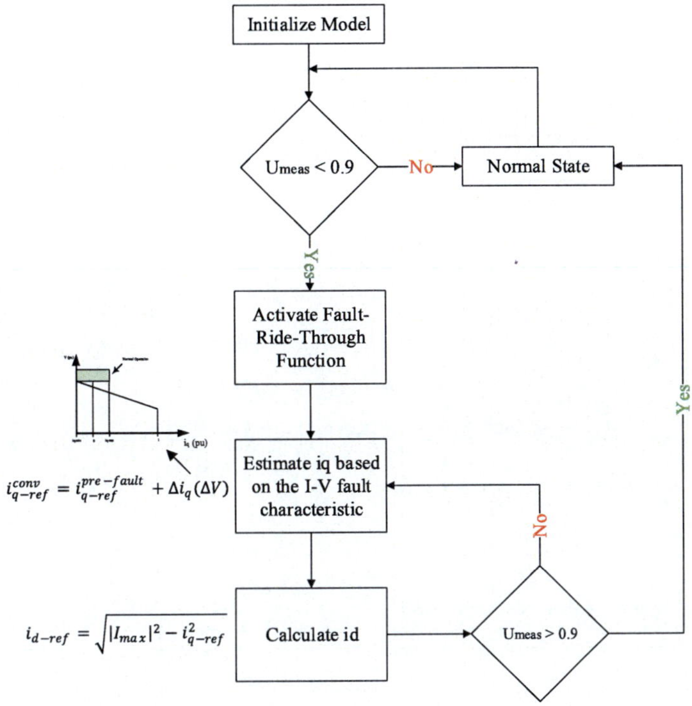

Fig. 7.15 Flow chart of the FRT function

$$K_{\text{frt}} = \frac{i_r - i_{q-\text{ind}-r}}{V_{\min} - V_{\text{threshold}}}. \tag{7.15}$$

The i_r current is the rated current of the converter (1.00 pu), $i_{q\text{-ind-}r}$ is the rated inductive current of the converter operating in normal mode, $V_{\min}$ is the voltage limit beyond which the reactive current injection requirement is 1.00 pu, and $V_{\text{threshold}}$ is the voltage at which the fault-ride-through mode is activated. For a voltage dip to 0.50 pu, we have chosen the reactive current injection to be equal to the maximum rated current of the converter (1.10 pu), and this current represents the maximum overcurrent that the converter can provide continuously.

In PowerFactory, the fault-ride-through algorithm is implemented by using the "*picdro*" special DSL function. The mode will activate once the voltage drops below the threshold (logical expression in the "*picdro*" function switches to true). The limits are updated depending on the mode of operation. For normal operation, the d- and q-axis current limits are set according to the active and reactive power rating of the battery system, while for the fault ride through mode the priority is given to the reactive power injection.

```
!Initial conditions
inc(uref)=uac
inc(frt)=0
inc(i_pre)=iqout
inc(iq_state)=iq_lim
inc(iq_ref_limit)= maxAbsCur*maxAbsCur-idin*idin
inc(id_ref_limit)= maxAbsCur*maxAbsCur-iqin*iqin

!Fault-ride-through flag variable
frt = picdro(abs(uac-uref)>=dU,0.001,0.01)

!Fault-ride-through current characteristic as per grid code
iq_lim=select(frt,lim(i_pre-3.625*((1-dU)-uac),-maxAbsCur,i_pre),i_pre)

!Current limit calculation
var_d=select(idin>idmax,idmax,select(idin<idmin,idmin,idin))
var_q=select(iqin>iqmax,iqmax,select(iqin<iqmin,select(frt,-iq_lim, iqmin),iqin))
lim_d=select(idin>0,idmax,idmin)
lim_q=select(iqin>0,iqmax,-lim(iq_lim,-iqmin,maxAbsCur))
!normal operation iq limit
iq_ref_limit=abs(lim(maxAbsCur*maxAbsCur-var_d*var_d,0,lim_q*lim_q))
!frt operation id limit
id_ref_limit=abs(lim(maxAbsCur*maxAbsCur-var_q*var_q,0,lim_d*lim_d))
!selecting limits based on operation mode
lim_sel=sqrt(select(frt,id_ref_limit,iq_ref_limit))

!Fault-Ride-Through current reference with low pass filter
iq_state.=(iq_lim-iq_state)/0.02

!Generating id and iq reference currents
idout=select(frt,lim(idin,-lim_sel,lim_sel),lim(idin,idmin,idmax))
iqout=select(frt,lim(iq_state,-maxAbsCur,maxAbsCur),lim(iqin,-lim_sel,lim_sel))
```

The output of the "*Charge-Discharge.ElmDsl*" common model blocks are the d-axis and *q*-axis currents, which serve as a reference for the power converter block. In PowerFactory, the power converter block has a built-in inner current controller, so it requires only the reference *d*- and *q*-axis currents, as well as the voltage phase angle in order to generate the reference voltage signal for the PWM.

7.4 Case Study: Testing the Battery Model and Control

In order to test the battery model and the implemented control strategies, we have used a simple network consisting of the battery system and a large synchronous machine rated at 2.55 GVA, 19 kV, 4 s inertia constant. This setup is used for the frequency support test, while for the voltage support test we have used a Thevenin equivalent to represent the grid (2800 MVA short circuit power, 33 kV voltage level, and 18.85 X/R ratio). The rated parameters of the battery system are given in Table 7.1.

7.4.1 *Frequency Control Functions*

As we previously discussed, there are three possible frequency support functions within the battery model. The droop control is based on the frequency deviation,

Table 7.1 Rated parameters of the battery pack

Single Cell Capacity	13 Ah
Nominal Cell Voltage (~ 50% SOC)	2.228
Number of Cells in a Single String	585
Number of Strings Connected in Parallel	197
Nominal Voltage of a Battery Pack (~ 50% SOC)	1.3 kV
Maximum Charging and Discharging Current	3C
End of Discharge Voltage ($\approx$ 0% SOC)	1.07 kV
Minimum SOC	10%
Maximum SOC	100%
Rated Power (Charging and Discharging)	10 MW
Stored Energy	3.34 MWh

where the controller injects active power in proportion to the level of frequency change. The second function is based on the rate of change of frequency (RoCoF). The controller acts in proportion to the rate of frequency deviation. The third function consists of using both control strategy to address the frequency disturbance at different time scales. The RoCoF function acts very fast following a disturbance and tries to counter the frequency change in a similar way to the inertial response of a synchronous machine.

The RoCoF function has an influence on the slope of frequency change and the resulting nadir. The frequency droop control has an impact on the frequency nadir and the steady-state. We have performed the simulation with four different cases in order to capture the effect of the function implementation, as well as the proportional coefficient (droop and RoCoF). For the base case, the battery system does not provide any frequency support, while for the other three cases, we changed the droop and RoCoF parameter within the range of 0.01–0.05 (0.5–2.5 Hz frequency deviation). The results of the simulations are shown in Fig. 7.16.

At 1.0 s a load of 200 MW is connected to the same bus as BESS. The frequency response of the system clearly demonstrates the effect of the control strategies. The RoCoF function acts fast to arrest the frequency drop, while the droop function response is slower, affecting the frequency nadir and especially the new steady state value of the frequency. The effect of both strategies can be observed in the bottom-right plot where we compare the different strategies with the base case (no frequency support) at the same droop/RoCoF coefficient.

7.4.2 *Fault-Ride-Through Capability*

The fault-ride-through algorithm is a requirement for most grid-connected generating units. The reactive current vs. voltage characteristics are included in most grid codes, and the requirement is to inject a certain level of reactive current during severe voltage

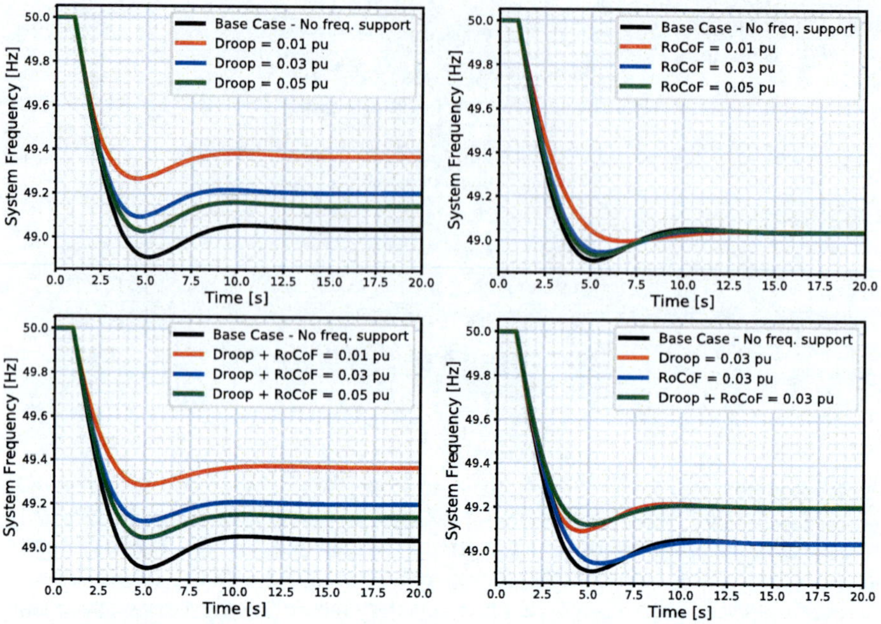

Fig. 7.16 Frequency response of BESS. Top left: Droop controller active; Top right: RoCoF controller active; Bottom left: Droop + RoCof controllers active; Bottom right: Comparison between the functions

drops (grid faults). In this example, we have tested the BESS response to a three-phase fault at the PCC. The results of the simulation are shown in Fig. 7.17. The fault is applied at 0.1 s, which results in a voltage drop to 0.5 pu. At the instance of the fault inception, the BESS was delivering full rated active power of 70 MW. Following the fault, the controller activates the fault-ride-through mode and injects the maximum permissible reactive current (1.1 pu, which includes 10% overcurrent) into the grid. The limiter acts to limit the active current so that the total injected current does not exceed the maximum overcurrent of the power converter.

The bottom left, and right plots show the reference currents before and after the limiter. The $i_{d\text{-ref-out}}$ and $i_{q\text{-ref-out}}$ are the reference currents that are sent to the converter model. As the injected reactive current matches the maximum overcurrent of the converter (in fault-ride-through mode priority given to the reactive current), the active current must be limited to zero.

7.4.3 Charging and Discharging Control

The active power control function is a simple control algorithm, where the system operator (or higher-level automatic control) sets the active power reference. This

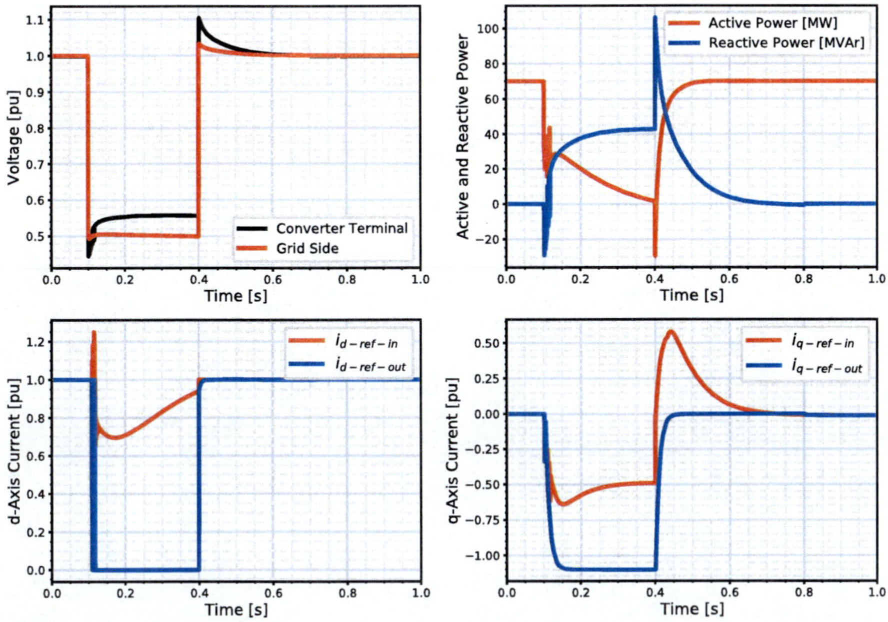

Fig. 7.17 BESS response in FRT mode of operation

function is used for scheduled charging and discharging of the battery. The top left and top right plots in Fig. 7.18 show the voltage and current of a single battery cell during 1.0 s of charging and 1.0 s of discharging. The rated charging and discharging currents for this battery cell are 3C or 39 A, which corresponds to the rated active power output of the entire battery system. The voltage of the battery cell exhibits a typical behaviour of a lithium-ion cell under load. It can be distinguished by the immediate voltage drop due to the R_0 resistance (Fig. 7.1), and the fast and slow dynamics exemplifying the two RC circuits.

The bottom left plot shows that the battery is being charged and discharged under equal power. However, the state of charge is not the same at the end of the simulation as it was at the beginning. The reason for this is that the discharging current will cause a voltage drop at the battery terminals (due to the internal resistances of the battery) and consequently the discharging current will increase in order to have a rated power output of 1.00 pu at the battery terminal. Similarly, the charging current will cause the voltage to rise and to cause a lower than nominal charging current for the rated power input. This phenomenon is the cause of SOC mismatch.

The charging/discharging controller of the battery system is constantly updated with the information about the state of charge. Most battery technologies have the longest lifetime if the state of charge is kept between 20% and 80%, which we have set as the minimum and maximum SOC for this example. Once the battery is fully charged or discharged, it is important not to interrupt the charging/discharging current instantaneously because this can cause unwanted transients and potential

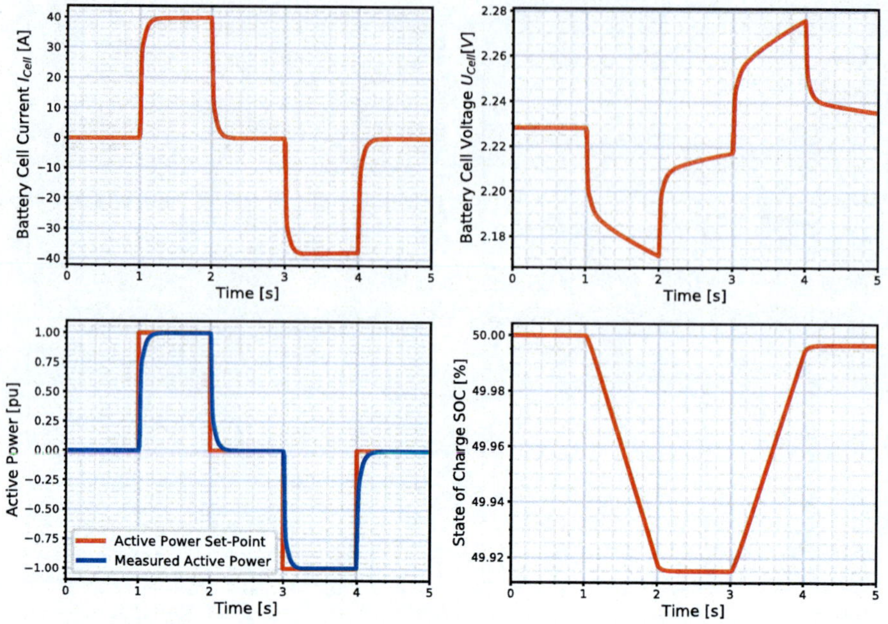

Fig. 7.18 Scheduled charging/discharging of BESS

instability in the grid. The user can set the time constant of the battery switching from charging/discharging mode to the standby mode (parameter T_e within the "*Charge-Discharge.ElmDsl*" common model). Figure 7.19 shows the battery response once the fully charged state has been reached.

Once the SOC reaches 80%, the charging current starts to drop. After approximately 5.0 s, the current is reduced to zero, and the battery is in the standby mode. The top- and bottom–left plot show that the active power and the current reference remain at 1.00 pu, while the actual measured power and current are zero. This indicates that the charging has stopped because the battery was fully charged. The top–right plot shows that SOC is slightly above 80%, and the reason behind this is that the battery was still charging after the standby mode was initiated. The charging current of a single battery cell (bottom–right plot) is not equal to 3C (even though the power reference was 1.00 pu). The reason for this is that when the battery is charging, the terminal voltage will be higher than in the case when the battery is discharging (even though the power exchange is the same). The other significant reason is that for higher SOC, the open-circuit voltage (OCV) of the battery will be high, and consequently the terminal voltage.

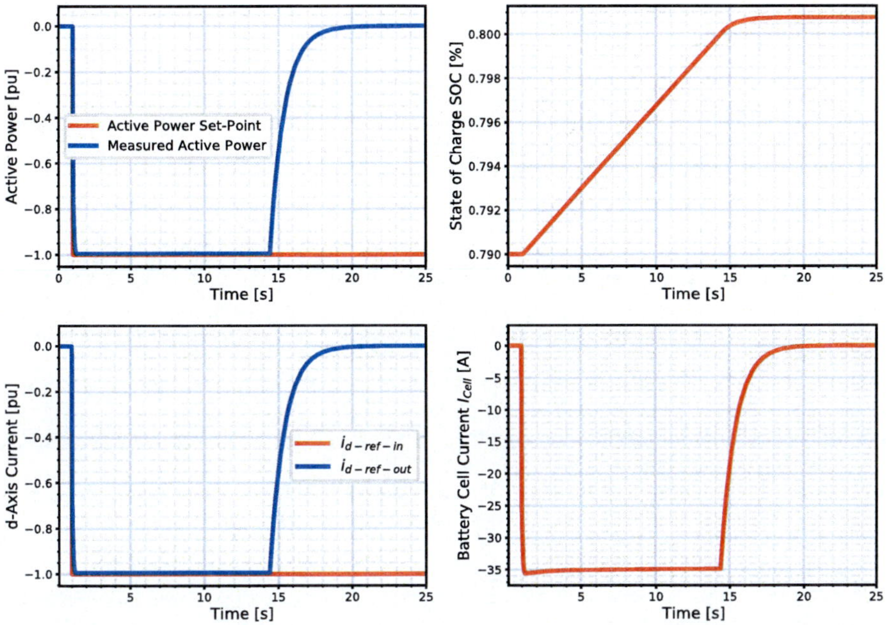

Fig. 7.19 Fully charged battery state

7.5 Conclusion

This chapter presented a procedure about the modelling of battery systems in DIgSILENT PowerFactory simulation software. The battery model is based on an equivalent electrical circuit, which captures the internal chemical processes that characterize the electrical performance of a battery cell. It has included two types of parameter representation, namely the approach based on a lookup table and the parametrized function approach. We have dedicated a significant space to control design and control strategies implementation. The main features of the model include frequency and voltage support, as well as the fault-ride-through function. The frequency control block includes two functions for frequency support, RoCoF and frequency droop, and the third function is the hybrid function, which combines the two into a singular frequency control function providing the frequency support at two different time scales. The voltage support consists of automatic voltage control for steady-state operation (voltages above 0.90 pu), and the fault-ride-through mode for operation under severe voltage dips. All the functions were tested under different scenarios in order to verify the validity of the model. The BESS model can be easily expanded to include the thermal model of the battery, as well as other ancillary services, such as oscillation damping, harmonic attenuation, etc.

References

1. M. Nuhic, G. Yang, A hybrid system consisting of synchronous condenser and battery - enhanced services for weak systems. *IEEE PES Innovative Smart Grid Technologies Europe (ISGT-Europe)*, (Bucharest, Romania, 2019), pp. 1–5
2. H. Jo, J. Choi, K. A. Agyeman, S. Han, Development of frequency control performance evaluation criteria of BESS for ancillary service: a case study of frequency regulation by KEPCO. in *IEEE Innovative Smart Grid Technologies—Asia (ISGT-Asia)*, (Auckland, 2017), pp. 1–5
3. P.V. Brogan, R.J. Best, D.J. Morrow, K. McKinley, M.L. Kubik, Effect of BESS response on frequency and rocof during underfrequency transients. IEEE Trans. Power Syst. **34**(1), 575–583 (2019)
4. C. Julien et al., Lithium batteries: science and technology, Springer International Publishing, 2015, pp. 1–619
5. D.W. Dees, V.S. Battaglia, A. Bellanger, Electrochemical modeling of lithium polymer batteries. J. Power. Sour. **110**(2), 310–320 (2002)
6. S. Li, B. Ke, Study of battery modeling using mathematical and circuit oriented approaches, in *Power and Energy Society General Meeting*, IEEE July 2011, pp. 1–8
7. H. Zhang, M.-Y. Chow, Comprehensive dynamic battery modeling for phev applications, in *Power and Energy Society General Meeting*, IEEE July 2010, pp. 1–6
8. M. Chen, G.A. Rincon-Mora, Accurate electrical battery model capable of predicting runtime and I-V performance. IEEE Trans. Energy Convers. **21**(2), 504–511 (2006)
9. E. Barsoukov, J.R. Macdonald, *Impedance Spectroscopy: Theory, Experiment, and Applications* (Wiley, 2005), pp. 1–595
10. W. Waag et al., Critical review of the methods for monitoring of lithium-ion batteries in electric and hybrid vehicles. J. Power. Sour. **258**, 321–339 (2014)
11. A.I. Stroe et al., Lithium-ion battery dynamic model for wide range of operating conditions, Proceedings—2017 International Conference on Optimization of Electrical and Electronic Equipment, Optim 2017 and 2017 Intl Aegean Conference on Electrical Machines and Power Electronics, Acemp 2017, Institute of Electrical and Electronics Engineers Inc., 2017, pp. 660–66
12. DIgSILENT, DIgSILENT PowerFactory 2018—User Manual, DIgSILENT GmbH, Gomaringen, Germany, June 2018
13. The Grid Code, National Grid UK, Issue 5, Revision 38, 4 Sept 2019

Chapter 8
A Benchmark Test System for the Power System Stability Assessment Considering Very High Penetration of Converter-Based Generation Units Including Grid Forming Converters

Abdul W. Korai, M. Ebrahim Adabi, E. Rakhshani, José Luis Rueda Torres, and Mart A. M. M. van der Meijden

Abstract In this book chapter, a benchmark test system has been studied for power system stability considering the high share of power electronic converter-based generation. Furthermore, both conventional PI controllers and grid forming control have been taken in to account in order to study the impact of the high penetration of power electronic converter on the dynamic response of the power system.

Keywords Wind park (WP) · Frequency control · Direct voltage control (DVC)

8.1 Introduction

In order to study the impact of the high penetration of power electronic converter on the dynamic response of power system, not only the share of power electronic converter generation but also other factors such as location, the topology of the network and the type and severity of the disturbance should be taken into account.

Electronic supplementary material The online version of this chapter (https://doi.org/10.1007/978-3-030-54124-8_8) contains supplementary material, which is available to authorized users.

A. W. Korai
DIgSILENT GmbH, Gomaringen, Baden-Württemberg, Germany

M. E. Adabi · E. Rakhshani · J. L. Rueda Torres (✉) · M. A. M. M. van der Meijden
Department of Electrical Sustainable Energy, Delft University of Technology, Delft, The Netherlands
e-mail: j.l.ruedatorres@tudelft.nl

M. A. M. M. van der Meijden
TenneT TSO B.V., Arnhem, The Netherlands

F. M. Gonzalez-Longatt and J. L. Rueda Torres (eds.), *Modelling and Simulation of Power Electronic Converter Dominated Power Systems in PowerFactory*, Power Systems,
https://doi.org/10.1007/978-3-030-54124-8_8

Before moving towards 100% power electronic generation-based grids, it is essential to determine the minimum share of conventional synchronous generators to provide stable operation of power system network [1–3].

The maximum penetration of power electronic-based generation unit with keeping stable operation of power system depends on different factors such as size of the power system, location of the power electronic converter-based generation units, location of the synchronous machines, the parameters of the controller of the power electronic converter and the dynamic and static behaviour of the loads.

In this book chapter, a benchmark test system for the power system stability assessment has been considered with high penetration of converter-based generation units, including grid forming converters. The remainder of this book chapter is structured as follows: Sect. 8.2 presents the test grid. Section 8.3 presents the model of wind turbine—Type 4 for grid integration studies. Section 8.4 describes the established grid codes for the frequency control. Section 8.5 describes the PI-based control, direct voltage control method (DVC) which is grid forming control and the simulation results. Section 8.6 summarizes the main conclusions.

8.2 Test System

Figure 8.1 displays the test system to study the minimum number of synchronous machines for stable operation of the power system. The aim is to study the transient voltage and frequency stability of network with high penetration of power electronic converters, by simulating high share of wind turbines (WTs).

Figure 8.2, shows the test grid of Fig. 8.1, implemented in PowerFactory. The model used in this chapter has been downloaded from [4].

All connections are 380 kV overhead lines with typical parameters, and the length of each line is 100 km. Two wind parks which are represented as WP1 and WP2 in Fig. 8.1 are the power in feed for two networks.

An equivalent transformer reactance of 30% has been assumed in order to consider the multiple step-up transformations from the wind turbine to the high-voltage bus. In this study, at first, there is no share of wind generation, and the total generation is through synchronous generators. Then, the share wind generation increased in line with decreasing conventional synchronous generators.

It should be noted that the total power remains constant, in all scenarios. To study the impact of the high share of power electronic converter generation unit, it was assumed that 44.5, and 25% of conventional units were kept in the system. In the end, a scenario has been considered with 98% of the share of power electronic converter-based generation unit. The synchronous generators marked as SG are shown in Fig. 8.1 on the right side. This unit assumes the task of primary control with 2% reserve power capacity in all cases under investigation [5].

This requirement leads to hysteresis behaviour, as can be seen in Fig. 8.2. According to offshore grid code represented in [6], the maximum rate of power

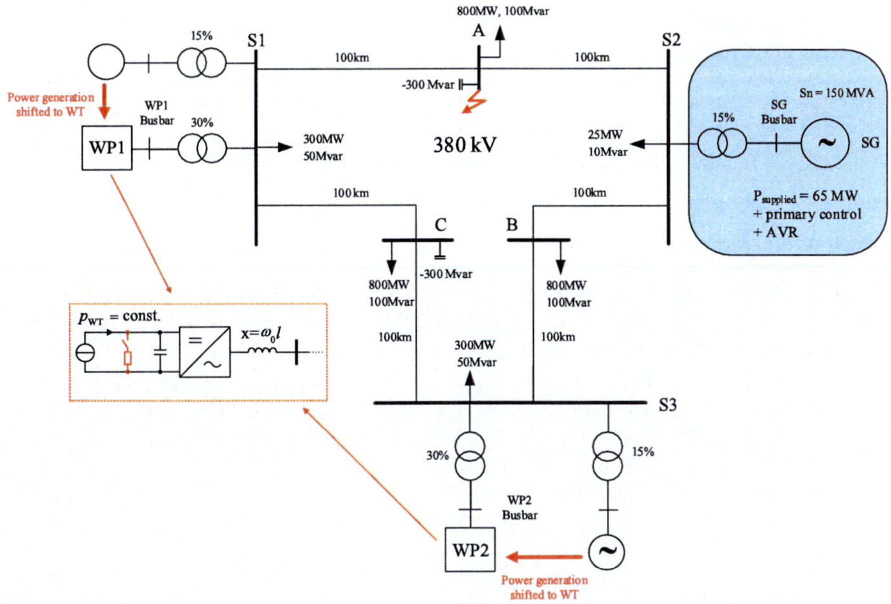

Fig. 8.1 Benchmark test system to study the minimum number of synchronous machines for stable operation of the power system

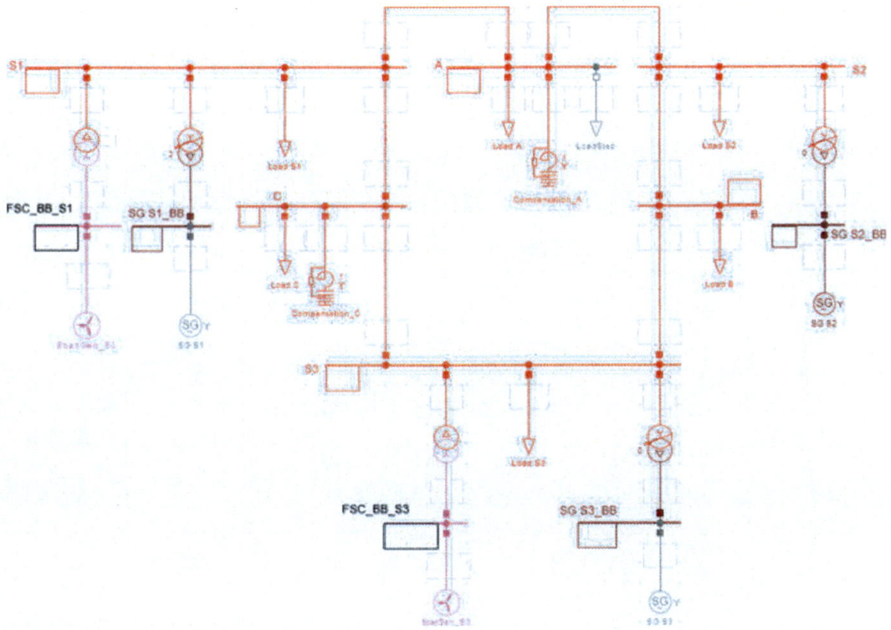

Fig. 8.2 Benchmark test system implemented in PowerFactory

change was also defined as 25% of the power corresponding to the operating point at the instant of the disturbance in both directions.

8.3 Classification of Modern WT Systems

Modern wind turbines are divided into four main types as shown in Fig. 8.3.

Type 1: Fixed speed induction generator (FSIG)-based WTs.
Type 2: Slip ring induction generator (SRIG) with variable rotor resistance-based WTs.

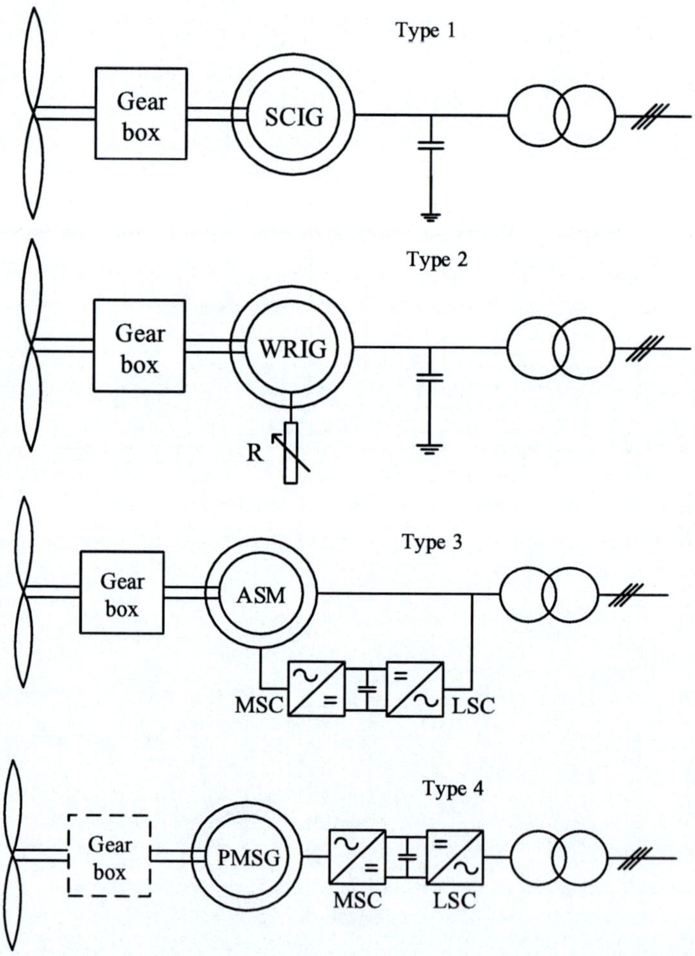

Fig. 8.3 Basic configuration of WT types

Type 3: Doubly-fed induction generator (DFIG)-based WTs.
Type 4: Full-scale converter generator (FSCG)-based WTs.

This book chapter focuses on the type 4-based WTs.

The speed of type 1-based WT cannot be controlled; hence, no optimal power extraction from the wind is possible. In such wind turbines, a capacitor bank is needed as the machine needs reactive power for the buildup of the magnetic field.

In type 2, the speed of the machine can be varied by the introduction of a variable resistor in the rotor windings, but the speed variation is not considerably high while the losses increase due to the variable resistance [7].

In types 1 and 2, the grid operators have no control over the operation of the wind turbines, and neither of these two types can be used to maximize the energy capture from the wind. Due to these reasons, the type 3 and type 4 have been installed extensively all over the world, and that is why the focus of this thesis will lie on type 4 of WTs.

8.3.1 Full-Scale Converter Generator (FSCG)-Based WTs—Type 4

Figure 8.4 shows the main configuration of the full converter-based WT.

It consists of the DC link capacitor and line side converter as well as machine side converter including the filters and power electronic converter protection circuit (chopper) along with the wind turbine transformer. An induction machine or synchronous machine can be used as a generator. The direct-drive permanent magnet synchronous generator (PMSG) is mainly used for this WT type due to its technical advantages. The power electronic converters have to fully rated as the rating of the WT in such configurations; hence, it is more costly than DFIG-based WT. The control

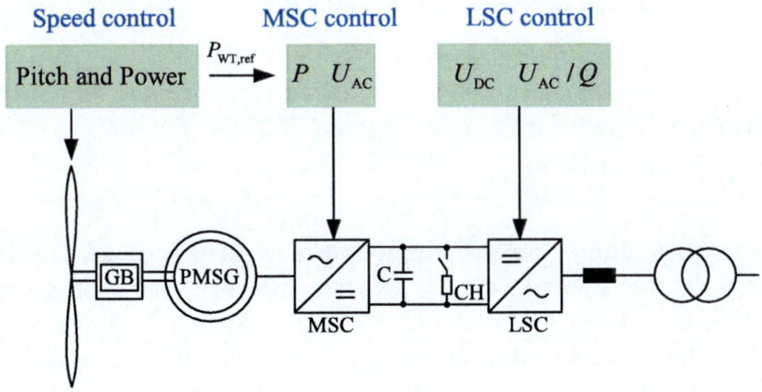

Fig. 8.4 Basic layout of the full-scale converter-based WT

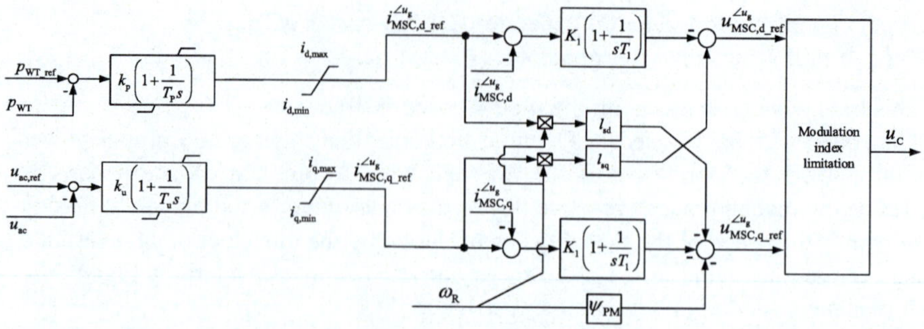

Fig. 8.5 Control of MSC of full-scale converter-based WT

of DC link voltage and the AC grid voltage is done by the LSC while the MSC usually controls the active power and AC voltage on the machine side.

8.3.2 *Machine Side Converter Control*

The MSC control loop is based on the following equations of the synchronous machine.

$$u_d = -r_s i_d + \omega_R \psi_{sq} \tag{8.1}$$

Figure 8.5 shows the control of the MSC of the FSCG-based WT. The MSC controls the active power and the AC voltage of the machine. The current controller is based on the equations given above neglecting the resistive terms.

Figure 8.6 shows the LSC side control of the FSCG. The LSC control controls the DC link voltage and the reactive power of the WT. During faults, the WT injects the reactive current into the network to support the grid voltages.

8.3.3 *Simplified Model of FSCG-Based WT for Grid Integration Studies*

The MSC and wind turbine generator parts are neglected because, for the grid dynamics, only the line side converter is relevant. However, the chopper and DC circuit have been included in the model of the WT so that the dynamics of the DC circuit can be observed and are well-represented. It is assumed that the machine side converter injects a constant active power into the DC circuit as shown in Fig. 8.7.

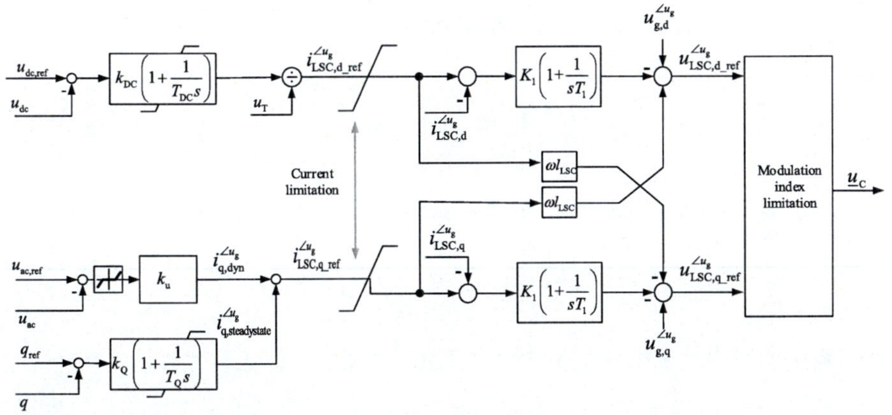

Fig. 8.6 Control of LSC of full-scale converter-based WT

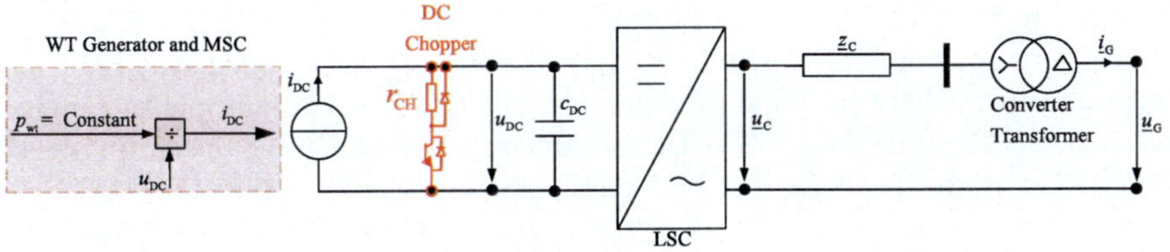

Fig. 8.7 Simplified Model of FSCG-based WT—Type 4 for grid integration studies

8.3.4 DC Circuit

The power balance between the MSC and LSC in FSCG-based WTs is governed by the Eq. (8.2). The chopper will dissipate the excess energy if the DC circuit voltage rises above the threshold value.

$$\begin{aligned}\frac{\mathrm{d}u_{\mathrm{DC}}}{\mathrm{d}t} &= \frac{1}{c_{\mathrm{DC}}}\cdot i_{\mathrm{DC}} = \frac{1}{c_{\mathrm{DC}}}\left(-i_{\mathrm{LSC,DC}} - i_{\mathrm{MSC,DC}} - i_{\mathrm{CH}}\right)\\ &= \frac{1}{c_{\mathrm{DC}}\cdot u_{\mathrm{DC}}}\left(-p_{\mathrm{LSC}} - p_{\mathrm{MSC}} - p_{\mathrm{CH}}\right)\end{aligned} \tag{8.2}$$

The time constant of the DC link is calculated by using Eq. 8.3. This equation describes the time, during which the DC capacitor can store rated power before it gets depleted during a disturbance. Whereas, C_{DC}, $U_{\mathrm{DC,N}}$ and S_{N} define the rated DC capacitance, nominal DC voltage and apparent power of the power electronic converter, respectively.

$$\tau = \frac{1}{2}\frac{C_{\mathrm{DC}}U_{\mathrm{DC,N}}^2}{S_{\mathrm{N}}} \tag{8.3}$$

The logic of the operation of the DC chopper is given in Eq. (8.4) [8].

$$p_{CH} = \begin{cases} 0, & \text{if chopper off} \\ \frac{u_{DC}^2}{r_{CH}}, & \text{if chopper on} \end{cases} \tag{8.4}$$

where u_{DC} and r_{CH} describe the rated DC voltage and chopper resistance in p.u. of the power electronic converter.

8.4 Established Grid Codes for the Frequency Control

Figure 8.8 shows the established grid codes for frequency control. In the German grid codes, it is necessary to decrease power generation when the frequency exceeds the value of 50.2 Hz. Also, there should be no increment in power until the frequency drops below 50.05 Hz.

This requirement leads to hysteresis behaviour as can be seen in Fig. 8.3. According to offshore grid code represented in [6], the maximum rate of power change was also defined as 25% of the power corresponding to the operating point at the instant of the disturbance in both directions.

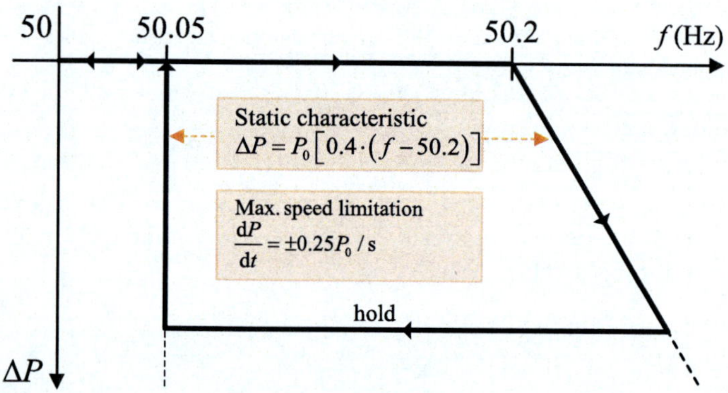

Fig. 8.8 Overfrequency power reduction with hysteresis

8.5 Control Strategies

8.5.1 PI-Based Current Control

In order to study the impact of penetration of power electronic converter-based generation, the share of the power electronic converter generation was increased stepwise, with the overall power remaining unchanged. The three-phase short circuit with a duration of 150 ms at bus A served as the trigger for the dynamic phenomenon. Figure 8.9 shows the voltage at bus A. Figure 8.10 demonstrates the speed response of the primary-controlled generator G1.

In the case of 25% synchronous generator which corresponds to 75% power electronic converter-based generation units, during the short-circuit, higher-frequency oscillations occur which are attributable to a certain instability of the power electronic converter control. Nevertheless, the system remains stable after fault clearing, but such oscillations are not desired during the fault. Generally, the voltage breaks down more during the short circuit in the network than in the network supplied solely by synchronous generators. Obviously, the voltage control by the power electronic converter is less effective than that of the synchronous machine. Both the 10% dead-band and the low reactive current gain of 2.0 p.u contribute to this.

Also, the voltage is only controlled indirectly via the reactive current, and the effect; thus, achievable depends on the effective equivalent impedance at the grid connection point. In a high-voltage network, the pure capacitive reactive current is almost optimal, but not in a distribution network with higher resistive components. By comparison, conventional synchronous generators always control the voltage

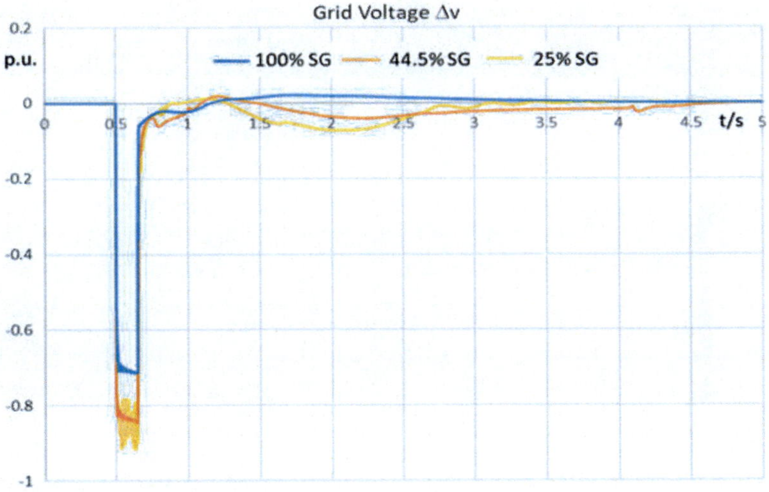

Fig. 8.9 Selected grid bus voltage

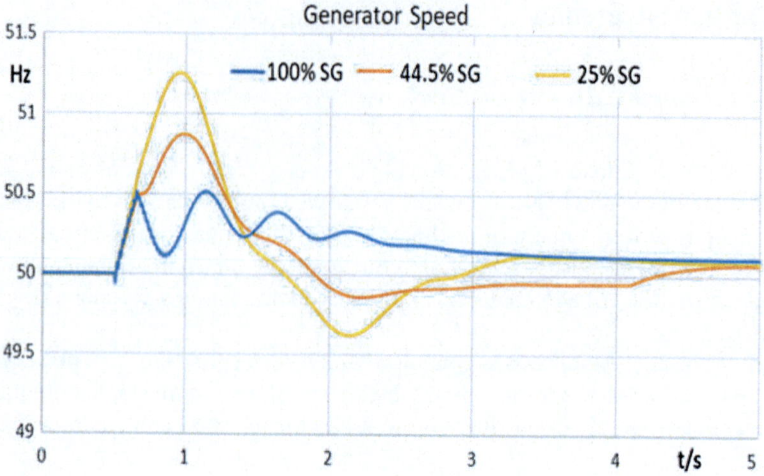

Fig. 8.10 Speed of synchronous generator (SG)

directly and the reactive current flow results from the required voltage boost and the characteristic of the network impedance.

Figure 8.11 shows the case with only 25% synchronous generation. Due to the increased grid frequency, the power electronic converter reduces the feed-in power

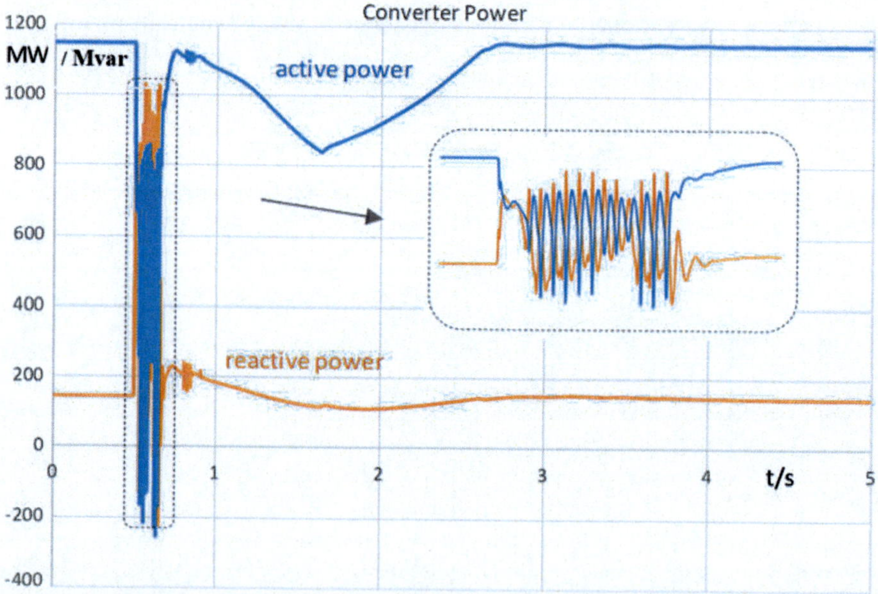

Fig. 8.11 Power electronic converter active and reactive power, the case with 25% residual synchronous generators (75% converter)

and raises it again only when the frequency falls below 50.05 Hz. The maximum rate of change is 25% per second. The hysteresis and the speed limitation lead to undesirable system behaviour. Although the frequency drops below 50 Hz, the power is still limited. At this point, direct and faster frequency control without a hysteresis characteristics would be much more suitable.

From the simulation results in this particular grid, the following conclusions can be deduced for the currently implemented power electronic converter control:

1. Downwards of about 30–40% conventional synchronous generation (70–60% power electronic converter), it seems that dynamic problems such as transient and voltage stability under the circumstances can no longer be controlled and may lead to instability (transient).
2. In cases of short circuits in the network, more profound and widespread voltage drops must be expected. The voltage–var control by power electronic converter is less effective compared to synchronous generators because of their sizing and limited rating as the power electronic converters are designed to provide maximum active power; hence, their capability to provide reactive power during steady state and during a short circuit is limited.
3. The power reduction above 50.2 Hz with hysteresis characteristic can lead to behaviours (frequency and short-term voltage stability), which are counterproductive for the operation of the network.
4. Due to the reduced grid inertia, the frequency response of the network depends largely on the frequency-power-control characteristic of the power electronic converters. The task of short-term frequency stabilization, which was inherently performed by the rotating inertia in the past, is replaced by the frequency-power-control behaviour of power electronic converters [5].

8.5.2 Grid Forming Control

The extended *direct voltage control* (DVC) strategy for reactive and active power are represented in Figs. 8.12 and 8.13, respectively.

Figure 8.13 also shows the voltage-dependent active power reduction, which is used as a way of feedforward and adjusts the power injection capability of the power electronic converter-based generation units. During a fault, the power electronic converters cannot inject power in the grid; hence, the VDAPR reduces the reference power set point of the power electronic converters immediately, thus improving the transient stability of the overall system.

The current limitation in the voltage source converters is a very important task as the overcurrent can damage the IGBTs very quickly. In case of an overcurrent, the power electronic converter is usually blocked as to avoid the damage to the IGBTs. Figure 8.14 shows the current limitation technique for the VSC. If there is no priority defined, the total magnitude of the current will be limited, irrespectively of the reactive or active current injection. In the case, that the current active priority is active, the reactive current will be reduced and limited to allow the maximum

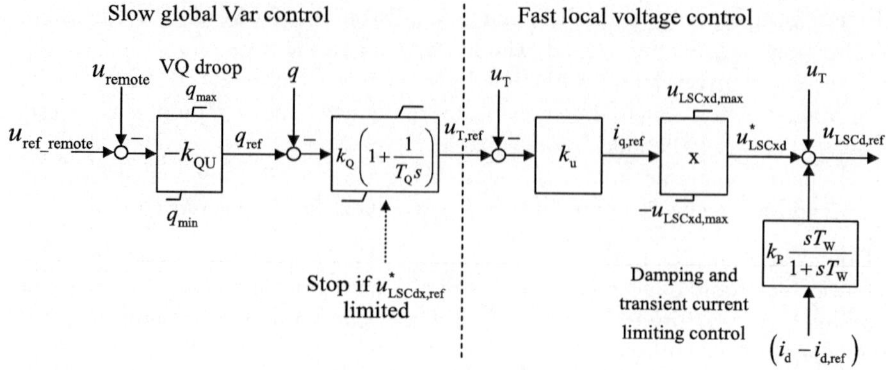

Fig. 8.12 Extended DVC reactive power control

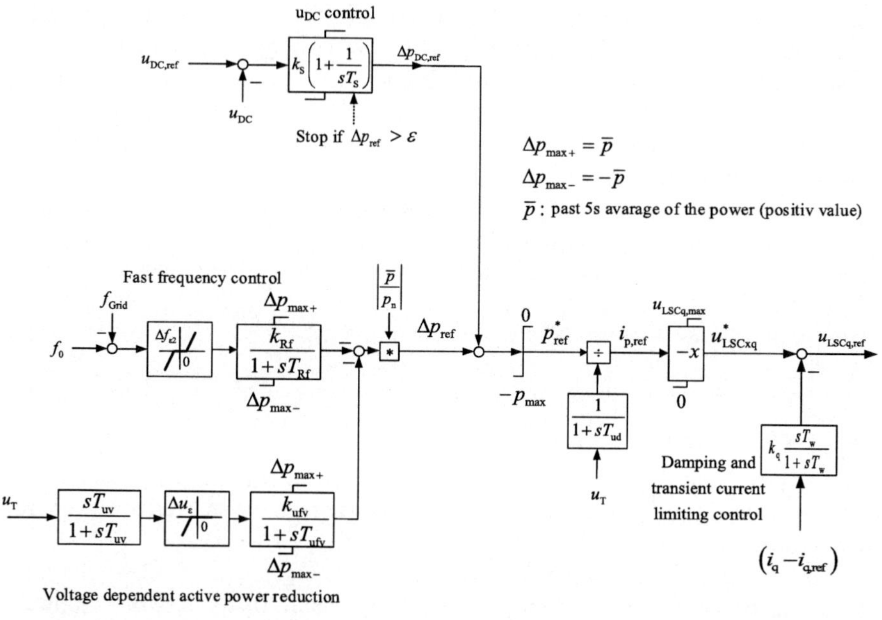

Fig. 8.13 Extended DVC active power control

active current injection possible, and in the case of the reactive current priority, the maximum reactive current will be injected while active current will be reduced so that the total current remains limited.

The active current priority is given by the following equation

$$i_{\text{p,ref}} = i^*_{\text{p,ref}}$$

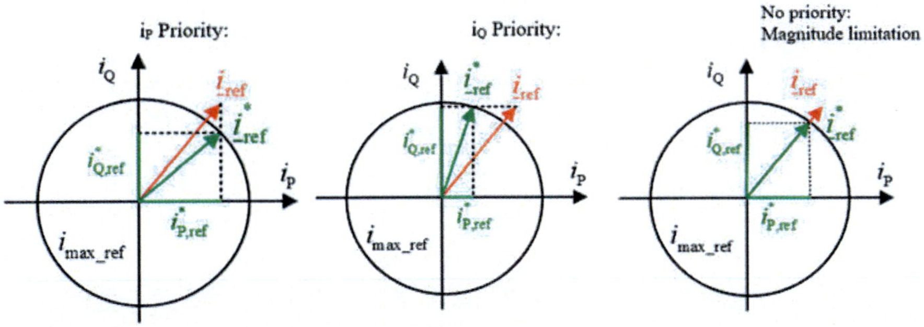

Fig. 8.14 Current limitation options for the VSC

$$i_{\mathrm{q,ref}} = \sqrt{\left(i_{\max\_\mathrm{ref}}\right)^2 - \left(i^*_{\mathrm{p,ref}}\right)^2} \tag{8.5}$$

The reactive current priority is given by the following equation

$$\begin{aligned} i_{\mathrm{q,ref}} &= i^*_{\mathrm{q,ref}} \\ i_{\mathrm{p,ref}} &= \sqrt{\left(i_{\max\_\mathrm{ref}}\right)^2 - \left(i^*_{\mathrm{q,ref}}\right)^2} \end{aligned} \tag{8.6}$$

Depending upon the set point for the voltage support, the active and reactive priorities are changed automatically by the control.

Figure 8.15 shows the current limiting method used in this extended controller. When the actual power electronic converter current exceeds the maximum current of the power electronic converter, a new maximum current value is calculated using the equation shown in Fig. 8.15. The new calculated maximum current value is used in the limiter block along with the angle of the grid impedance seen by the power electronic converter. The reference current, i.e. limits of the d- and q-converter control voltages, is limited based on the impedance of the grid and new calculated maximum current value. This means that the power electronic converter adjusts the reference currents based on the impedance seen by it, and thus, it provides the best possible voltage support. This idea is again depicted with the help of the phasor and circle diagram in Fig. 8.15.

Using the above-mentioned power electronic converter control, the same disturbance as with the PI-based current control was simulated. It was noticed that none of the previously observed critical phenomena occurred. Instead, an improvement of the dynamic behaviour was observed, which compares favourably even with the case with exclusively synchronous generator-based supply. In order to demonstrate how powerful the suggested power electronic converter control can perform and to show how even strongly power electronic converter-dominated network can be made to cope with the disturbance, the in-feed via power electronic converters was

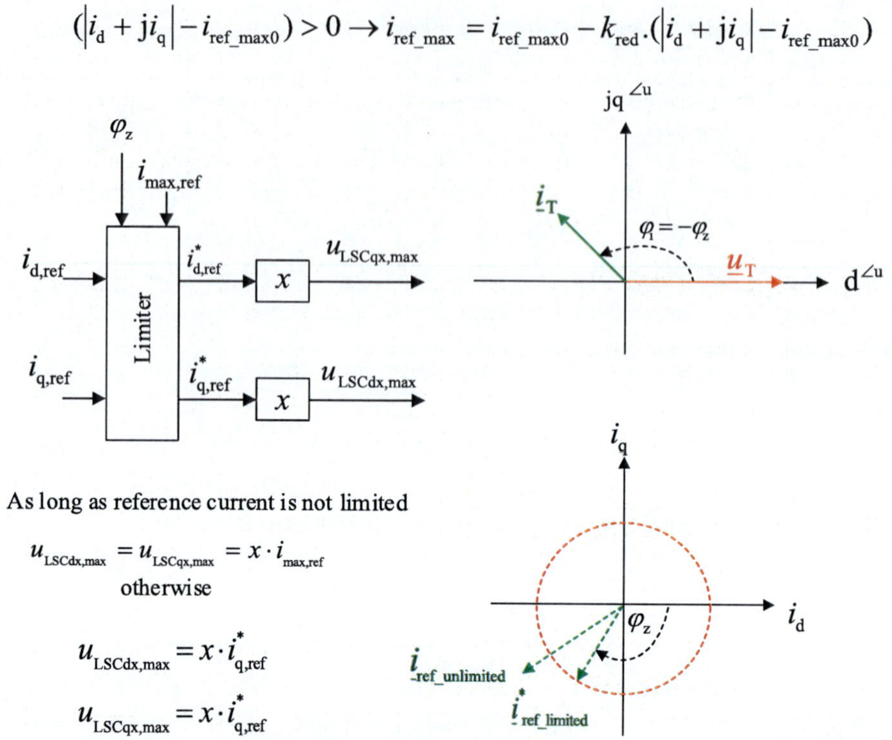

Fig. 8.15 Algorithm for current limitation of DVC

increased to 98%. Only the synchronous generator G1 representing 2% of the generation has been retained to reproduce the primary control similar to the previous cases. The diagrams in Fig. 8.16 demonstrate that even in such an extreme scenario, the network can be operated in a stable manner.

The same disturbance (the three-phase short circuit with a duration of 150 ms) as considered for PI-based current control was simulated for DVC strategy.

It is observed that none of the critical phenomena with the previous case happened. Furthermore, the dynamic behaviour has improved, which seems favourable even compared with the case when the network supplied solely by the conventional synchronous generator.

Figure 8.10 shows the response to grid short circuit by using DVC and only 2% conventional generation. The share of power electronic converter-based generation increased to 98%, in order to prove the strong performance of the proposed power electronic converter control method and also to show that even strongly power electronic converter—the dominated network can be made to cope with the disturbance.

The plots in Fig. 8.10 represent the stable operation of the network, even in such an extreme scenario.

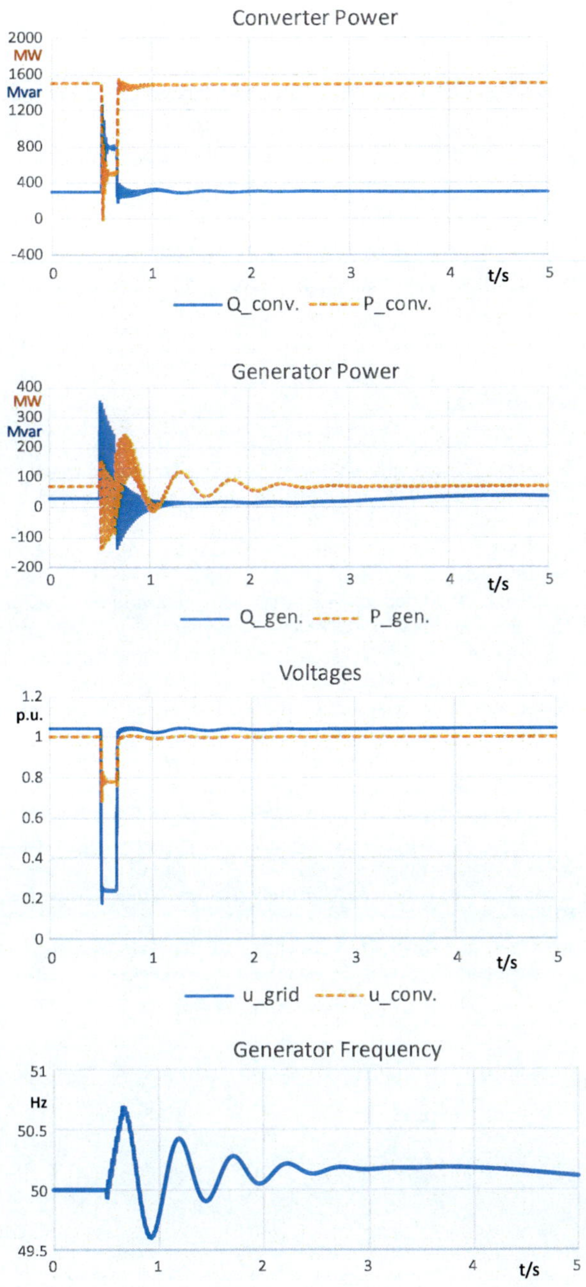

Fig. 8.16 Response to grid short circuit by using DVC and only 2% conventional generation

8.6 Conclusion

In this book chapter, a benchmark test system has been for the power system stability assessment considering very high penetration of converter-based generation units, including grid forming converters. Both conventional PI-based control strategy and grid forming control strategy have been considered. The simulation results show that the prosed control strategy can provide the stable operation and improvement in the dynamic behaviour of the power system, even in a scenario with extremely high penetration of power electronic converter-based generation unit.

References

1. G. Denis, "From grid-following to grid-forming: the new strategy to build 100 % power-electronics interfaced transmission system with enhanced transient behaviour. Ph.D. Thesis, University Lille, France, Nov 2017
2. Q. Zhong, Virtual synchronous machines: a unified interface for grid integration. IEEE Power Electron. Mag. **3**(4), 18–27 (2016)
3. D. Daniel, W. Friedrich, G. Christoph, *Operational Behaviour of the Virtual Synchronous Machine, Energiewende in Der Stromversorgung – Systemstabilität Und Systemsicherheit - 12* (ETG/GMA-Tagung, Netzregelung Und Systemführung, Berlin, 2017)
4. The link: https://www.digsilent.de/en/faq-reader-powerfactory/do-you-have-an-example-of-a-fault-tolerant-power-system-with-98-share-of-renewables/searchfaq/98.html
5. I. Erlich, A. Korai, F. Shewarega, *Study on the Minimum Share of Conventional Generation Units Required for Stable Operation of Future Converter-Dominated Grids* (IEEE Power & Energy Society General Meeting, Portland, 2018)
6. Grid codes, high and extra- high voltage, TenneT TSO GmbH, 1st November 2015. Available at: https://www.tennet.eu/fileadmin/user_upload/The_Electricity_Market/German_Market/Grid_customers/tennet-NAR2015eng.pdf
7. J. Machowski, J. Bialek, J. Bumby, *Power System Dynamic: Stability and Control*, 2nd edn. (Wiley, 2008)
8. A. Korai, I. Erlich, Modelling of Full Converter-based Wind Turbine for the Simulation Environment DIgSILENT PowerFactory. Project report (2017)

Chapter 9
System Protection Schemes as a Way to Prevent Bottlenecks of the Power System Considering the Integration of Offshore and Onshore Wind Turbines and HVDC Link

Abdul W. Korai, M. Ebrahim Adabi, E. Rakhshani, José Luis Rueda Torres, and Mart A. M. M. van der Meijden

Abstract In this book chapter, innovative protection schemes have been suggested to prevent bottlenecks of the power system considering the integration of offshore and onshore wind turbines and HVDC link. Four different countermeasures are proposed and investigated. Their effect on the system overloading and stability is also taken into account. The models for the simulation have been implemented in PowerFactory.

Keywords Offshore wind turbine (OWP) · Onshore wind turbine · HVDC · Special system protection (SPS)

9.1 Introduction

Operational bottlenecks such as overloading of the electrical grid which originates from the integration of power electronic converter-based generation unit such as integration of offshore and onshore wind turbines and HVDC link lead to re-dispatching of the electrical power. It should be noted that the re-dispatching of the electrical power puts a burden of millions of euros on the consumers [1].

Electronic supplementary material The online version of this chapter (https://doi.org/10.1007/978-3-030-54124-8_9) contains supplementary material, which is available to authorized users.

A. W. Korai
DIgSILENT GmbH, Gomaringen, Baden-Württemberg, Germany

M. E. Adabi · E. Rakhshani · J. L. Rueda Torres (✉) · M. A. M. M. van der Meijden
Department of Electrical Sustainable Energy, Delft University of Technology, Delft, The Netherlands
e-mail: j.l.ruedatorres@tudelft.nl

M. A. M. M. van der Meijden
TenneT TSO BV, Arnhem, The Netherlands

F. M. Gonzalez-Longatt and J. L. Rueda Torres (eds.), *Modelling and Simulation of Power Electronic Converter Dominated Power Systems in PowerFactory*, Power Systems,
https://doi.org/10.1007/978-3-030-54124-8_9

It is required to propose a creative solution to prevent the bottlenecks of the electrical grid, which in turn decreases the cost regarding re-dispatching of the electrical power and also makes the integration of power electronic converter-based generation unit to the electrical power network, easier.

In this book chapter, the test grid has been represented in Sect. 9.2 to analyse the effect of the operational bottlenecks on the power system stability. Four countermeasures are suggested and investigated in order to prevent grid bottlenecks, considering the high share of power electronic converter-based generation units. Their effect on the system overloading and stability is also analysed. These countermeasures will be discussed one by one in Sects. 9.3–9.6, respectively.

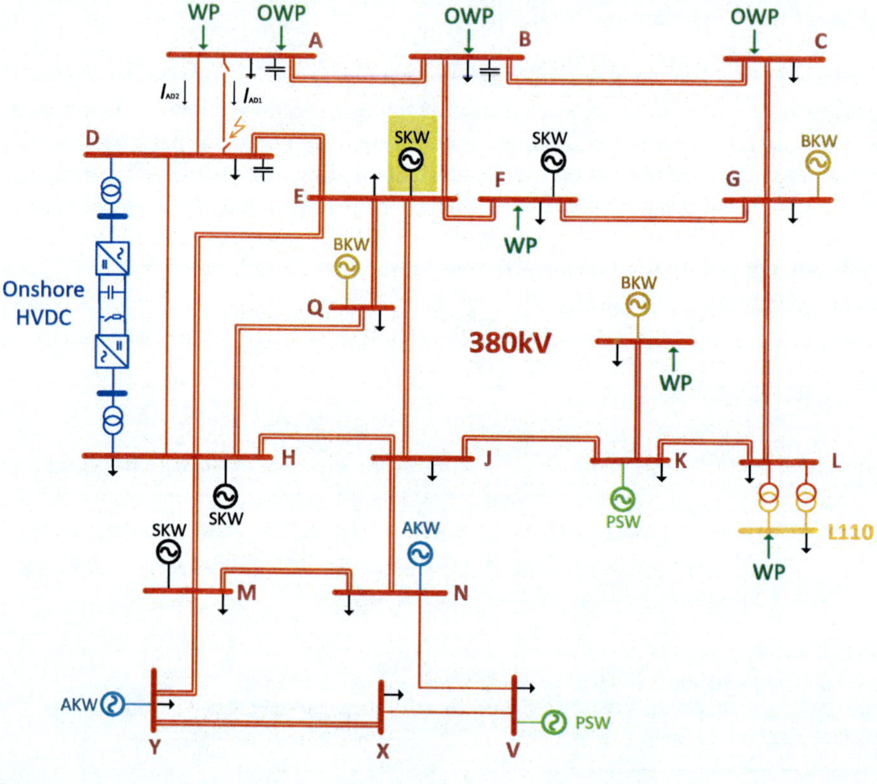

Fig. 9.1 Test grid for bottleneck prevention

9.2 Test System

Figure 9.1 shows the test grid that has been modelled to study the impact of operational bottlenecks on the power system stability in order to suggest a creative solution to prevent such obstacles.

The total generation is 205,32 MW in the test grid. The share of power electronic converter-based generation unit is about 50% of total generation. 7000 MW of electrical power comes from three offshore wind parks (OWPs) in the north, 2500 MW of total power comes from the three onshore wind parks in the north, middle and south of the grid.

The controllers of the HVDC, OWPs and WPs are according to the existing grid codes [2]. As it can be observed from Fig. 9.1, the test grid consists of one onshore HVDC connection. The HVDC transfers 2000 MW of electrical power from north to south.

There are eleven conventional power plants categorized into four types, namely

1. SKW: Coal-fired power plant
2. BKW: Lignite-fired power plant
3. PSW: Pump-storage power plant
4. AKW: Nuclear power plant.

The primary control is performed by one of the coal-fired power plants that have been marked in yellow, as it can be seen from Fig. 9.1. The reserve capacity of this generator is around 500 MW, which is 2.5% of the total load of the test grid.

Only one disturbance will be considered and investigated in this study. One of the two overhead lines connecting bus A to bus D will be disconnected after 150 ms during a three-phase fault near bus D to isolate the fault, as it can be observed from Fig. 9.1. This scenario is a classical n -1 scenario.

Figure 9.2 shows the results when the line A-D1 is switched off during the short circuit. It can be observed from Fig. 9.2a that the frequency goes up to 51.3 Hz despite the fact that the generator responsible for the primary control reduces its power almost to zero as can be seen in Fig. 9.2b.

The wind parks do not contribute towards the improvement of frequency as well as the voltage of the grid. The active power from the wind parks remains at the pre-fault value as shown in Fig. 9.2c, whereas the fast voltage support is inactive due to the dead band in the voltage control; hence, the voltages in the network go down as low as 0.9 p.u. The remaining line, which connects bus A to bus D, is overloaded to up to 120% as shown in Fig. 9.2d. This scenario will probably result in system collapse if no re-dispatch takes place.

Several countermeasures are suggested and investigated. Their effect on the system overloading and stability is also analysed. These countermeasures will be discussed one by one.

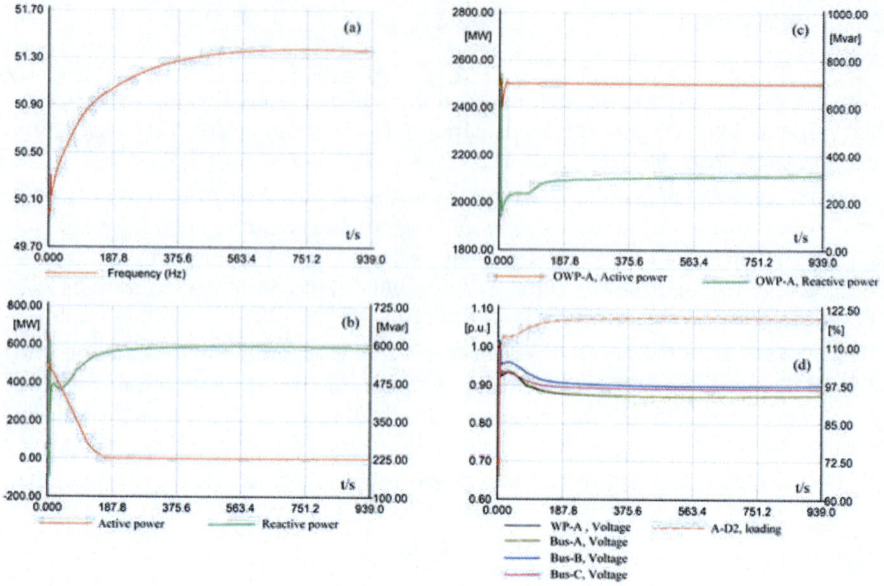

Fig. 9.2 Results in case of a line switching off

9.3 Countermeasure No: One

1. Removal of the dead band on the OWP controllers, i.e. onshore HVDC systems which connect the OWPs to the mainland grid.
2. Increase of the reference voltage up to 0.07 p.u. in the OWP controllers, i.e. onshore HVDC systems which connect OWPs to the mainland grid.

Figure 9.3 shows the results with the countermeasure one. It can be seen in Fig. 9.3a that the frequency can be controlled around 50 Hz. The loading of the line is reduced from 120 to 110% as well as the voltages in the network have improved Fig. 9.3d. The system seems to be stable; however, the continuous overloading of the line will increase the temperature of the line and will lead either to the line disconnecting or in re-dispatching. The outage of the line will result in system collapse while the re-dispatch will incur additional costs.

9.4 Countermeasure No: Two

1. Removal of the dead band on the OWP controllers, i.e. onshore HVDC systems which connect the OWPs to the mainland grid.
2. Reduction of the active power from the OWPs when the frequency goes above 50.2 Hz.

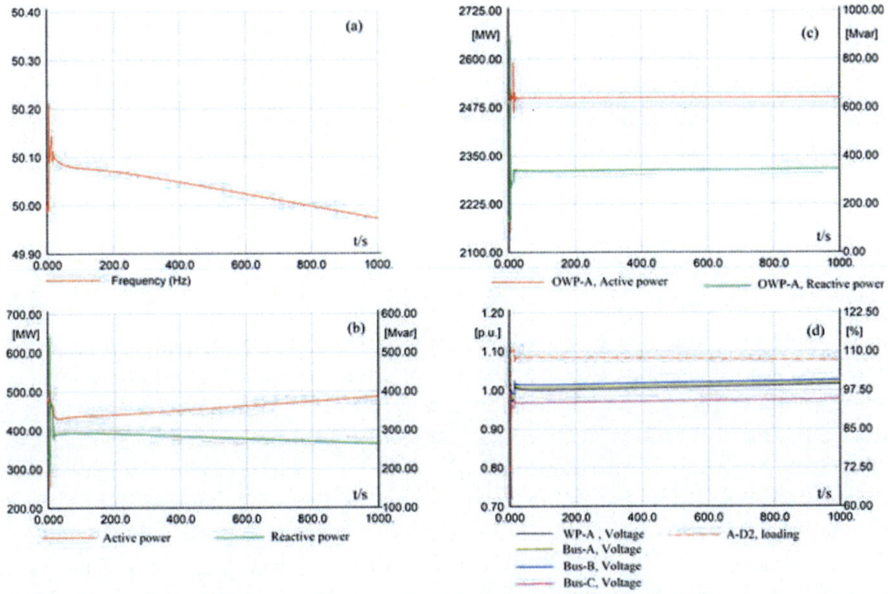

Fig. 9.3 Results of countermeasure no: one

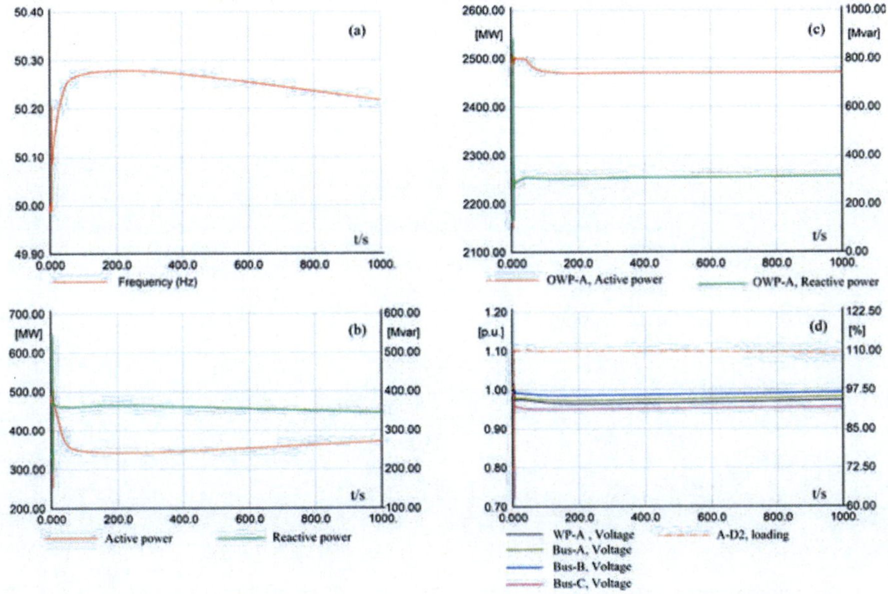

Fig. 9.4 Results of countermeasure no: two

Figure 9.4 shows the results with the countermeasure two. Due to the reduction of the active power in the three OWPs, the frequency is controlled around 50.2 Hz as shown in Fig. 9.4a. However, the line overloading still exists as can be seen in Fig. 9.4d, and as explained before, this can lead to an outage of the line and system collapse or re-dispatching.

9.5 Countermeasure No: Three

Special protection schemes are preventive measures that can be activated in case of pre-defined disturbances so that the system collapse can be prevented or the overloading of lines, transformers and cables can be avoided [3]. Figure 9.5 shows the implementation of special protection schemes (SPS). The SPS can be used to control WP's active and reactive power depending upon the temperature of electrical line and weather conditions (wind and ambient temperature). Depending on these factors, the active and reactive power of WP's can be controlled in order to prevent overloading of line or the re-dispatch of active power in this particular example. The SPS can be used for a variety of disturbances and corrective measures. In this study, however, only one disturbance is investigated.

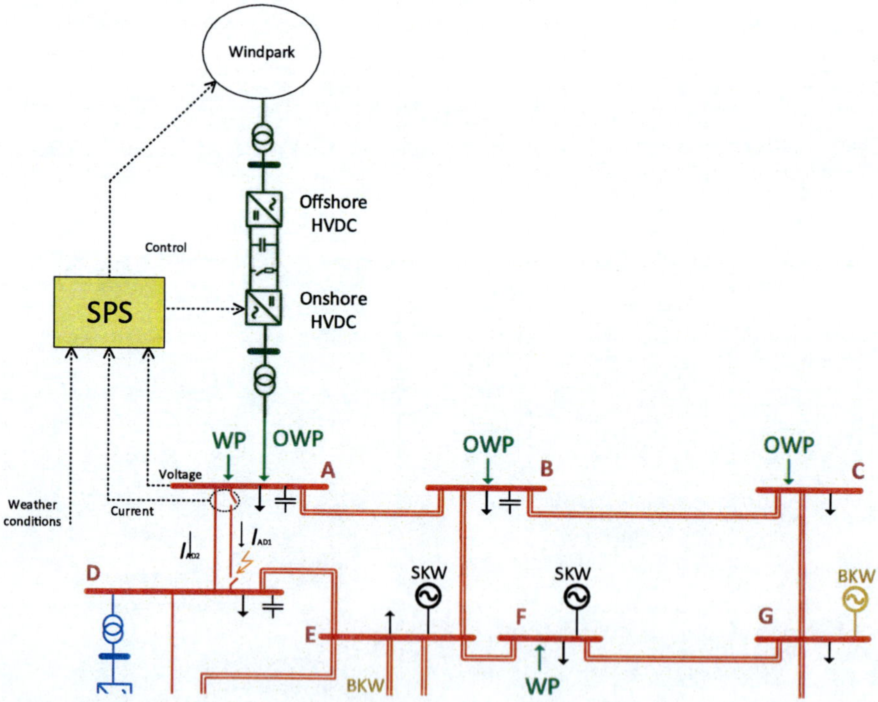

Fig. 9.5 Application of SPS on the OWP

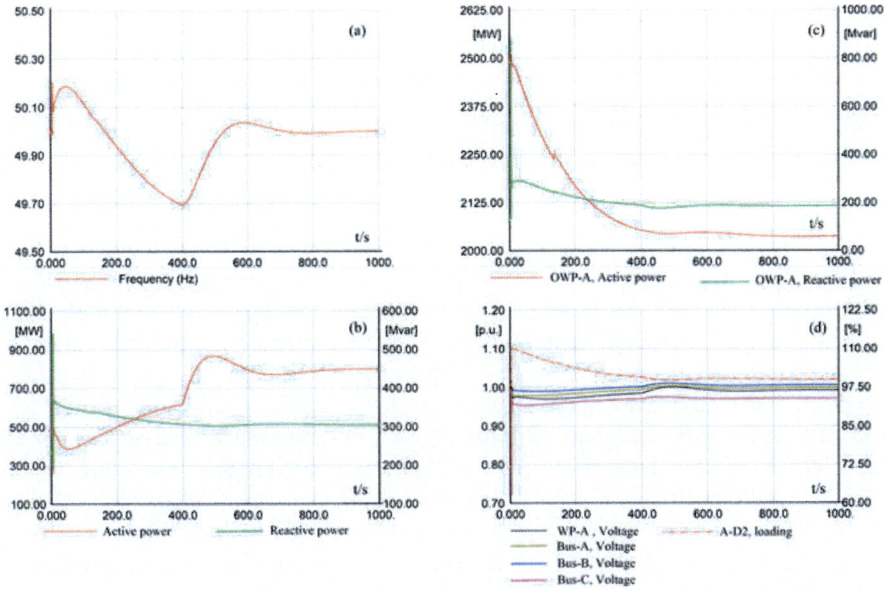

Fig. 9.6 Results of countermeasure no: three

1. Removal of the dead band on the OWP controllers, i.e. onshore HVDC systems which connect the OWPs to the mainland grid.
2. Reduction of the active power from the OWPs when the frequency goes above 50.2 Hz.
3. Reduction of the active power from the OWPs based on the overloading on the line A-D2. The reduction follows the integral characteristic.

Figure 9.6 shows the results of the usage of SPS for the reduction of overloading of the line and possibly the avoidance of re-dispatch of electrical power in the grid. As can be seen in Fig. 9.6b, the SPS sends a signal to OWP-A to reduce the active power as long as the overloading on overhead line A-D2 is above 100%. In a more realistic scenario, the temperature of the line can be estimated, which depends on the wind velocity and ambient temperature as well as loading of the line. For the case of simplicity, only loading is considered here. Figure 9.6c shows that the reduction of the active power from OWP-A causes the frequency to drop, which is then stabilized by the primary control of the generator. After some time, the secondary control is activated, which brings the frequency back to 50 Hz shown in Fig. 9.6b. The reduction of the active power follows an integral characteristic, which means that the active power reduction on OWP-A side can be reduced as long as the temperature or in this example loading of the line is below a pre-defined value.

The voltages of the grid seem to be also in the acceptable limits, as shown in Fig. 9.6d. The use of the SPS, as shown by the results here, can prevent the re-dispatch measures.

9.6 Countermeasure No: Four

1. Removal of the dead band on the OWP controllers, i.e. onshore HVDC systems which connect the OWPs to the mainland grid.
2. Reduction of the active power from the OWPs when the frequency goes above 50.2 Hz.
3. Shifting of the electrical power from line A-D2 to line A-B using the phase-shifting transformer with the help of SPS.

Additional usage of the special protection schemes (SPS) to regulate the phase-shifting transformer installed between bus A and B as shown in Fig. 9.7. The phase-shifting transformer is installed between bus A and B, in order to ensure that the power can be diverted through this line so that the line A-D2 can be unloaded. It is necessary that the overhead line A-B must not be heavily loaded in normal operation and should be capable of transferring diverted power without being overloaded. This also takes into consideration that the diversion might overload other lines in the network. In this example, however, the system is designed such that the diversion does not create any bottlenecks. The installation of the phase-shifting transformer, however, costs additional investment.

Figure 9.8 shows the result of countermeasure number four. The frequency is controlled as shown in Fig. 9.8a by reducing the active power injection from OWPs as shown in Fig. 9.8c and after 200 s (in this example), the phase-shifting transformer is activated by the SPS, which diverts the power through line A–B and unloads the overloaded line immediately as shown in Fig. 9.8d. In this particular example, the activation of the transformer tap changer is carried out considering the loading of the line, but in practice, another criterion can also be used to regulate the transformer by SPS.

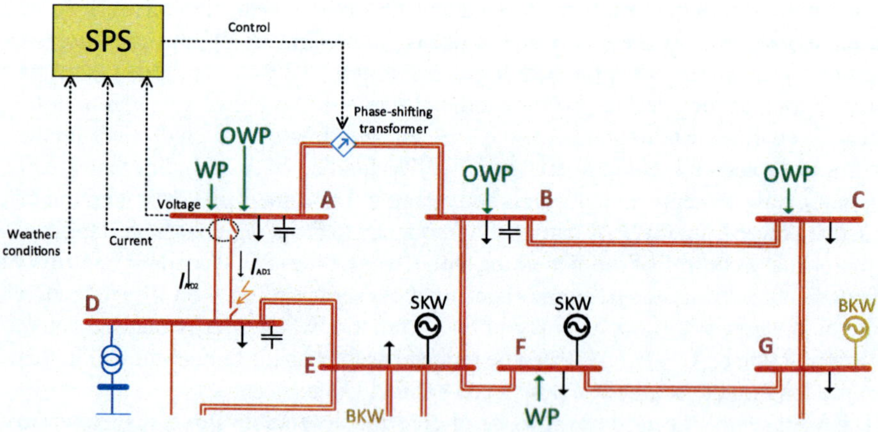

Fig. 9.7 Application of SPS on the phase-shifting transformer

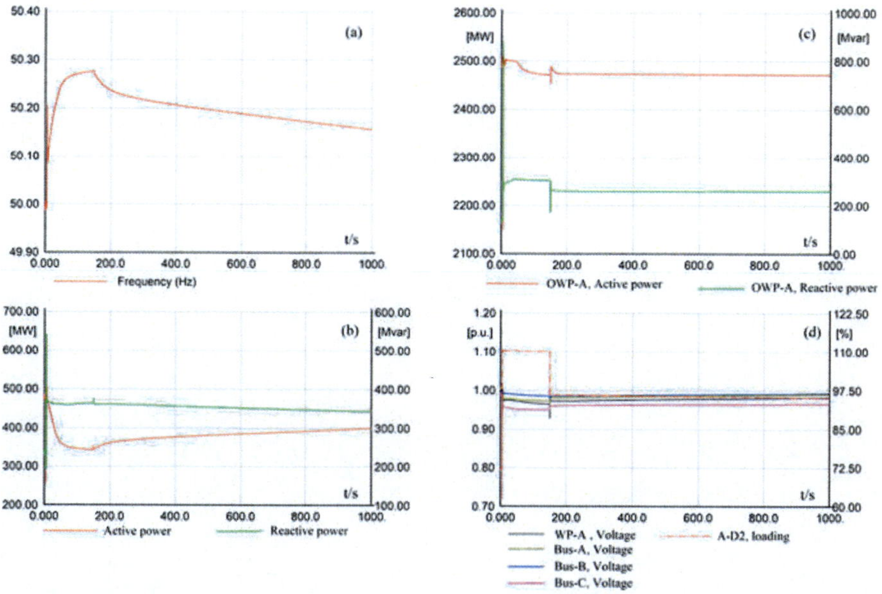

Fig. 9.8 Results of countermeasure no: four

9.7 Conclusion

In this book chapter, the special protection system has been suggested to prevent the bottlenecks of power system with considering the integration of power electronic converter-based generation unit. Different countermeasures have been taken in to account to study the impact of protection schemes on overloading and power system stability. The simulation results prove the performance of the protection schemes to keep the bottlenecks bellow pre-defined value.

References

1. Was ist eigentlich ein "Netzengpass. Available at, https://www.bmwi-energiewende.de/EWD/Redaktion/Newsletter/2018/03/Meldung/direkt-erklaert.html
2. Grid codes, high and extra high voltage, TenneT TSO GmbH, 1st November 2015. Available at :https://www.tennet.eu/fileadmin/user_upload/The_Electricity_Market/German_Market/Grid_customers/tennet-NAR2015eng.pdf
3. ENTSO-E, Establishing a guidelines on electricity transmission system operation. Available at: https://eur-lex.europa.eu/legal-content/EN/TXT/?uri=uriserv:OJ.L_.2017.220.01.0001.01.ENG&toc=OJ:L:2017:220

Chapter 10
Implementation and Performance Comparison of Derivative and Virtual Synchronous Power Methods for Enhancement of System Frequency Stability

E. Rakhshani, M. Ebrahim Adabi, José Luis Rueda Torres, Mart A. M. M. van der Meijden, and Francisco M. Gonzalez-Longatt

Abstract This chapter is dedicated to present some control mechanism to cope with the challenges due to the growth of the penetration level of the power electronic interfaced generation (PEIG) in sustainable interconnected energy systems. Specifically, this chapter presents different forms of fast active power injection (FAPI) control schemes for the analysis and development of different mitigation measures to address the frequency stability problem. Among the considered FAPI control schemes are the traditional droop-based scheme, and two propositions implemented in the form of a derivative-based control and a second-order virtual synchronous power (VSP)-based control. All the detailed explanation, DSL-based control is presented for the simulations in DIgSILENT software. Simulation results show that thanks to proposed FAPI controllers, it is possible to increase the maximum share of wind power generation without violating the threshold limits for frequency stability problem in low-inertia systems.

Keywords DIgSILENT RMS simulation · DSL programming · Control · Frequency control · Virtual inertia · Wind turbine

Electronic supplementary material The online version of this chapter (https://doi.org/10.1007/978-3-030-54124-8_10) contains supplementary material, which is available to authorized users.

E. Rakhshani · M. E. Adabi · J. L. Rueda Torres (✉) · M. A. M. M. van der Meijden
Department of Electrical Sustainable Energy, Delft University of Technology, Delft, The Netherlands
e-mail: j.l.ruedatorres@tudelft.nl

M. A. M. M. van der Meijden
TenneT TSO B.V, Arnhem, The Netherlands

F. M. Gonzalez-Longatt
Department of Electrical Engineering, Information Technology and Cybernetics, University of South-Eastern Norway, Notodden, Norway

F. M. Gonzalez-Longatt and J. L. Rueda Torres (eds.), *Modelling and Simulation of Power Electronic Converter Dominated Power Systems in PowerFactory*, Power Systems,
https://doi.org/10.1007/978-3-030-54124-8_10

10.1 Introduction

Modern power systems are evolving and changing in many ways; one of those changes is the massive penetration of power electronic converters (PEC). As the power electronics is penetrating the power systems novel mechanism of control are required in order to cope with the progressive substitution of the traditional synchronous generators and all the primary services provided by them. One critical aspect is the total system rotational inertia reduction.

Systems with low inertia are the consequence of phase-out of conventional power plants with synchronous generators due to increasing the share of PEIG such as solar photovoltaic systems and wind power plants. The main concern with power electronic-based generation units is that they cannot completely replace the inertial response of conventional generators. This is due to their limited capability of storing energy which makes their contribution to the overall system inertia very small. Therefore, this results in system frequency being more volatile.

This chapter is looking into the fast active power injection (FAPI) control schemes for the analysis and development of different mitigation measures to address the frequency stability problem.

The proposition of different FAPI controllers with the application of DIgSILENT PowerFactory for analysing and comparisons in low-inertia power systems with PEC-based generation like WT generation considering the fact that future wind power will be required to participate on inertia emulation capabilities with active and reactive power management systems is the main motivation of presented materials in this chapter.

A well-designed inertia emulation (IE) controller empowers the wind generator, or storage element, to release the stored energy for arresting the frequency drop within 10.0 s [1, 2]. Fast frequency controllers can be classified into three main families, namely

(i) droop-based controllers (or proportional controllers) [3–5],
(ii) derivative-based controllers [6–9] and
(iii) other approaches which are usually based on the swing equation of conventional SG, thus attempting to represent a virtual synchronous machine (VSM) for IE [10, 11].

FAPI enables the controlled source to modulate its output power within 10–30 s to arrest the frequency deviation. Depending on the system requirements, the recovery time may be bounded to be within a relatively short period. In this chapter, different forms of FAPI control schemes manly derivative-based and a second-order VSP-based controller are discussed and compared.

This chapter is organised as follows: In Sect. 10.2, the system modelling with the wind turbine generator (WTG) control schematic with its integrated structure is presented. Details of the proposed FAPI controller and implementation of different control laws for FAPI in DIgSILENT PowerFactory software are presented and discussed in Sect. 10.3. In Sect. 10.4, the use of different FAPI controllers on the test system is studied and compared. Finally, the conclusions are presented in Sects. 10.5.

10.2 Wind Power Integration and Control

The WTG model is implemented according to the standards of IEC 61,400–27 series [5]. It is clarified that, due to confidentiality agreements, the pdf file of the implemented controller with the wind turbine model in PowerFactory cannot be provided to the readers. For this chapter, a simple illustrative example for one of the presented controller is provided. Following the explained details in this chapter, the readers are encouraged to build their own model for this purpose.

As shown in Fig. 10.1, different blocks that represent modifications in respect to the IEC standard with a brief explanation of their roles are presented [5]. The key aspects of the studied model are explained as follows:

- In the measurement section of this model, the measurement blocks for frequency, power and voltage have a direct connection to the WG terminals. These measurement and also the currents which are coming from generator blocks are necessary for initialisation of the used model.
- The generator block consists of the "static generator" from PowerFactory elements, and it will work as a controlled current source. The details of the generator block are based on the IEC 61400-27-1 standard [12].
- The aerodynamic block is used for the representation of the mechanical parts of the wind generation unit. In this block, the mechanical power of the turbine is calculated. It is a single-mass component following the IEC standard [13].
- The input block contains wind speed and power inputs. The wind speed is provided by using an external file with wind speed measurements in metre per second (m/s). The power input is an alternative input which contains the maximal available power.
- The *P* control and pitch angle controller blocks are built according to [13]. The "FAPI controller" block, shown in light grey in Fig. 10.2, is an additional block to the developed wind generation model. Depending on the applied control strategy within the FAPI control clock, its input signal can be the measured power or frequency, while the output signal is the supplementary reference to the active power control block for fast active power injection capabilities.

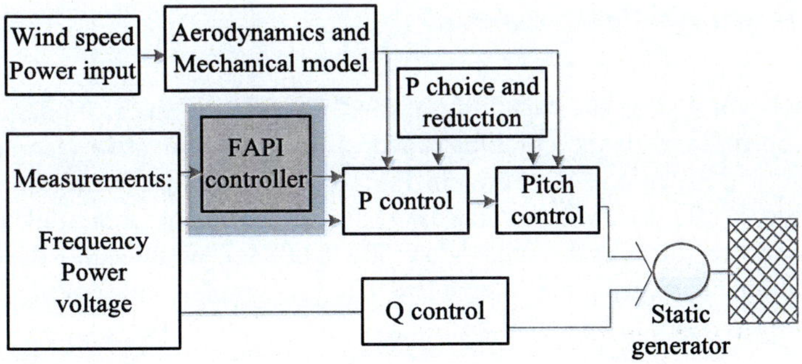

Fig. 10.1 Control structure for the type-4 wind generator

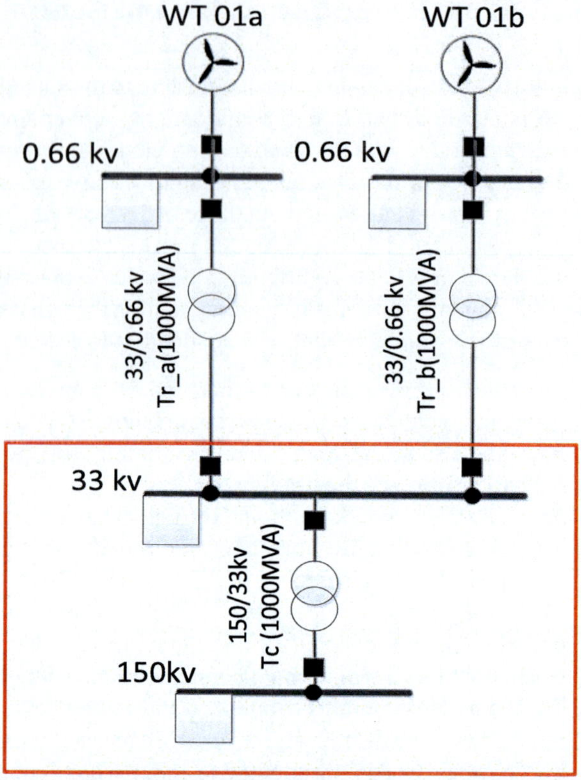

Fig. 10.2 Grid interface of a wind park with two types of WGs

As shown in Fig. 10.2, a wind park is implemented which can be used for representing the connection of various feeders with wind generations.

10.3 FAPI Controllers

10.3.1 Derivative-Based Control

In this section, for adding the capability of inertia emulation to the model of WG, a detailed controller with the combination of droop and derivative techniques is designed and implemented. As shown in Fig. 10.3, the droop control loop injection/extraction of the active power can be changed according to the deviation of measured frequency from its nominal value (50 Hz/60 Hz). While within the derivative control loop, considering the control law in (1), a complementary action for fast active power injection can be performed.

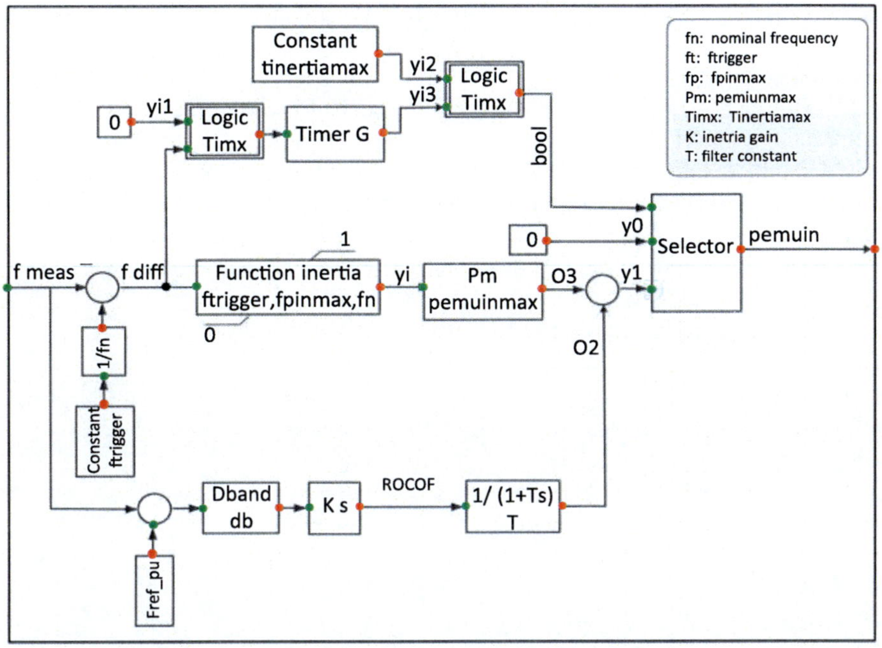

Fig. 10.3 Diagram of the derivative plus droop-based FAPI in PowerFactory

$$\Delta P_{\text{emuin}} = -k\frac{d(\Delta f)}{dt} \tag{10.1}$$

The input signal of the proposed FAPI controller is the frequency deviation, and its output is the additional power that is added to the active power reference of the wind turbine controller. The required energy for this supplementary power is taken from the rotating masses of the WG or from the embedded additional energy in the DC link.

The main control parameters of this FAPI controller that affect the dynamic response during activation of the controller are: the activation threshold for the FAPI controller (f_{th}), the maximum duration of FAPI activation (T_{imx}), the threshold for maximum emulated power (f_p), the allowable additional power output (K_{pm}) and parameters of the derivative loop: the derivative gain (K) and the filter time constant (T). The value of f_{th} is assumed to be equal to 49.85 Hz. In case the frequency reaches an even lower value, corresponding with a second threshold f_p, then the maximal allowed power through emulated inertia is released.

From Fig. 10.3, (cf. control loop in the middle of the block diagram), it can be inferred that there is a linear dependence between frequency and the additional power signal (P_{emuin}), when the frequency deviation is between 1 and 0. During this interval, the following formula is used:

$$P_{\text{emuin}}(t) = \frac{f_{th} - f_{\text{meas}}(t)}{f_{th} - f_p} K_{pm} \tag{2}$$

The value of K_{pm} is assumed to be set between 10 to 25% of the nominal wind turbine active power [14, 15].

Correct initialisation of a model in a power system simulation tool avoids fictitious electrical transients and makes it possible to evaluate the real dynamic performance of the system correctly. Therefore, the initialisation equations of important states and variables of this composite model (shown in Fig. 10.3) are presented below:

```
Vardef (ftrigger) ='Hz';'Frequency in Hz below which emulated
inertia will be activated'
Vardef (fpinmax)  ='Hz';'Frequency in Hz at and below which the
maximum emulatad inertia power will be released'
vardef(fn)           ='Hz';'Nominal frequency'
vardef(pemuinmax)    ='Pn';'Maximum allowed additional power output
from emulated inertia in p.u. '
vardef(tinertiamax) ='s';'Maximum duration of inertia emulation'
inc(xtimer)=0
inc (x)=0
inc (x1)=0
```

10.3.2 Second-Order VSP-Based FAPI Control

Figure 10.4 presents the general structure of the synchronous controller, which aims at controlling the VSC, for emulating inertial droop response and damping of a normal synchronous generator (SG). The structure of this controller is coming from suitable programming of the electrical performance of a conventional SG on a digital basis that is taking care of controlling the VSC converter. Depending on the converter technology, an additional source of energy, like storage elements, can be added into the DC side of a VSC station. The VSP is located in the outer-loop control of the VSC, providing a power reference for the inner loop (current) controller.

Figure 10.5 shows the main electromechanical loop of the VSP strategy. This figure implies a control approach, in which differences between the measured output power of the converter (P_o) and the input (reference) power (P_{in}) are managed with a block called power loop controller (PWRLC) for setting a virtual frequency which has to be added to the grid's frequency, for creating the rotating frequency of a virtual rotor. The integration of such virtual frequency will give the angular position of the virtual rotor, which corresponds to the phase angle of the induced *emf* in the virtual stator, which enables power injection like in the inertial response of an SG.

The representation of the PWRLC in the VSP controller has a second-order characteristic, which makes it possible to simultaneously impact the damping (k) and the

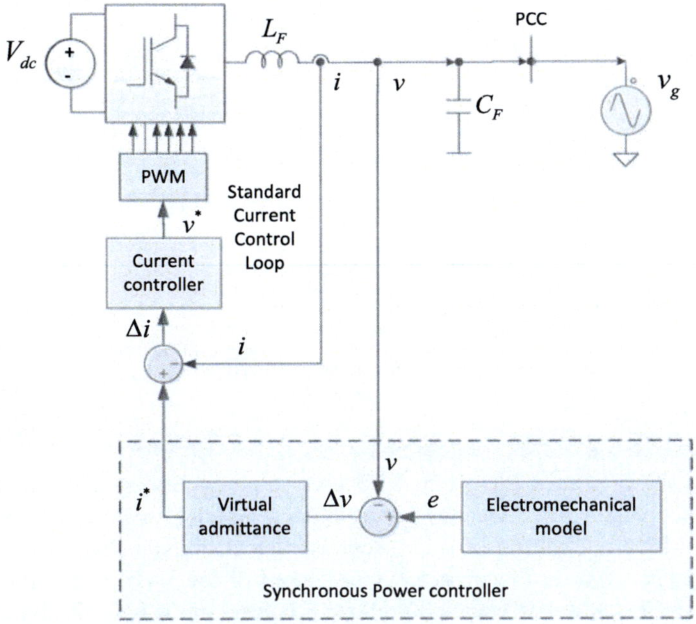

Fig. 10.4 Control structure of VSP-based controller for a VSC

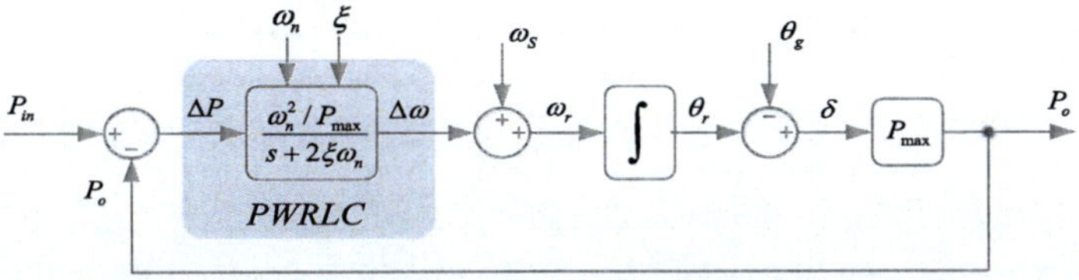

Fig. 10.5 Electromechanical representation of the VSP control

inertia (J) of the system. The dynamic relations between the input and the output power of the presented VSP can be lead to the following second-order equation:

$$\frac{P_o}{P_{in}} = \frac{\omega_n^2}{s^2 + 2\zeta\omega_n s + \omega_n^2} = \frac{\frac{P_{max}}{J.\omega_s}}{s^2 + \frac{k}{J.\omega_s}s + \frac{P_{max}}{J.\omega_s}} \tag{3}$$

where ζ and ω_n are the damping factor and natural frequency, respectively. While the P_{max} is the maximum delivered power, for implementing this second-order function in DIgSILENT PowerFactory the following equation is used:

$$\frac{P_o}{P_{in}} = \frac{1}{1 + K_z s + K_w s^2} = \frac{\frac{1}{K_w}}{s^2 + \frac{K_z}{K_w}s + \frac{1}{K_w}} \tag{4}$$

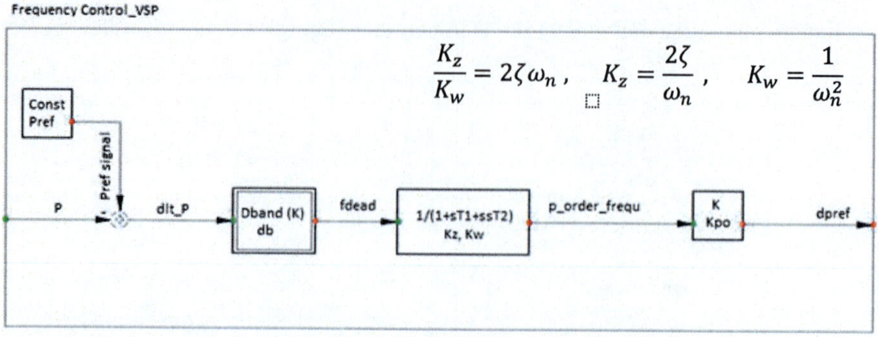

Fig. 10.6 Block definition of the second-order control in PowerFactory

where $1/K_w = \omega_n^2$ and $K_z/K_w = 2\zeta\omega_n$.

Based on the presented idea from VSP concept, a generic second-order-based FAPI is implemented for frequency control functionalities in PowerFactory. The implemented block definition of such second-order virtual synchronous power is presented in Fig. 10.6. In PowerFactory, the output of the VSP control block will provide the surplus power reference (dpref) for P control of the unit. The input of the VSP frequency control block is also the measured power (from "PQ-Measurement" block in DIgSILENT) coming from the main connected bus.

It should be noted that, in real practice, there might be a slight delay due to filtering the measured power, i.e. through a low-pass filter implemented in the electronic analogue measurements of currents and voltages. If the input signal contains high-frequency harmonics, the digital controller implemented in a microcontroller will be able to reject it by proper sampling method of the equivalent analog-to-igital (AD) conversion module. In case, other noise sources are identified the usage of digital notch filters which can be used as a supplement to provide high impedance for the measurement and reject undesirable harmonics.

It is clarified that, due to confidentiality agreements, the complete.pdf file of this controller implemented in PowerFactory with full-order converter cannot be provided to the readers. For this controller, an illustrative example for the presented second-order electromechanical block for a generic energy storage system is provided as supplementary file. Thus, the initialisation equations of important states and variables for such a composite model are presented below:

```
inc(x1)=0;  'state variable of the second-order block'
inc(x2)=0;  'state variable of the second-order block'
inc0(dpref)=0;  'defined ad the output power'
vardef(db)='pu'; 'deadband for frequ. control'
```

10.4 Results and Comparisons in DIgSILENT PowerFactory

In this section, the implementation of the models and the benchmark system and the time-domain simulations in DIgSILENT PowerFactory is presented. Several scripts were developed in Python 3.4 and MATLAB R2016b to automate the FAPI parameter variations and the simulation of different operational scenarios and network topologies.

The single line diagram of the used three-area test system (reported in Sect. 3.2.3 of D1.2 [14] is shown in Fig. 10.7 highlighting the added wind power plants, which are used to create different levels of a share of PEIG in the grid. A worst case of active power imbalance is created by considering the outage of the biggest synchronous generator (A1aG, which entails the loss of 1000 MW) for the winter load profile at $t = 5$ s. The total demand for the system is 15,480 MW.

As the base scenario, it is considered that 50% of the demand is supplied by wind power plants. Details of different operational scenarios, dispatches and system data

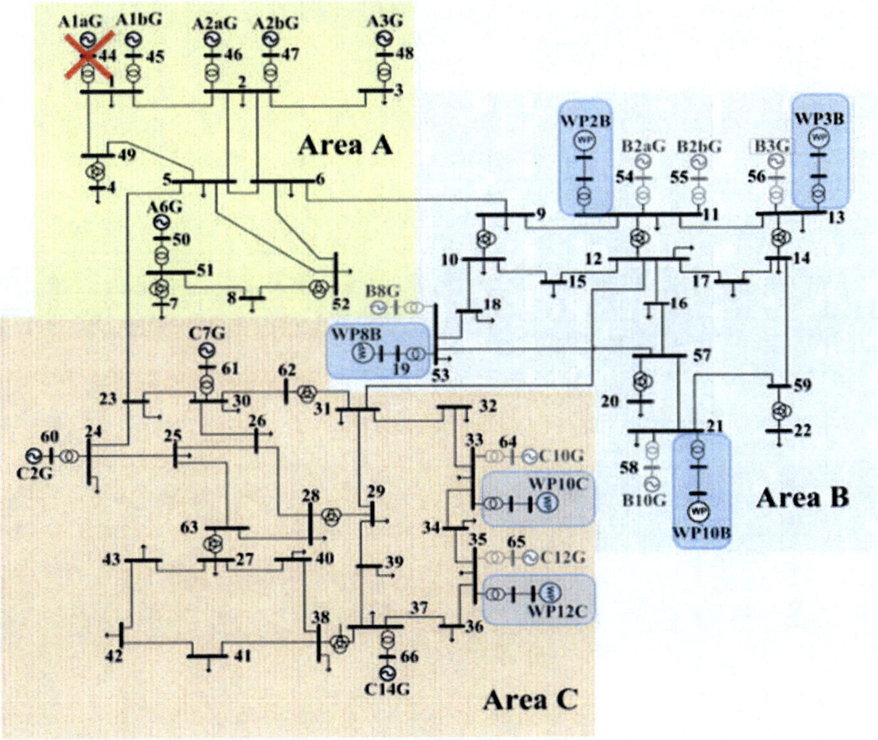

Fig. 10.7 Three-area benchmark with WP integrations

Table 10.1 Parameters used for the FAPI controller

Selected control parameters	Value
f_t[Hz]	49.95
f_P[Hz]	49.75
K_p [pu]	0.25
T_{imx}[s]	15
f_n[Hz]	50
K_d [pu]	10
T [s]	0.25
K_z [pu]	0.07
K_w [pu]	0.3

are reported in D1.2 of the MIGRATE project [14]. The parameters used for this study are given in Table 10.1.

10.4.1 Results of the Derivative-Based FAPI Controller in Three-Area System

The derivative-based FAPI controller (explained in Sect. 10.3.1) is analysed with the generic three-area test system. The impact of the derivative-based FAPI controller for different values of the derivative control gain (K) is shown in Figs. 10.8, 10.9, which indicates the mitigation of the frequency deviation after the occurrence of an active power imbalance. It should be noted that PLL and device measurements might introduce a time delay into the control loop, which makes it difficult to affect the RoCoF immediately from the instant of occurrence of an imbalance. This is an example of impacting the RoCoF in practice.

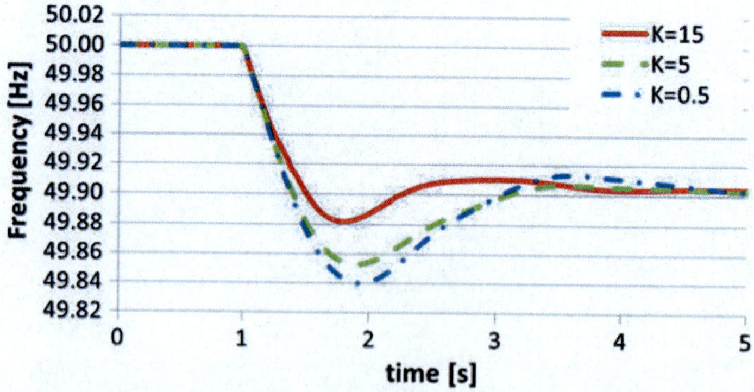

Fig. 10.8 Frequency response for derivative-based FAPI controller

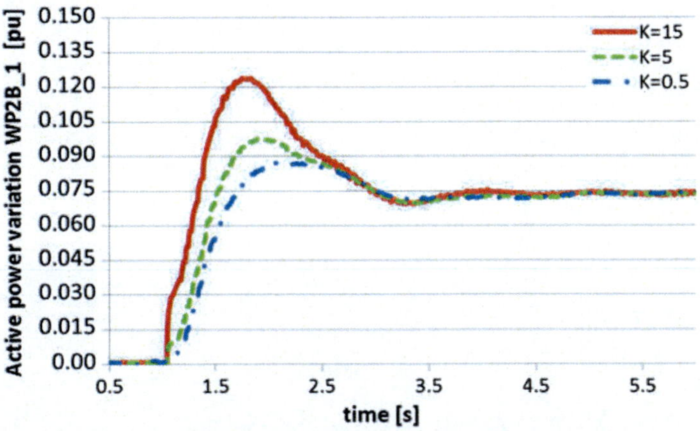

Fig. 10.9 Variation of wind output power during derivative-based FAPI activation (unit 1 of WP2B)

As reported in [16], two sections of the curve of the frequency signal (measured in the period of primary frequency control) can be considered for the evaluation of the RoCoF. The first-time window is around 0.5 s from the time of occurrence of an active power imbalance, whereas the second time window lies between 0.5–2 s from the time of occurrence of an active power imbalance.

As shown in Fig. 10.8, the increase of the derivative gain entails an improvement of the frequency response, with a slight improvement of RoCoF in the first-time window of evaluation, followed by more appreciable improvement in the second time window of evaluation of RoCoF and Nadir. For the sake of illustration, Fig. 10.9 shows the shape of the active power injection due to the action of the derivative-based FAPI controller attached to the largest wind generator of the system. From Fig. 10.8, it can be noted that a higher derivative gain causes a faster active power injection which is leading to a stronger influence on the frequency response. It should be noted that since in this case study the wind generators operate very close to their nominal output if the values of derivative gain became very high (comparing Figs. 10.8 and 10.9) might lead to exceeding the allowed 10% margin above the nominal power. Hence, this margin should be considered as a bound for choosing a value of the derivative gain.

10.4.2 Results of the Second-Order VSP FAPI Controller

The impact of the second-order VSP-based FAPI controller, as an alternative mitigation measure, is analysed with the three-area test system shown in Fig. 10.10. The controller is implemented by considering the addition of battery energy storage at the same location of the wind power plant. Due to the boundaries of the wind generator type-4 model used for RMS simulations, the battery storage was connected in the

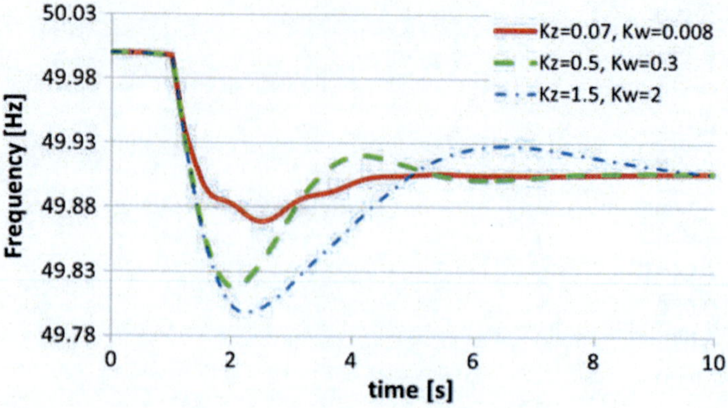

Fig. 10.10 Frequency response for second-order VSP FAPI controller

system as an additional component while for the EMT simulation the battery can be seen as an embedded component through the DC link. The studied operational scenario and disturbance (occurs at 1 s) are the same as indicated in Sect. 10.4.1. The results of the application of the VSP-FAPI controller are presented in Figs. 10.10 and 10.11. It is important to the point that, for the sake of quantitative comparison, it was assumed that the amount of the active power required from the battery storage during the activation of VSP-FAPI controller was nearly the same as the amount of active power required during activation of derivative-based FAPI controller (i.e. approximately 10% of the nominal power of the wind power plant).

The values of Kz and Kw (cf. (3), and (4)) were determined via parametric sensitivity approach [17]. Figure 10.10 shows three example values for combinations of K_z and K_w. Note that both parameters should have relatively low values to achieve a faster active injection (defined by K_w) with high damping (defined by K_z), which leads to improvement of the frequency performance (cf. Figs 10.10 and 10.11). The

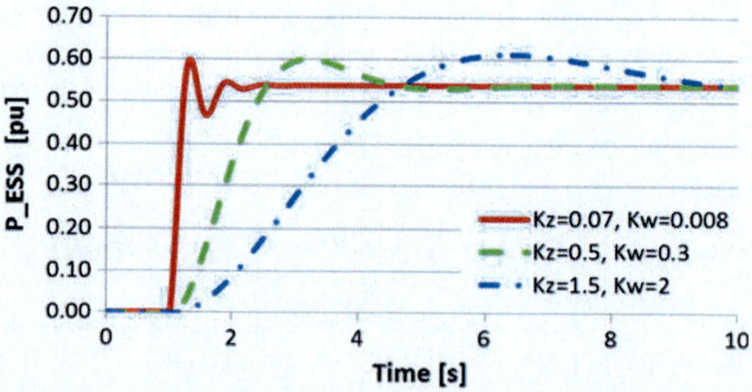

Fig. 10.11 Output power of the storage element in the VSP-FAPI controller case1: $K_z = 0.07$, $K_w = 0.008$ case2: $K_z = 0.5$, $K_w = 0.3$, case3: $K_z = 1.5$, $K_w = 2$

degree of improvement by the VSP-FAPI controller is similar to the one achieved by using derivative-based FAPI controller.

Nevertheless, the VSP-FAPI controller is more attractive since it can be tuned to simultaneously fasten the rate of active power injection and supporting the damping of electromechanical oscillations. Besides, the VSP-FAPI controller does not need PLL (i.e. free from limitations of PLL performance), which is needed in other methods for tracking the frequency. But in the studied application of the proposed FAPI, a PLL might be needed in the inter control loop of the converter to which the storage element is connected. Finally, it is important to point out that the physical connection of the battery will influence the tuning of K_z and K_w. For instance, if the VSP-FAPI controller with battery is connected to the DC link of the wind generator Type-4, then the tuning is bounded by the physical limit of the converter. By contrast, if the VSP-based FAPI controller is implemented with the battery as an independent component, then, depending on the size and technology of the battery, a faster and higher contribution for fast active power injection may be pursued.

10.5 Sensitivity-Based Approach for FAPI Tuning

According to the materials presented in previous subsections, three main classes of FAPI controller for frequency stability issues are covered and discussed. These FAPI controllers are droop-based, derivative-based and VSP-based FAPI controllers. Each of these approaches is analysed and discussed in order to increase the share of PEIG in the grid without jeopardising the frequency stability of a power system. These controls are explained in the subsequent sub-sections.

Figure 10.12 shows the proposed approach for sensitivity analysis to tune and determines the maximum share of wind power generation as a function of the FAPI controller's parameters located in selected wind power plants. The procedure considers a given power system model with a number of wind generations associated with a given level, an initial set of FAPI control parameters, a selected disturbance (e.g. generator outage), and a selected operational scenario (e.g. peak load demand and corresponding generation dispatch and network topology) as inputs. Next, for each wind power plant equipped with FAPI, an iterative process (implemented in a Python script) is performed to sweep over each FAPI control parameter, while keeping the other parameters fixed. The min–max values of the FAPI parameters can be selected according to the physical limits of the control unit and the system operator priorities. Next, time-domain (RMS) simulations are executed for the selected contingency and specific operational scenarios in PowerFactory. A MATLAB script is then executed to extract the time data series of relevant measurements (e.g. grid frequency) to assess the performance of the system (e.g. calculation of Nadir).

The objective is to identify the value of each FAPI control parameter, which enhances the fast frequency response of the wind power plant. The iterative process is applied to individually tune each single wind power plant. Afterwards, the share of wind power generation can be increased, and by applying the same procedure,

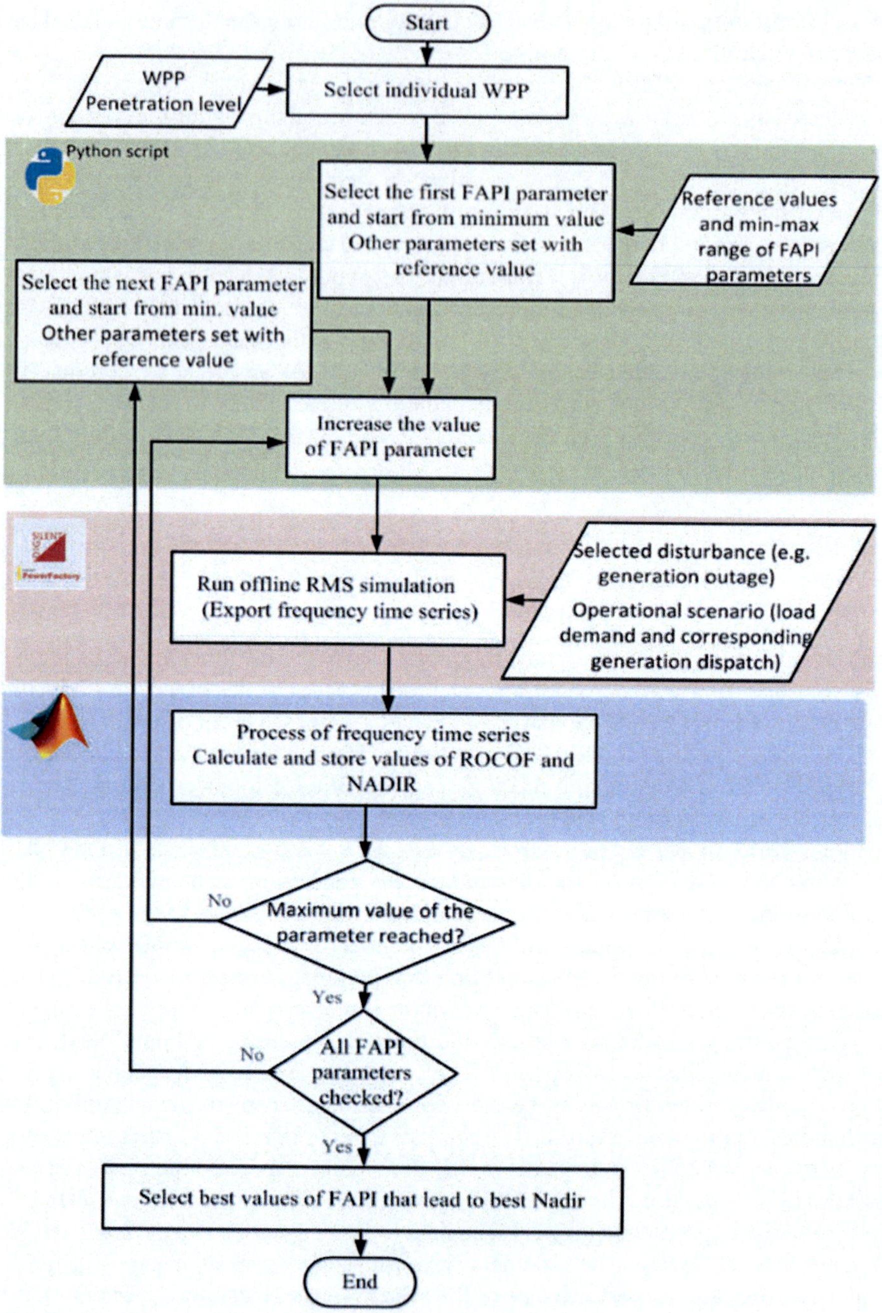

Fig. 10.12 Procedure for tuning of FAPI control parameters

different combinations of wind power plants (each with its own FAPI parameters) with FAPI activated are assessed to determine the best combination that entails satisfactory values of Nadir. The maximum share of wind power generation is found when the Nadir threshold cannot be met.

This procedure is applied to the presented three-area system shown in Fig. 10.7. In order to analyse the impact of FAPI controller in a system configuration with a high share of wind power generation, a given different operational scenario is considered. Operational scenarios for the three-area system are taken from D1.2 [14]. The system with a 50% share of wind power generation was chosen as the base case. During the sensitivity-based tuning approach, it is assumed that each parameter can sweep from its min to max values, whereas the rest of the parameters have the values shown in Table 10.1.

10.6 Combination of Multiple WP with FAPI Controller

Ideally, it could be assumed that all wind power plants of a power system have an active FAPI controller. However, depending on the system characteristics, it might not be necessary to have all WPs with FAPI controller. The size of the WP is the main key factor for implementing the FAPI. FAPI should be implemented in WPs which are large enough to facilitate the most significant possible volume reserves for FAPI controller to impact the grid. It is worth mentioning that, in addition to the size, the location of wind turbines with FAPI might be an alternative criterion for impacting the dynamic performance of the system. It is better to have the PEIG units in proximity to low-inertia regions and far from the centre of inertia in the system.

Furthermore, when the wind speed is low, the contribution of FAPI controller (time duration and proportional gains) should be accordingly reduced in coherence with the available kinetic energy, and if the speed is very low below rated operation, FAPI should not be activated to avoid stalling.

Thus, it is worth evaluating if a minimum subset (combination) of wind power plants with FAPI can entail the satisfactory frequency performance as in the case when all wind power plants perform with FAPI. Indeed, as shown in Fig. 10.13, the combination of WP2B, WP3B, WP12C has the highest and the most similar performance compared to the case that all the wind power plants actively perform FAPI.

The parameters of the FAPI of the three selected wind power plants are given in Table 10.1. This combination is kept for further analysis in the remainder of this section in terms of increasing the share of wind power generation in the whole grid.

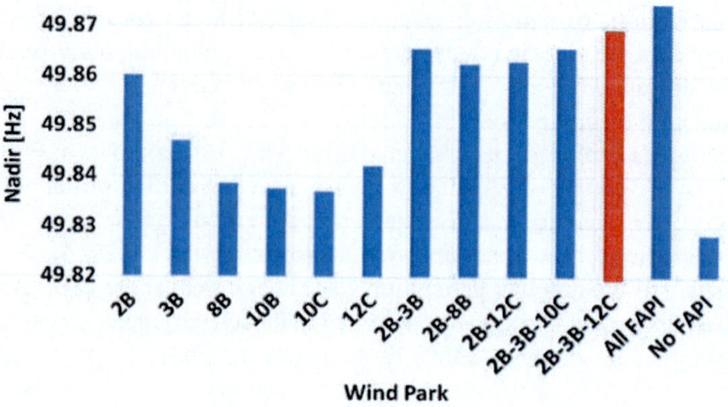

Fig. 10.13 Nadir for different combinations of WP with FAPI controller

10.6.1 Maximum Share of PEIG

In this section, the effectiveness of FAPI controller when increasing the share of wind power is evaluated. According to the ENTSO-E evaluation criteria, stationary values for Nadir between 49.8 and 50.2 Hz can be considered as the limits after the occurrence of an active power imbalance. As shown in Fig. 10.14, within the existing controls, the maximum reachable share of PEIG level without FAPI controller is around 50% while after FAPI activation, with the best combination shown in Table 10.2, it can be reached up to 57% for the three-area test system. It has to be noted that the selected results shown in Fig. 10.14 correspond with the worst-case scenario (an outage of the largest generation unit in a heavily loaded condition). In each studied method for each penetration level, the amounts of available reserves for primary frequency control are the same. It is concluded that by means of properly tuning of FAPI control in the key (selected) wind power plants, it is possible to ensure compliance of Nadir limit when increasing the share of wind power generation (which is a maximum 57% in case of the genetic test case 1).

It should be noted that the frequency threshold (e.g. Nadir) for defining the maximum penetration level in the studied three-area test system has been considered according to continental Europe limits, while this can vary in different systems as

Table 10.2 Parameters of droop-based FAPI for different wind power plants

Selected parameters	WP2B	WP3B	WP12C
f_t[Hz]	49.95	49.95	49.95
f_p[Hz]	49.75	49.75	49.75
P_m[pu]	0.25	0.25	0.25
T_{imx}[s]	15	15	15
f_n[Hz]	50	50	50

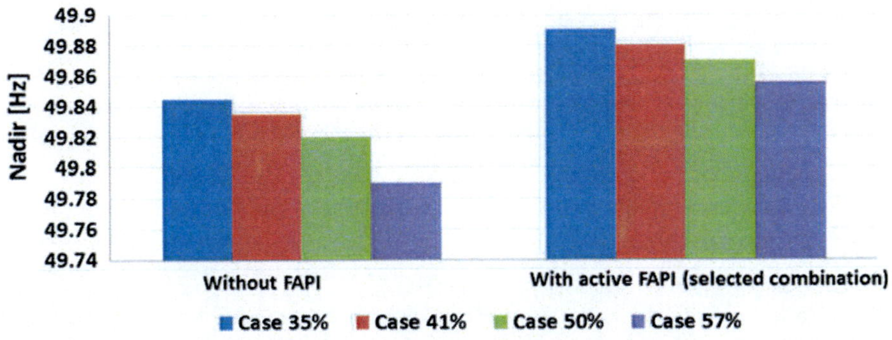

Fig. 10.14 Nadir for different shares of wind power generation with and without droop-based FAPI controllers

reflected on their national grid codes. In this study, the penetration is considered as the amount of total demand covered by total generated wind power.

10.7 Conclusive Remarks

Different forms of mitigation measures, namely droop-based FAPI, derivative-based FAPI and VSP-based FAPI controllers, for enhancing the dynamics of the frequency of a low-inertia system in DIgSILENT PowerFactory have been presented and later discussed. The control implementation and its initialisation in PowerFactory were also explained.

Based on the obtained results, it becomes clear that the fast delivery of a large amount of power is essential to quickly arrest the frequency deviations due to a large active power imbalance. This goal can be achieved by using the derivative or the VSP-based FAPI controls.

The practical implementation of this method will depend on the availability of the energy source like battery storage systems. The size of the battery storage system determines the degree of influence of the VSP, especially if the battery is connected as an independent component.

References

1. M. Dreidy, H. Mokhlis, S. Mekhilef, Inertia response and frequency control techniques for renewable energy sources: A review. Renew. Sustain. Energ. Rev. **69**(July 2016), 144–155 (2017)
2. F. Ha, A. Abdennour, Optimal use of kinetic energy for the inertial support from variable speed wind turbines. Renew. Energ. **80**, 629–643 (2015)

3. S. Mishra, P.P. Zarina, A novel controller for frequency regulation in a hybrid system with high pv penetration, in *2013 IEEE Power and Energy Society General Meeting*, (Vancouver, BC, 2013), pp. 1–5
4. W. Yao, K.Y. Lee, A control configuration of wind farm for load-following and frequency support by considering the inertia issue, in *IEEE Power and Energy Society General Meeting*, (San Diego, CA, 2011) pp. 1–6
5. S. Engelken, A. Mendonca, M. Fischer, Inertial response with improved variable recovery behaviour provided by type 4 WTs. IET Renew. Power Gener. Spec. **11**(3), 195–201 (2017)
6. E. Rakhshani, P. Rodriguez, Inertia emulation in AC/DC interconnected power systems using derivative technique considering frequency measurement effects. IEEE Trans. Power Syst. **32**(5), 3338–3351 (2017)
7. E. Rakhshani, D. Remon, A. Mir Cantarellas, P. Rodriguez, Analysis of derivative control based virtual inertia in multi-area high-voltage direct current interconnected power systems. IET Gener. Transm. Distrib **10**(6), 1458–1469, (21 Apr 2016)
8. F. Gonzalez-Longatt, E. Chikuni, E. Rashayi, Effects of the synthetic inertia from wind power on the total system inertia after a frequency disturbance, in *IEEE International Conference on Industrial Technology (ICIT), Cape Town*, (2013) pp. 826–832
9. T. Ackermann (ed.), *Wind Power in Power Systems*, vol. 140, (2005), p. 1
10. E. Rakhshani, D. Remon, A.M. Cantarellas, J.M. Garcia, P. Rodriguez, Virtual synchronous power strategy for multiple power systems. IEEE Trans. Power Syst. **32**(3), 1665–1677 (2017)
11. Q. Zhong, S. Member, G. Weiss, Synchronverters: Inverters that mimic synchronous generators. IEEE Trans. Ind. Electron. **58**(4), 1259–1267 (2011)
12. IEC, IEC standard 61400-27-1, Wind turbines—Part 27–1: Electrical simulation models wind turbines, 2nd edn. (2017)
13. Energynautics, MIGRATE Project, Type-3 and Type-4 EMT—Model Documentation. (Germany, 2017)
14. MIGRATE Work package 1, *MIGRATE Deliverable D1.2: Report on Power System Analysis and Key Performance Indicators*. (MIGRATE consortium, 2018) www.h2020-migrate.eu
15. Z. Miao, S. Member, L. Fan, S. Member, D. Osborn, Wind farms with HVdc delivery in inertial response and primary frequency control. IEEE Trans. Energy Convers. **25**(4), 1171–1178 (2010)
16. ENTSO-E, *Rate of Change of Frequency (ROCOF) Withstand Capability*. (Brussels, 2018). [online]. Available at: https://docstore.entsoe.eu/Documents/Network%20codes%20documents/NC%20RfG/IGD_RoCoF_withstand_capability_final.pdf
17. MIGRATE Work package 1, *MIGRATE Deliverable D1.5: Report on Power System Risk Analysis and Mitigation Measures*. (MIGRATE consortium, 2019) www.h2020-migrate.eu

Chapter 11
Modelling and Simulation of Wind Turbines with Grid Forming Direct Voltage Control and Black-Start Capability

Abdul W. Korai, E. Rakhshani, M. Ebrahim Adabi, José Luis Rueda Torres, and Mart A. M. M. van der Meijden

Abstract In this chapter, a grid forming control approach called direct voltage control (DVC) for wind turbine control with restoration capability of power system with a high share of power electronic-based generation units is presented and discussed. All the detailed explanation, DSL-based control is presented for dynamic simulations in DIgSILENT software.

Keywords DIgSILENT simulation · DSL programming · Wind turbine control · Grid forming control · Direct voltage control · Restoration

11.1 Introduction

Within strong gird, the current injection control or grid following control approaches will not lead to any significant problems because they are located in a mostly strong grid that is capable of absorbing the injected current without any difficulty. In isolated networks predominantly supplied through a power electronic converter interface, however, the situation is different. If a network is connected to the grid mostly through power electronic converters and has little or no load of its own, as is the case in an offshore grid, the current injection-based controllers cannot physically work.

Electronic supplementary material The online version of this chapter (https://doi.org/10.1007/978-3-030-54124-8_11) contains supplementary material, which is available to authorized users.

A. W. Korai
DIgSILENT GmbH. Gomaringen, Baden-Württemberg, Germany

E. Rakhshani · M. E. Adabi · J. L. Rueda Torres (✉) · M. A. M. M. van der Meijden
Department of Electrical Sustainable Energy, Delft University of Technology, Delft, The Netherlands
e-mail: j.l.ruedatorres@tudelft.nl

M. A. M. M. van der Meijden
TenneT TSO B.V, Arnhem, The Netherlands

F. M. Gonzalez-Longatt and J. L. Rueda Torres (eds.), *Modelling and Simulation of Power Electronic Converter Dominated Power Systems in PowerFactory*, Power Systems,
https://doi.org/10.1007/978-3-030-54124-8_11

This chapter presented a novel grid forming control called direct voltage control (DVC), and it is modeled and tested by using time-domain simulations. The control is developed in DIgSILENT and is tested for restoration and black-start capability in a large-scale power system.

The rest of this chapter is organized as follows: In Sect. 11.2, the direct voltage control approach is explained while its DSP implementation in PowerFactory is briefly presented in Sect. 11.3. Testing of the proposed control with its modification for black-start capabilities and power system restoration is also presented in Sect. 11.4. While different simulation test for DVC application is also presented in Sects. 11.5–11.7, respectively. Summary and the main conclusions of this chapter are presented in Sect. 11.8.

11.2 Direct Voltage Control Approach

The current injection control did not lead to any significant problems thus far since there were mostly strong grids, which were capable of absorbing the injected current without any difficulty. In isolated networks predominantly supplied through a power electronic converter interface, however, the situation is different. If a network is connected to the grid mostly through power electronic converters and has little or no load of its own, as is the case in an offshore grid, the current injection cannot physically work. Figure 11.1 demonstrates the fundamental concept behind the proposed controller.

While the integral component of the PI current control has been discarded, the proportional component is merely shifted toward the output terminal, which entails only graphical change without implications in terms of performance. Without the integral component in the current controller, the setpoints and actual values of the power electronic converter currents will not be the same. This has no significance here since the setpoints are used for limiting the power electronic converter currents and the power electronic converter current will be adjusted based on the grid situation, whereas the power balance will be controlled by the outer loop controller, which employs a PI-based current control. In order to limit the effect of this part on the dynamic behavior (similar to the case in the conventional PI controller), this component uses a high-pass washout filter in this approach. This expansion also opens up the possibility for frequency-selective damping if these terms at the output of the current controller are augmented by appropriately designed band-pass filters.

In the following, additional explanations regarding the components used at the output are also presented. One of the objectives of the scheme is to ensure adequate damping of transient processes. This can be achieved by, for example, emulating a series resistor in the power electronic converter circuit, as shown in Fig. 11.2.

Physically, the power loss on a resistor clearly represents damping. However, inserting an actual resistance in the primary circuit means active power losses and the accompanying operational disadvantages. However, if the controller is designed in such a way that the power electronic converter emulates the voltage drop generated

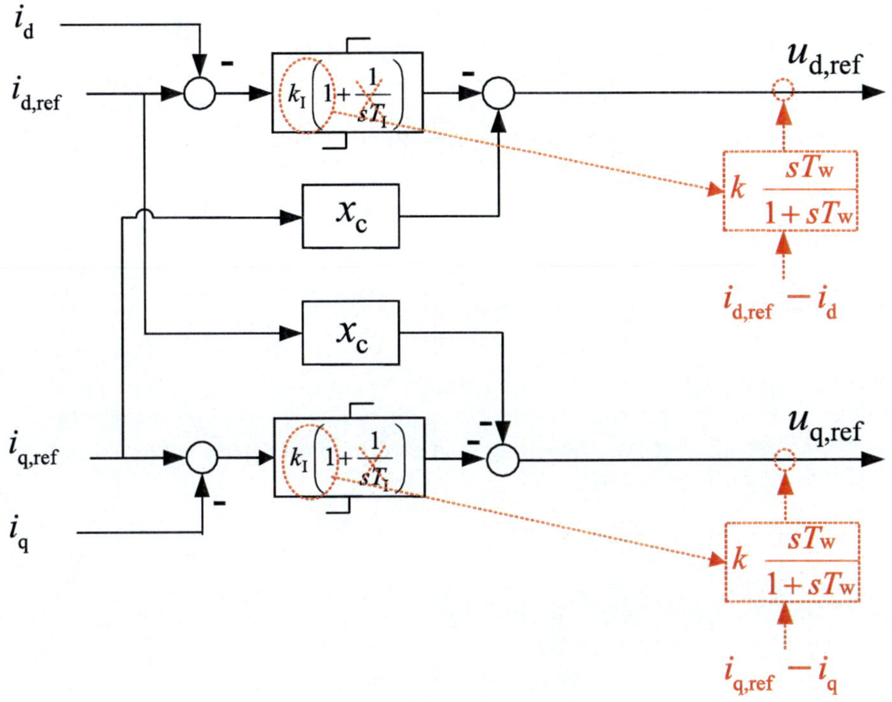

Fig. 11.1 Development of the DVC by modification of the PI-based current controller

Fig. 11.2 Equivalence of damping terms with physical resistive damping

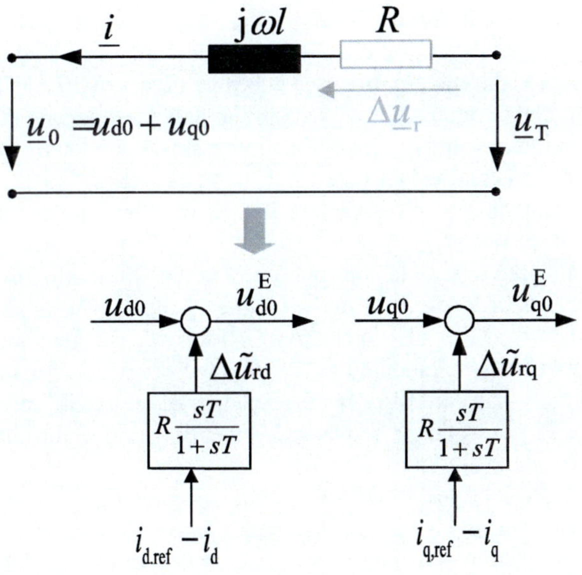

by the virtual resistor, the same damping effect (but with no accompanying power loss) can be achieved. The voltage drop on the resistor is given by:

$$\Delta \underline{u}_r = R \cdot \underline{i} \tag{11.1}$$

If the power electronic converter voltage is to emulate this voltage drop, the following current-dependent term must be superimposed on the power electronic converter voltage:

$$\underline{u}_0^E = \underline{u}_0 + R \cdot \underline{i} \tag{11.2}$$

where the superscript, E indicates extension by the emulation term. It is practical to limit the effect of the resistance emulation on the transient currents only. The high-pass filter (washout filter) was used for this reason. The current in the voltage-oriented coordinate system then becomes:

$$\Delta \tilde{\underline{i}} = \Delta \tilde{i}_\mathrm{d} + \mathrm{j}\Delta \tilde{i}_\mathrm{q} \tag{11.3}$$

Alternatively, in notation using components:

$$\begin{aligned} u_{d0}^E &= u_{d0} + R \cdot \Delta \tilde{i}_d \\ u_{q0}^E &= u_{q0} + R \cdot \Delta \tilde{i}_q \end{aligned} \tag{11.4}$$

This equation corresponds exactly with the current-dependent terms added to the current controller as shown in Fig. 11.1.

The analysis of the islanding phenomenon revealed that the cause of the voltage rise was the customarily used control structure for the WT based on current injection. Its integral characteristic causes the power electronic converter voltage to reach its maximum value in a short time interval following a mismatch between the reference and the actual values [1].

The direct voltage controller in the final form is shown in Figs. 11.3 and 11.4, respectively, including the modification and extension. The significant change compared to simple current control is the elimination of the integrators but keeping the proportional terms in the former PI block. The main control task is done by the feedforward terms, which are sufficient because there are already integral controllers upstream, implemented as DC voltage controller for the active current and in the voltage–var controller for the reactive current. The remaining proportional part of the PI controller is replaced by damping and transient current limitation control extension [1].

As can be seen, reactive current injection requirement is no longer defined in this schema in favor of direct voltage control. The current adjusts itself optimally for the objective of voltage control according to the network conditions in response to a changing power electronic converter voltage. Therefore, the scheme is more like conventional voltage control in synchronous generators. In the case of islanding,

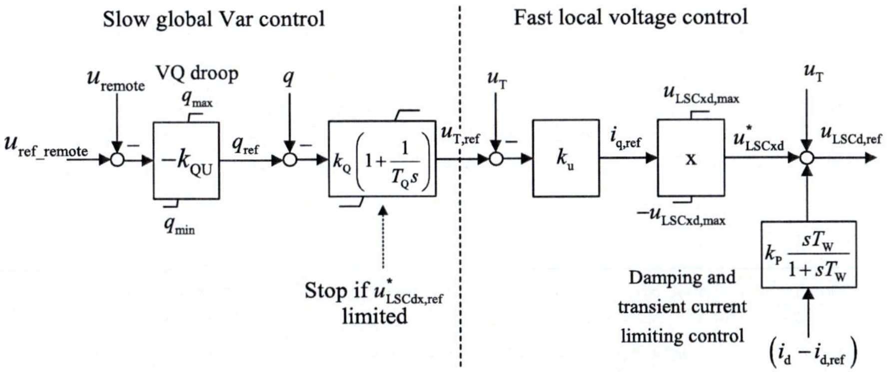

Fig. 11.3 Reactive power control

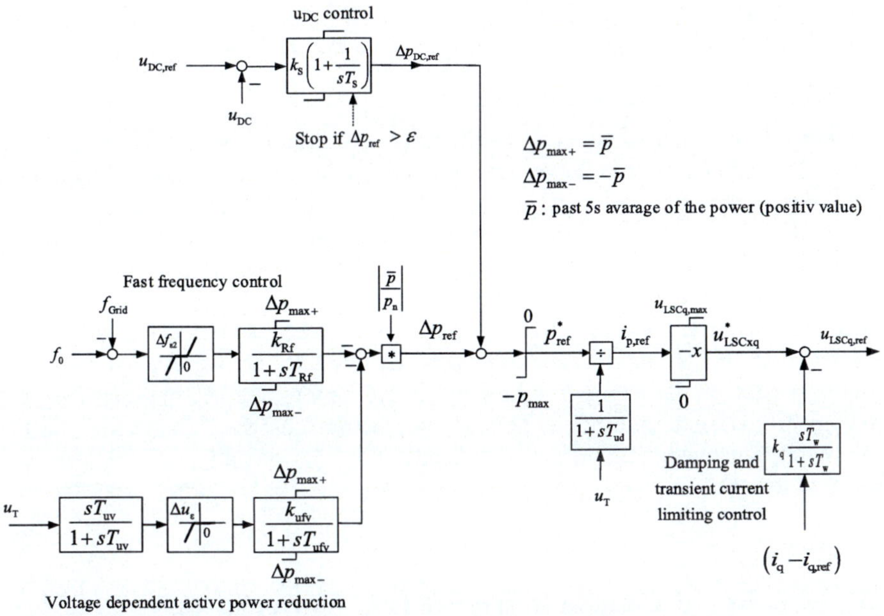

Fig. 11.4 DVC scheme in active power control

there is no risk of integrator windup, as will be shown in the simulation example in the next section.

The active power control channel with suggested frequency control strategies is shown in Fig. 11.4. The structure of the DC voltage controller with the PI characteristic is the same as in the PI-based current control schemes. The output of the DC voltage controller is the active power injected into the network. The active power injection can be calculated as:

$$p = u_T i_d = -u_T \frac{u_{C_q}}{x} \tag{11.5}$$

The x as can been seen in the active power channel in Fig. 11.4 is the reactance of the power electronic converter and presents the feedforward action of the power electronic converter. The output voltage, which results from multiplying the current with the x, is the set voltage of the power electronic converter in steady-state, which in turn corresponds with the supplied active and reactive powers from the power electronic converter.

For the balance between the power injected into the machine side and that delivered to the network, the DC link controller with a PI characteristic is responsible.

Equation (11.5) describes that the active power can be controlled using the q-component of the power electronic converter voltage. As the active power is controlled through the DC link capacitor and in the absence of the integral term in the current control, there will be a steady-state error between the reference currents and measured currents.

Frequency control is explained below and shown in Fig. 11.4.

Over-frequency and under frequency control

The control is activated when the frequency exceeds a preset threshold value, e.g., 50.2 Hz. The gain k_R defines the frequency deviation at which the power reduction corresponds to the total power p_{ref} (e.g., 51.5 Hz). The time constant T_R is small totally with the fast response time of power electronic converters. One can also define in the delay block a limitation of the rate of change, the effect of which will be demonstrated in the next section [2].

A voltage-dependent active power reduction can be implemented concurrently, as suggested in [2]. For the inertia emulation, however, the MSC converter must provide the energy or the excess energy required by the DC link to emulate the inertia or it must come from an external source. This, in turn, requires that the DC capacitor must be oversized. The oversizing, however, depends on the design and inertia emulation capacity of the WT.

11.3 DSL-Based Control in DIgSILENT PowerFactory

Figure 11.5 shows the model of the grid side converter, which was built as a composite model and contains the following components: (a) converter used in wind farm model which includes the PWM converter (ElmVsc); (b) DC voltage and power controller (Udc_PQ Controller, ElmDsl); (c) current controller (ElmDsl); (d) DC model (ElmDsl); (e) AC voltage measurement (StaVmea); (f) AC current measurement (StaImea); (g) power calculation block (ElmDsl); (h) Park transform (ElmDsl), model of current source used in simplified WT (ElmDsl), PLL: angle measurement (ElmPhi_pll); and (v) Mod Limiter (ElmDsl).

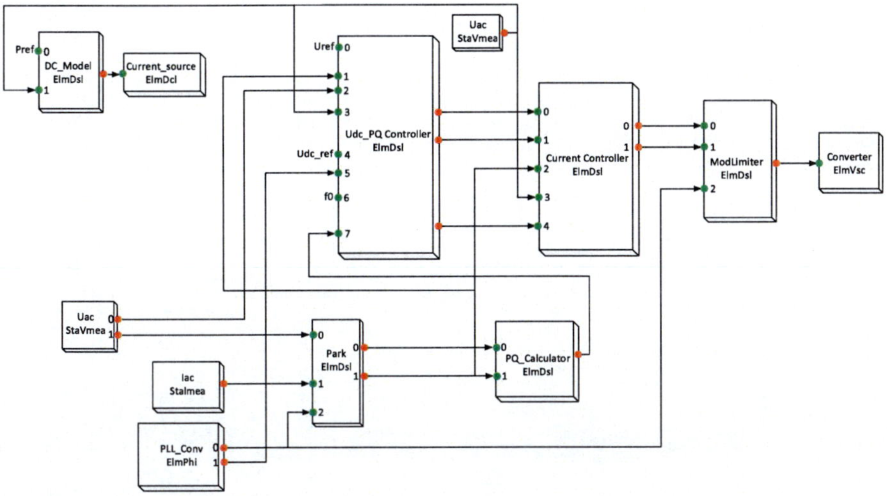

Fig. 11.5 Structure of the grid side converter model in PowerFactory

Correct initialization of a model in a power system simulation-tool avoids fictitious electrical transients and makes it possible to evaluate correctly the real dynamic performance of the system. Therefore, the initialization equations of the important dynamic block of this composite model are presented below:

DC Model initialization:

```
vardef(VdcN)='kV';'Nominal DC Voltage'
vardef(Vdcmax)='kV';'Max DC Voltage'
vardef(Pnom)='MW';'Nominal active power of the converter'

inc(xr)=1
inc(vdc)=1
inc(pref)=-1
```

Mod Limitation initialization:

```
inc(ud_ref)= ur*cos(phi) + ui*sin(phi)
inc(uq_ref)=-ur*sin(phi) + ui*cos(phi)
inc(ur)=Pmr /0.612
inc(ui)=Pmi /0.612
```

Current controller initialization:

```
vardef(Kw) = '-';'Washout out filter gain'
vardef(Tw) = 's';'Washout filter time constant'
vardef(Un) = 'V' ; 'Nominal AC voltage of converter'
vardef(UDCn) = 'V' ; 'Nominal DC voltage of converter'
vardef(ud_min) = 'pu' ; 'Minimum modulation index d-channel'
vardef(uq_min) = 'pu' ; 'Minimum modulation index q-channel'
vardef(ud_max) = 'pu' ; 'Maximum modulation index d-channel'
vardef(uq_max) = 'pu' ; 'Maximum modulation index q-channel'

inc(uq_set)=0
inc(xdw)  =0
inc(xqw)  =0
inc(ulsc_d)=1

inc(id_ref_)=id
inc(iq_ref)=iq
Ksin = 2*sqrt(2)*Un/(sqrt(3)*UDCn)
```

Udc_PQ Controller initialization:

```
vardef(Kv) = '-' ; 'Global Var control P.constant'
vardef(Tv) = 's' ; 'Global Var control Integral time constant'
vardef(Kdc) = '-' ; 'DC link control P.constant'
vardef(Tdc) = 's' ; 'DC link control Integral time constant'
vardef(uDC_CHon) = 'pu' ; 'Chopper activation voltage level'
vardef(uDC_CHoff) = 'pu' ; 'Chopper deactivation voltage level'
vardef(iChopper) = '-' ; 'Chopper active [1]- inactive[0]'
vardef(Ku) = 'pu' ; 'Fast voltage control P.constant'
vardef(db_spannung_VDACR) = 'pu' ; 'deadband for VDAPR'
vardef(irmax) = 'pu' ; 'Maximum converter current'
vardef(Tuu) = 's' ; 'Voltage measurement delay'
vardef(T_rate_limit_freq) = 's' ; 'Frequency control time delay'
vardef(l) = 'pu' ; 'Converter reactance'
vardef(T) = 's' ; 'VDAPR washout filter time constant'
vardef(qmax) = 'pu' ; 'Maximum reactive power in steady state'
vardef(Kvdapr) = '-' ; 'Gain of the VDAPR'
vardef(Tvdapr) = 's' ; 'Time delay of the PT1 filter of VDAPR'
vardef(Tp_aver) = 's' ; 'Time constant for the calcualtion of p average'
vardef(Kqu) = '-' ; 'Static gain of the reactive power control'

inc(ulsc_d)=1
inc(xv)= 1.01
inc(qref)=q
inc(xu)=ulsc_d
inc(id_ref)=-0.95/us
inc(idref)=-0.95/us
inc(uref)=us
inc(xdc)=id
inc(udc_ref)=1
inc(xtst)=-0.95
inc(xuuu)=0
inc(xuu)=us
inc(x)=us
inc(f0)=50
inc(xp_avg)=p
inc(xf)=0
```

11.4 Grid Restoration by WT

With the increase of the share of power electronic converter-based generation units and the replacement of the synchronous machine, there is a need for these units to be able to do the grid restoration and pick up the loads so that the grid can be formed again. This task was previously assigned to the synchronous machines and especially the gas turbine drives ones, which were responsible for the grid restoration and load peak.

The current German grid codes do not specify any requirement for the grid restoration by wind turbines. Therefore, there are no grid restoration mechanisms in place that enables the power electronic converter-based generation units especially wind turbines to do the grid restoration.

This section introduces a new approach based on the DVC, which can enable the grid restoration and pick up of the active as well as reactive loads.

In this simulation, only one WT with a rated power of 3.6 MW has been used for the load pickup and grid restoration scenario.

11.4.1 Modification of Control for Grid Restoration and Simulation Results

Figure 11.6 shows the test grid and the control modification that has been proposed for the WTs to do the grid restoration. Modified frequency control for grid restoration is also presented in Fig. 11.7.

The operation of the WT can be divided into two parts:

- Startup mode.
- Normal mode.

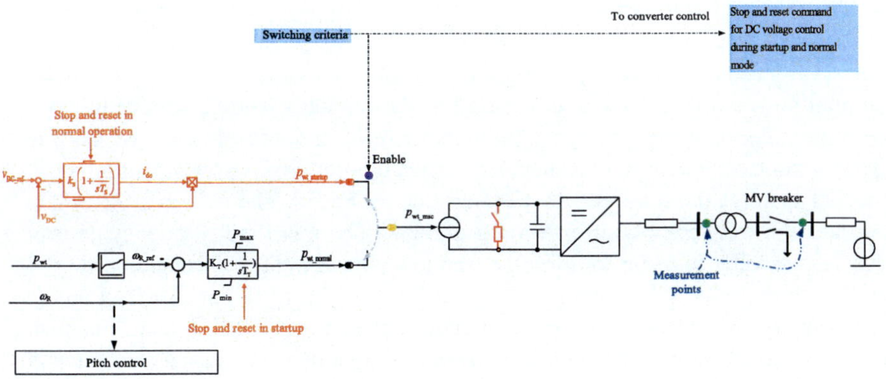

Fig. 11.6 Controller modification for grid restoration

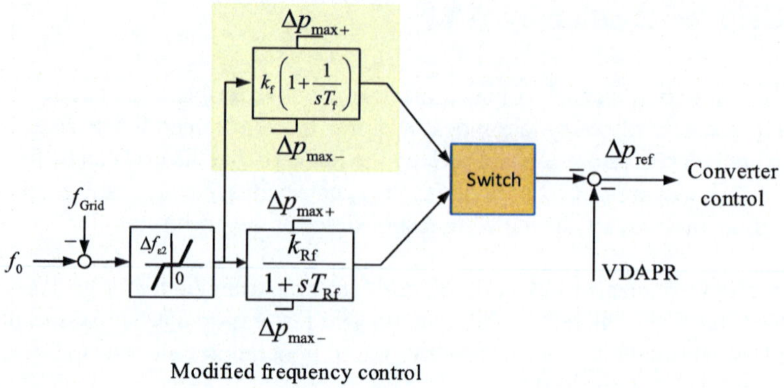

Fig. 11.7 Modified frequency control for grid restoration

In startup mode, the lookup table for the speed–power characteristic is disabled and DC voltage control in the LSC is disabled. The control as shown in Fig. 11.6 of the DC link is done by an additional DC power controller, which regulates the active power of the WT based on the measured DC voltage; moreover, the PLL used in the normal mode uses fixed frequency so that the power electronic converter control can be established.

In normal mode, the additional DC power controller is disabled and DC link control in LSC, as well as power tracking characteristic of the WT, is enabled. Moreover, the PLL with PI characteristic is enabled, and fixed frequency PLL is disabled. Through these steps, the WT can be synchronized with the grid.

This switching between the two modes depends on the availability of the medium voltage. During the load pick and grid restoration, it is assumed that enough and the constant wind is there so that the WT can supply the required power to the load.

11.4.1.1 Active Power Pickup

Figure 11.8 shows the active power pickup results. At time t = 2 s, a load with a nominal power of 0.85 p.u. is connected to the medium voltage side of the WT transformer. The frequency drops rapidly as there is no inertia in the grid but the frequency control, which is modified for the islanding case as shown in Fig. 11.7, of the WT restores the frequency back to around 49.85 Hz. The difference to 50 Hz is because the frequency control has a deadband. The modified frequency control uses a PI controller in order to bring the frequency back to 50 Hz. It is important to note that the usage of the PI controller on the frequency control has no effect on the power delivery as the load is supplied immediately. The modified frequency control only changes the q-component of the converter voltage to establish the frequency at 50 Hz. The switching between the PI controller and the first-order delay block occurs when WT is synchronized with the grid.

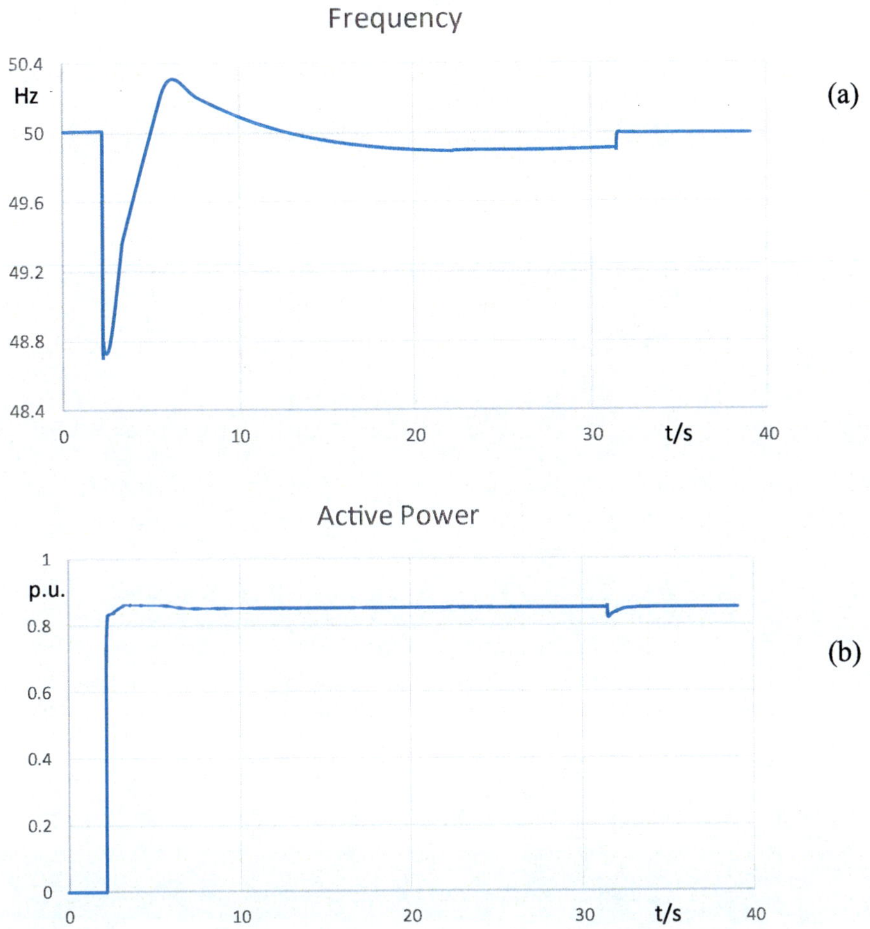

Fig. 11.8 Active power pick up results

The synchronization with the main grid happens at time t = 32 s. It can also be observed in Fig. 11.8c that the voltage at the WT before the load pickup is kept higher because, after the load connection, the grid voltage can be lower than the allowable operating voltage.

After the load pickup, the pitch angle of the WT is reduced to accommodate for the supplied power from the WT. The speed of the wind turbine after load pick up is also controlled to the reference speed which is generated by the speed-power characteristic curve of the FSCG-based WT as shown in Fig. 11.8f,g, respectively.

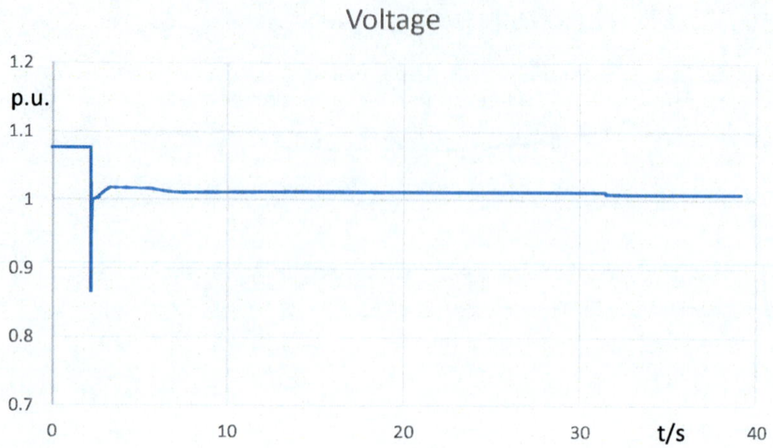

(c)

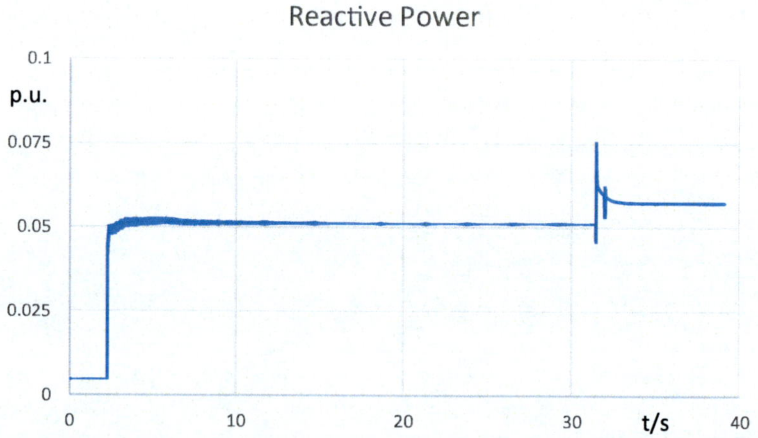

(d)

Fig. 11.8 (continued)

11.4.1.2 Capacitive Reactive Power Pickup

Figure 11.9 shows the capacitive reactive power pickup. The capacitive reactive power pickup is important in the sense that the cables generally have high capacitances and in case of no active load and no grid connection, the WTs should be able to supply the corresponding reactive power. Figure 11.9c shows that the WT is able to supply until 0.25 p.u. of capacitive reactive power.

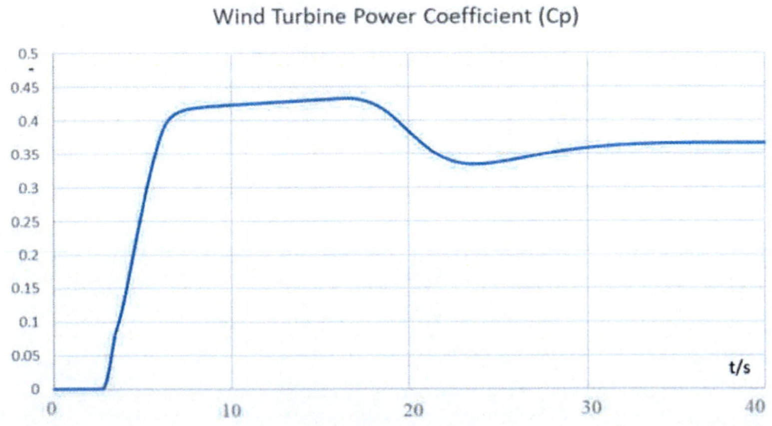

(e)

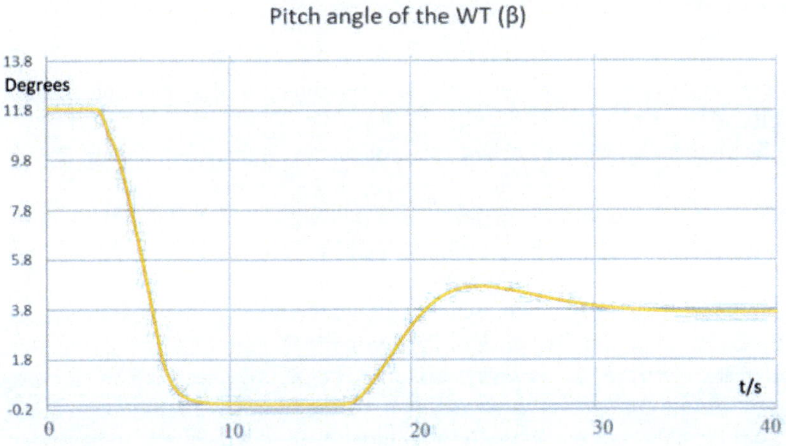

(f)

Fig. 11.8 (continued)

11.5 Simulation of Islanding by Using the Direct Voltage Control

Figure 11.10 shows the simulated behavior of the grid the following islanding by using the suggested direct voltage control-based WT control scheme. The current limiting and damping blocks are defined to pass frequencies about 50 Hz. As can be seen from Fig. 11.10a, the direct voltage control does not lead to voltage rise. The *d*- and *q*-components of the outputs of the controller remain within limits. For comparison, the same values obtained using the PI-based current control controller are shown in Fig. 11.10b–d. The direct voltage controller limits the frequency rise at

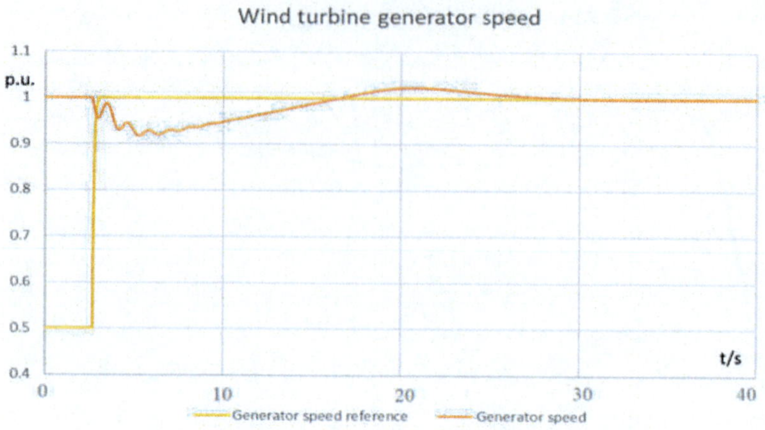

(g)

Fig. 11.8 (continued)

51 Hz by decreasing the reference value of the *q*-component of the power electronic converter voltage.

One of the major challenges of WT control is the fault ride-through (FRT) behavior. Therefore, a grid three-phase fault has been simulated in the 155-kV network with PI-based current control and compared with the DVC approach with connection to the grid remaining intact. Figure 11.11a,b shows that the response of the DVC is very similar to that of the PI-based current control. The current reaches in both cases a peak of about 1.75 p.u., which may be too high for the IGBT. Nevertheless, it should be noted that this case of deep voltage drop in offshore grid is not common and in any case, the WT as well as offshore HVDC converter will be blocked instantly but for the simulation and comparison case, the blocking has not been done during and after the short circuit. It can be concluded from the results that the direct voltage control will not change the established FRT characteristic. Right after the voltage drop a transient oscillation in the voltage is triggered with frequency around 1000 Hz (in abc-components). It represents the resonance frequency between the positive sequence inductances and capacitances of the grid as shown in Fig. 11.11c.

11.6 Effect of the Grid Short Circuit on the Direct Voltage Controller Stability

To evaluate the stability of the direct voltage control approach with changing grid impedance, a small test network is modeled as shown in Fig. 11.12. Three wind turbines are connected in parallel to a voltage source. The wind turbine models are linearized, and eigenvalues are calculated for each grid impedance to evaluate the stability limits of the DVC.

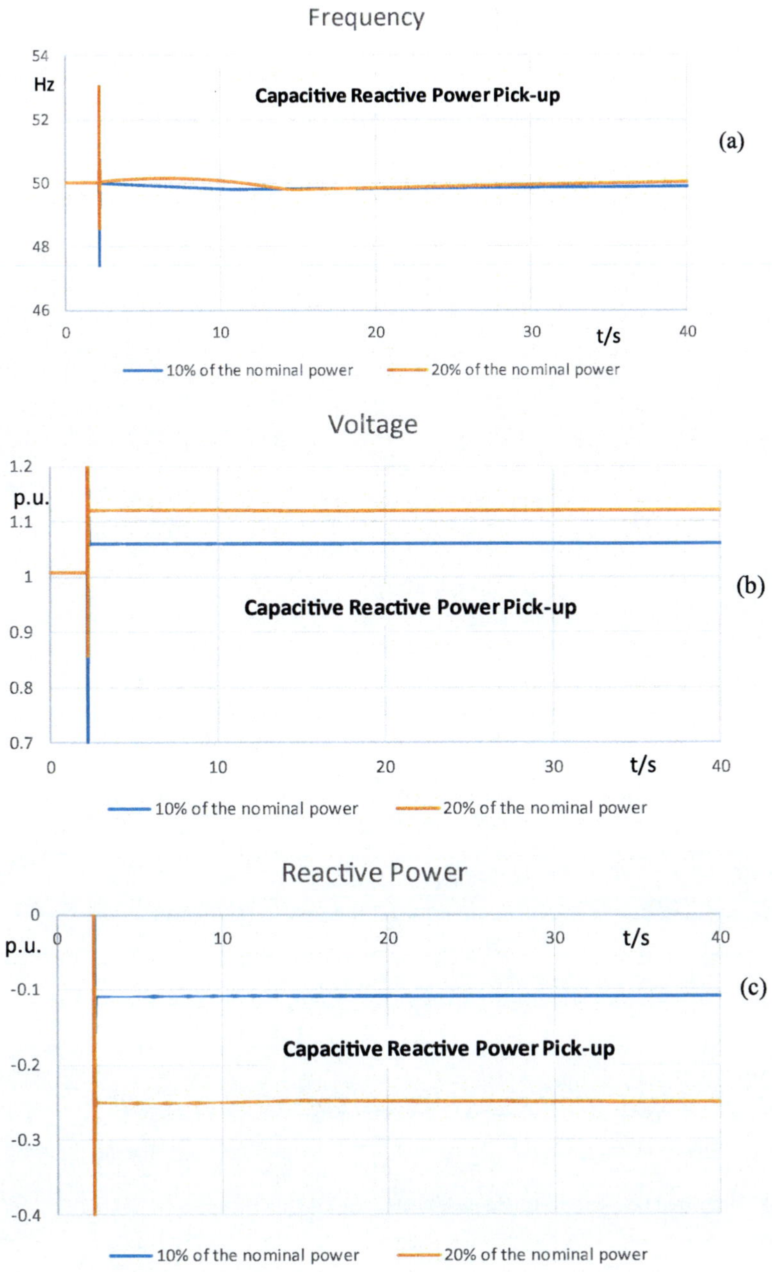

Fig. 11.9 Reactive power pickup

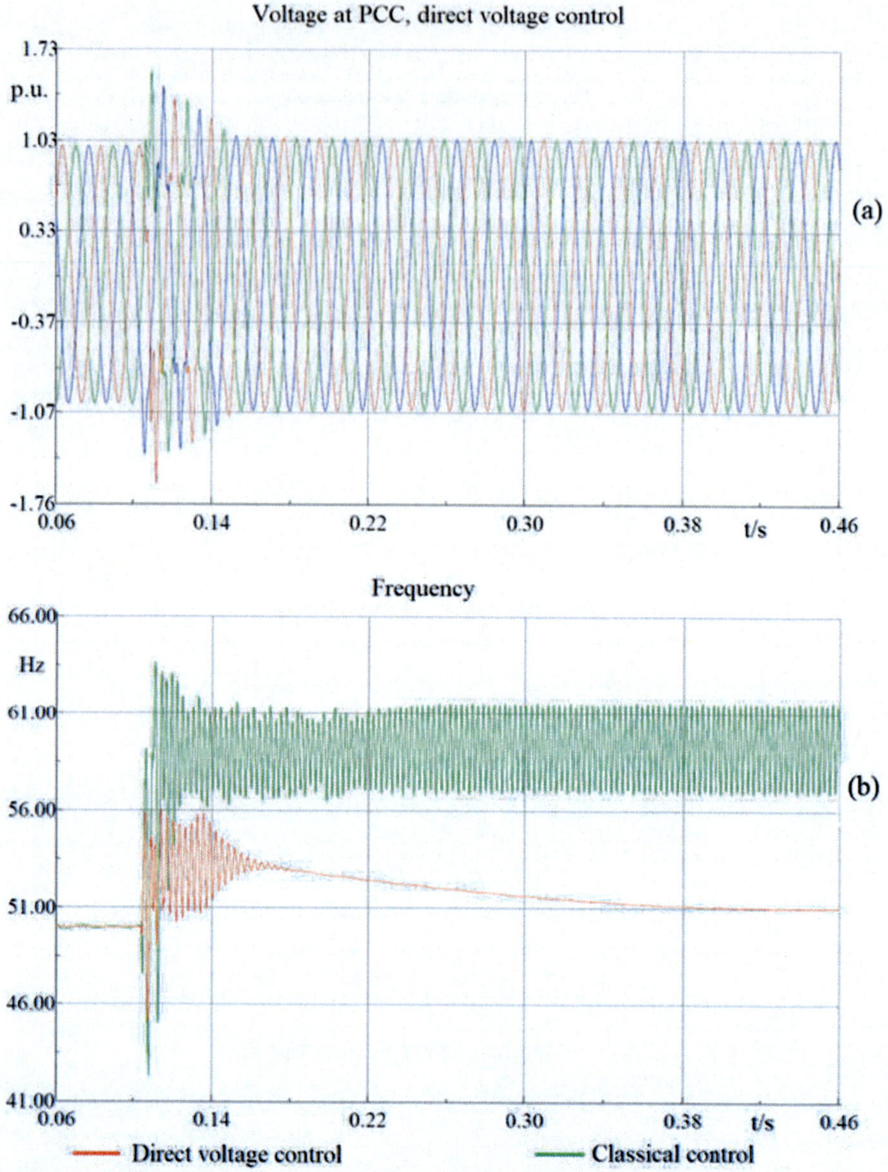

Fig. 11.10 Simulated behavior of islanding by using the DVC

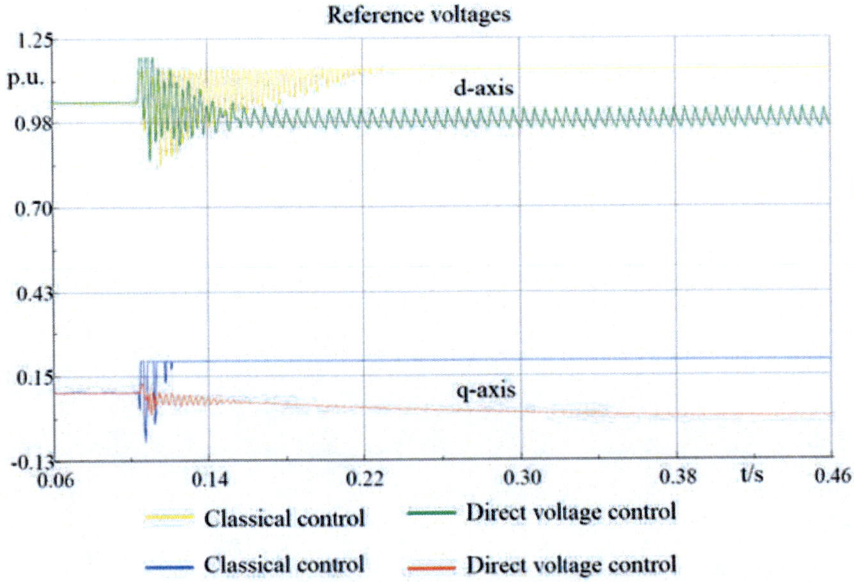

(c)

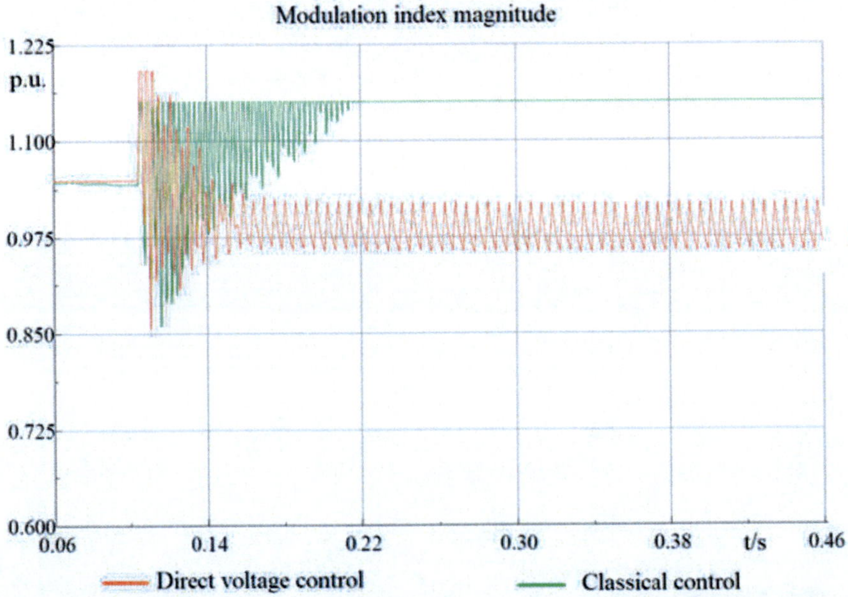

(d)

Fig. 11.10 (continued)

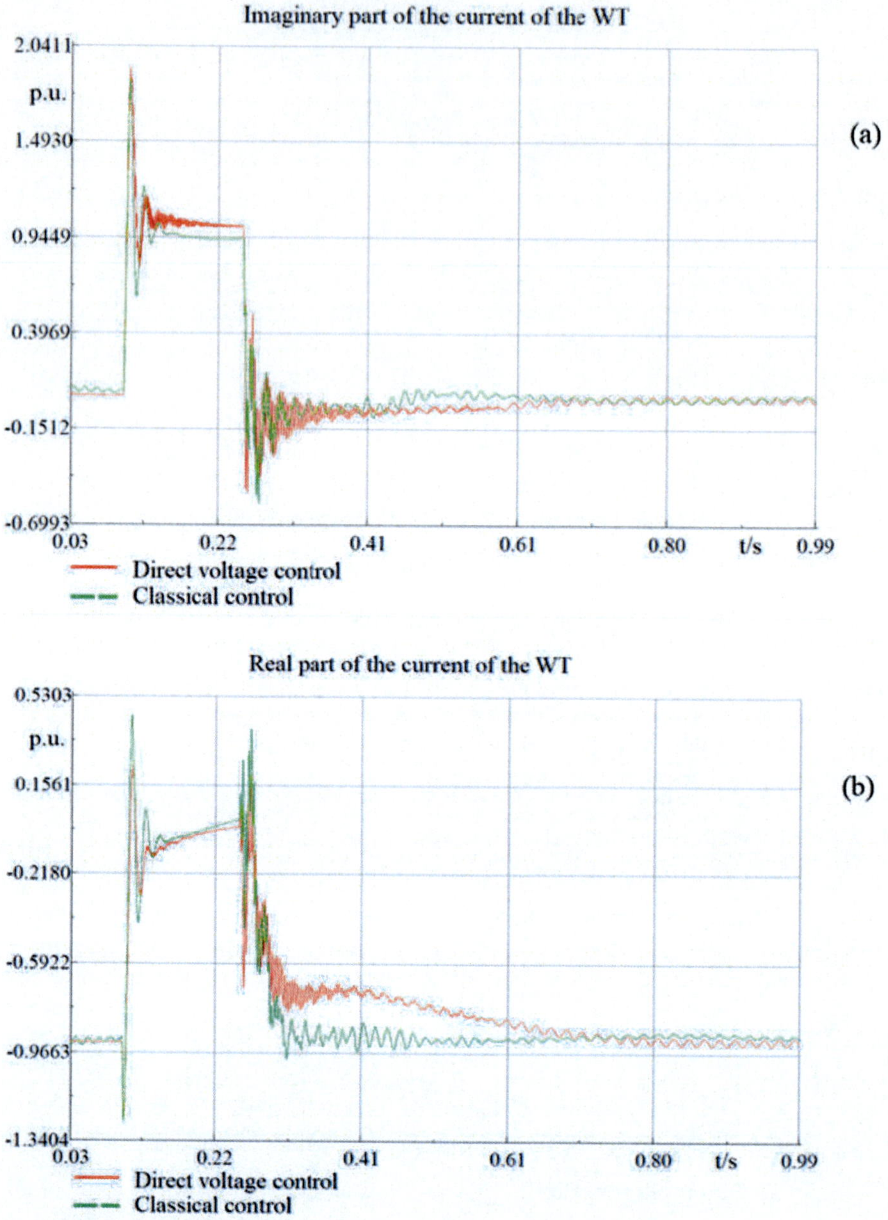

Fig. 11.11 FRT behavior following three-phase grid fault

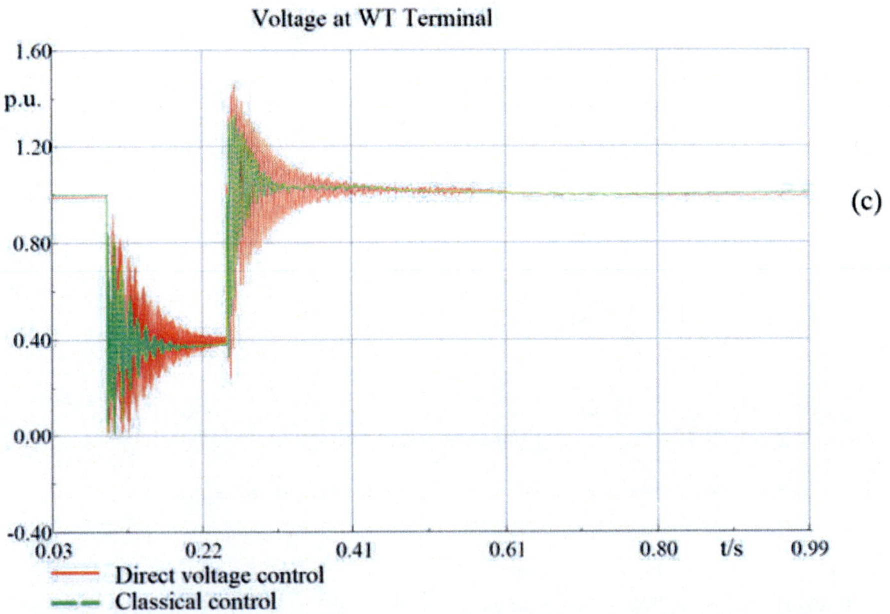

Fig. 11.11 (continued)

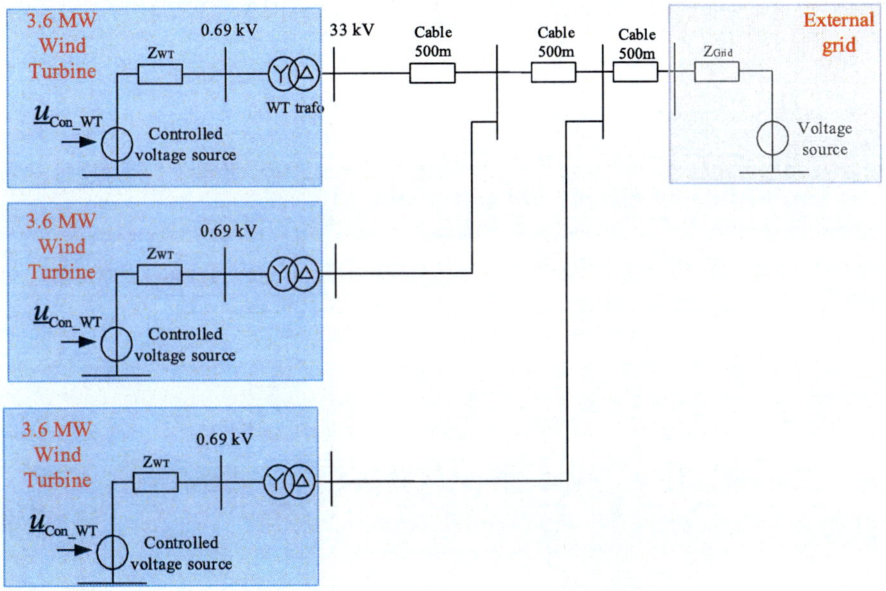

Fig. 11.12 Grid used to for linearization

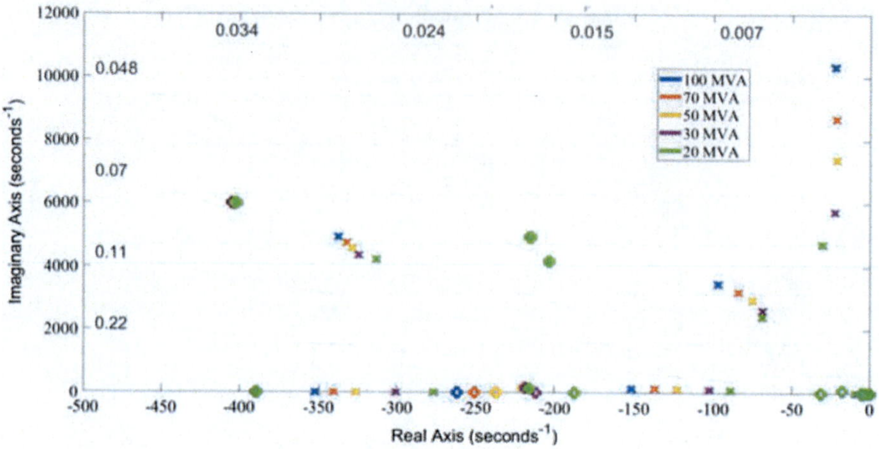

Fig. 11.13 Eigenvalues of the system with different short circuit ratios of the grid

Figure 11.13 shows the eigenvalues of the closed-loop system. It became evident that changing the impedance (lowering the MVA rating of the voltage source, thus increasing the effective impedance seen by the WTs), decreases the damping of the system and decreases the swing modes. It can be concluded from the analysis that the weak grids might impose challenge to the operation of power electronic converter-based grids.

11.7 Reference Tracking and Disturbance Rejection Capability of Direct Voltage Control

Figure 11.14 shows the SISO model which will be used to analyze the disturbance rejection and reference tracking capability of the direct voltage controller.

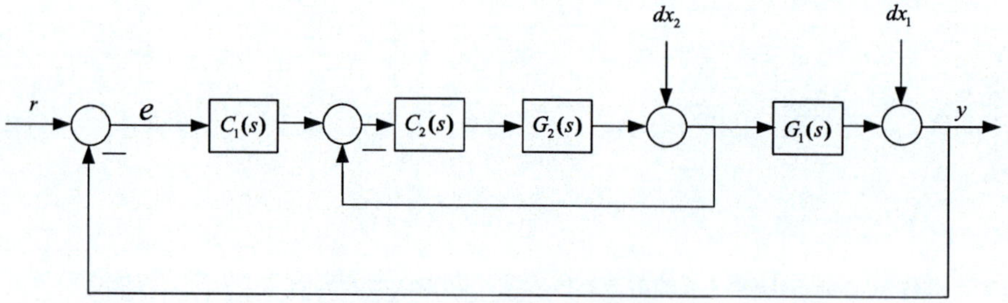

Fig. 11.14 SISO feedback control system

11.7.1 Loop Transfer Function

Open-loop gain of SISO system shown in Fig. 11.14 is given by Eq. (11.6)

$$L(s) = G_1(s)C_1(s)G_2(s)C_2(s) \tag{11.6}$$

Equation (11.7) shows that the external disturbance dx_2 is completely rejected when open-loop gain is infinite.

$$\frac{y}{dx_2} = G_1(s)/(1 + G_2(s)C_2(s) + G_1(s)G_2(s)C_1(s)C_2(s)) \tag{11.7}$$

The two control options are compared with each other, namely

1. PI-based current control (classical/conventional controller).
2. Direct voltage controller (washout with gain).

considering the disturbance rejection and reference tracking capabilities. Equation (11.8) shows the transfer functions of each of these control configurations.

The outer loop transfer function denoted by $C_1(s)$ is same in both the control concepts as it represents the DC link voltage controller, which is based on PI control. $G_1(s)$ and $G_2(s)$ represent the transfer functions of the DC capacitor and the power electronic converter reactance and resistance, respectively.

$$\begin{aligned} C_1 &= K_1 + \frac{1}{T_1 s} \\ C_{2\_\mathrm{PI}} &= K_{2\_\mathrm{PI}} + \frac{1}{T_{2\_\mathrm{PI}} s} \\ C_{2\_\mathrm{DVC}} &= K_\mathrm{P} \frac{T_{ws} s}{1 + T_{ws} s} \\ G_1 &= \frac{1}{C_f s} \\ G_2 &= \frac{1}{R_f + L_f s} \end{aligned} \tag{11.8}$$

Figure 11.15 shows the Bode plot of the Eq. (11.6). As it is clearly seen, the DC gain of the PI-based current controller is a little bit higher than the one from direct voltage controller and hence both the disturbance rejection and reference tracking with PI-based current control will be a little bit better.

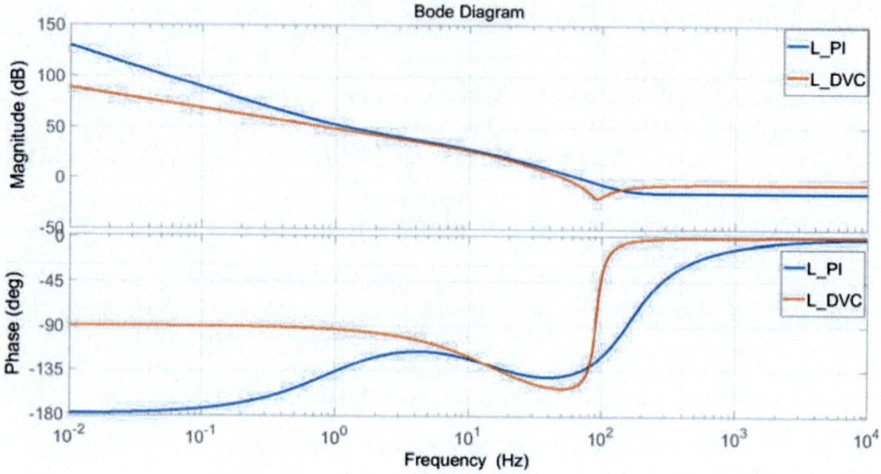

Fig. 11.15 Open-loop gain of two control configurations

11.7.2 *Disturbance Rejection and Reference Tracking*

Figure 11.16 shows and compares the disturbance rejection capability of PI-based current controller and direct voltage controller. It can be seen that the direct voltage controller cannot reject the disturbance completely due to the low DC gain. On the other hand, the PI-based current controller has excellent disturbance rejection capability.

Figure 11.17 shows the compares the reference tracking capability of the two controllers. The direct voltage controller has no integral part in the current control

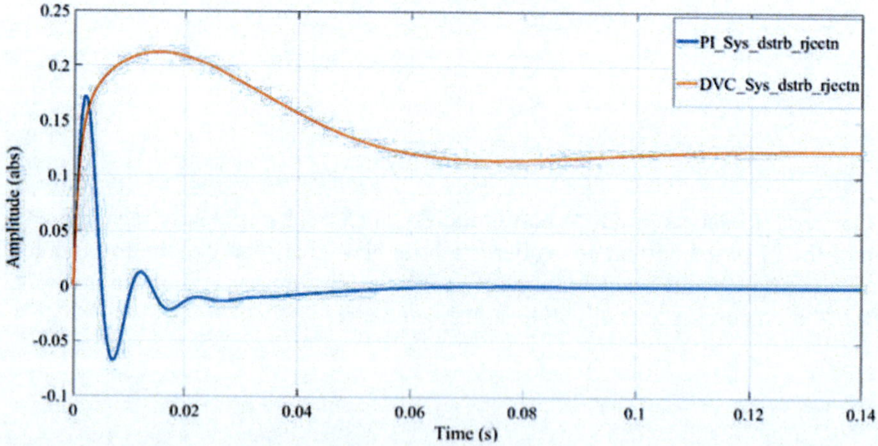

Fig. 11.16 Disturbance rejection capability of the two controllers

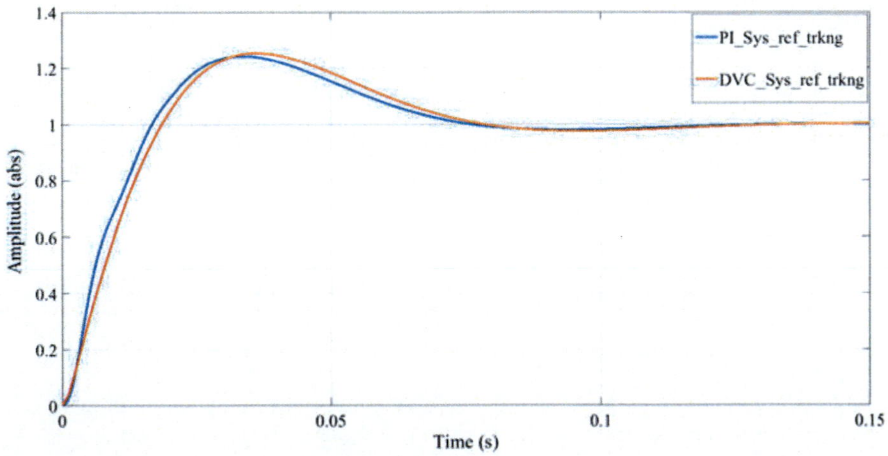

Fig.11.17 Reference tracking capability of the two controllers

loop; nevertheless, the reference tracking is done by the outer loop of the control, which employs a PI controller. It can also be seen that the direct voltage controller is slow compared to the PI-based current controller due to the missing integral in the current loop and low DC gain.

From these results above, it can be concluded that the direct voltage controller has no issue when it comes to tracking the references, whereas the disturbance rejection capability of such a controller is not as good as compared to PI-based current controller. Additional measures can be taken to improve the disturbance rejection capability of the direct voltage controller such as measurement filters, increasing the proportional gain of the washout filter and additional lead–lag compensators but such considerations, however, are not part of this chapter.

11.8 Conclusions

In this chapter, a grid forming wind turbine control principal based on DVC has been presented and discussed. The control implementation and its initialization in PowerFactory were also explained.

Numerical simulations show that the electrical torque can be controlled according to the intermediate circuit voltage, and, then, the pitch control stabilizes the wind turbine by adapting the mechanical torque according to the mechanical speed, resulting in a stationary operating point. Black-start and restoration capabilities of the proposed controller for wind turbine were also presented and discussed.

References

1. I. Erlich et al., New control of wind turbines ensuring stable and secure operation following islanding of wind farms. IEEE Trans. Energ. Convers. **32**, 1263–1271 (2017)
2. I. Erlich, A. Korai, F. Shewarega, *Control Challenges in Power Systems Dominated by Converter Interfaced Generation and Transmission Technologies* (IEEE Power and Energy Society General Meeting, Chicago, IL, 2017)

Chapter 12
Generic Modelling of PEM Technologies for Power System Stability Studies Based on PowerFactory

Feras Alshehri, Víctor García Suárez, Arcadio Perilla, M. Ebrahim Adabi, José Luis Rueda Torres, and Mart A. M. M. van der Meijden

Abstract In this chapter, a generic model of fuel cells and electrolysers suitable for power system stability studies has been developed in PowerFactory. Both theoretical modelling background and software implementation of fuel cells and electrolysers are detailed. Furthermore, a case study based on a three area test system has been performed, which provides valuable insight into the benefits that the synergy between the electricity and hydrogen sectors can bring to power system stability.

Keywords Ancillary services · Frequency stability · Fuel cell · Electrolyser

12.1 Introduction

The reduction of CO_2 emissions in the power system is being achieved by increasingly substituting conventional thermal synchronous generators with renewable energy resources. Although this new scenario leads to more complex system operation, it is still essential to maintain current robustness and reliability standards.

Flexibility plays an essential role in the success of the energy transition toward a more sustainable energy supply. Flexibility in a generation, demand-side technologies, and multi-energy sector couplings are key enablers to guarantee that the full potential of the future power system can be reached.

Electronic Supplementary Material The online version of this chapter (https://doi.org/10.1007/978-3-030-54124-8_12) contains supplementary material, which is available to authorized users.

F. Alshehri · V. G. Suárez · A. Perilla · M. E. Adabi · J. L. Rueda Torres (✉) ·
M. A. M. M. van der Meijden
Department of Electrical Sustainable Energy, Delft University of Technology, Delft, The Netherlands
e-mail: j.l.ruedatorres@tudelft.nl

M. A. M. M. van der Meijden
TenneT TSO B.V., Arnhem, The Netherlands

F. M. Gonzalez-Longatt and J. L. Rueda Torres (eds.), *Modelling and Simulation of Power Electronic Converter Dominated Power Systems in PowerFactory*, Power Systems,
https://doi.org/10.1007/978-3-030-54124-8_12

The provision of ancillary services such as frequency balancing, voltage support, and congestion management ensure a successful real-time power system operation. Since conventional generators have been the main provider of such services, it is essential to begin searching for other alternatives. Hydrogen-based technologies, namely fuel cells and electrolysers, can be categorized as promising alternatives for such purpose [1, 2].

In this chapter, a generic model of fuel cells and electrolysers suitable for power system stability studies has been developed in PowerFactory. The remainder of this chapter is structured as follows: Sect. 12.2 discusses the theoretical modelling background of fuel cells and electrolysers. Section 12.3 describes the implementation of the fuel cell and electrolyser model in PowerFactory. Section 12.4 presents the simulation results, and Sect. 12.5 summarizes the main conclusions.

12.2 Theoretical Background of the Model

12.2.1 Fuel Cell

The dynamic model of the PEM fuel cell provides the instantaneous power output of the fuel cell. The input for the dynamic model is the current drawn by the load, and the output is the stack voltage. The dynamic model relates these two values to establish the output power of the stack. While the current reacts instantaneously to the load change, the voltage response is slower due to the slower thermodynamics of the fuel cell. This results in an asymptotic exponential response to a step change in the load. The model developed here is based on the 1.2 kW Nexa PEM fuel cell semi-empirical model developed in previous research [3].

In developing this model, it is assumed that the supplied hydrogen and air are ideal and uniformly distributed gases supplied at constant pressure, the ambient temperature is constant at 25 °C, the thermodynamic properties are evaluated at the average stack temperature, temperature variations across the stack are neglected, the overall specific heat capacity of the stack is assumed to be a constant, the parameters for individual cells can be lumped together to represent a fuel cell stack, and the individual fuel cell stacks can be lumped together to represent the fuel cell array.

The stack voltage is related to the drawn current as follows:

$$V = E_0 - IR - A\ln\left(\frac{I}{I_{\text{ex}}}\right) \tag{12.1}$$

where $\boldsymbol{E_0}$ here is the Nernst potential, $\boldsymbol{R}$ is the resistance in ohms, $\boldsymbol{A}$ is the Tafel Slope in Volts, and $\boldsymbol{I_{ex}}$ is the exchange current which is considered a constant.

The Nernst potential, and it is calculated as follows [4]:

$$E_0 = 47 \times \left[1.482 - 0.000845\,T_K + 0.0000431\,T_K \ln\left(p_{\text{H}_2} p_{\text{O}_2}^{0.5}\right)\right] \tag{12.2}$$

Here, the multiplication by 47 is to account for the 47 individual cells within the stack, $\boldsymbol{T_K}$ is the stack temperature in Kelvin, and $\boldsymbol{p_{H_2}}, \boldsymbol{p_{O_2}}$ are the hydrogen and oxygen pressures in atm, respectively.

The resistance $\boldsymbol{R}$ and the Tafel Slope $\boldsymbol{A}$ are dependent on the temperature and are determined empirically using the following:

$$R(T) = R_0 \times \exp\left(\frac{E_{a,R}}{R_g T_K}\right) \tag{12.3}$$

$$A(T) = A_0 \times \exp\left(\frac{E_{a,A}}{R_g T_K}\right) \tag{12.4}$$

where $\boldsymbol{R_0}$ (Ω) and $\boldsymbol{A_0}$(V) are pre-exponential factor, $\boldsymbol{E_{a,R}}$ (J/mol) is the activation energy, $\boldsymbol{E_{a,A}}$ (J/mol) is the activation energy, and $\boldsymbol{R_g}$ is the universal gas constant?

The dynamic model of the fuel cell stack temperature is defined by the heat generation and dissipation. The temperature increases with heat generation and decreases with dissipation as defined in the following equation

$$\mathrm{mc}_p \frac{\mathrm{d}T}{\mathrm{d}t} = \dot{Q} - H_t(T - T_\infty) \tag{12.5}$$

Here, $\mathbf{mc_p}$ is the thermal capacitance (J/°C), $\dot{\boldsymbol{Q}}$ is the heat generation rate, and $\boldsymbol{H_t}$ is the heat transfer coefficient (W/°C).

The heat generation within the fuel cell stack is defined by

$$\dot{Q} = (E_h - V)I \tag{12.6}$$

where $\boldsymbol{E_h}$ an imaginary potential (V) obtained by converting all the enthalpy of water into electricity. It can be considered as a constant that is calculated by the following equation.

$$E_h = \frac{\Delta h^{\mathrm{o}}_{\mathrm{f,H_2O}}}{2F} \times 47 \tag{12.7}$$

Here, $\Delta \boldsymbol{h}^{\mathbf{o}}_{\mathbf{f,H_2O}}$ is the formation of enthalpy of water vapor at 25 °C and $\boldsymbol{F}$ is Faraday number.

It can be assumed that the current will be steady long enough that the temperature does not change which means that $\frac{\mathbf{d}T}{\mathbf{d}t}$ becomes zero which simplifies Eq. (12.5) to become

$$H_t = \frac{(E_h - V)I}{T - T_\infty} \tag{12.8}$$

Consequently, the dynamic temperature model can be defined by the simplified equation

Table 12.1 Constants for PEM fuel cell dynamic model

Parameter	Value	Equations
I_{ex}	1×10^{-6} A	(12.1)
R_0	0.1537 Ω	(12.3)
$E_{a,R}$	1800 J/mol	(12.3)
A_0	0.1591 V	(12.4)
$E_{a,A}$	5344 J/mol	(12.4)
mc_p	4304 J/°C	(12.5)
H_t	$15.07 \times I^{0.2358}$ W/°C	(12.5)
$\Delta h^{o}_{f,H_2O}$	−285.83 kJ/mol @ 25 °C	(12.7)
F	96,654 C/mol	(12.7)

$$T(t) = T_2 + (T_1 - T_2) \times \exp\left(-\frac{H_t}{mc_p}t\right) \tag{12.9}$$

Here, $\boldsymbol{T}_1$ is the initial temperature, $\boldsymbol{T}_2$ is the final steady-state asymptotic temperature.

Table 12.1 gives the values for the constants used in the model equations.

12.2.2 Electrolyser

Note: For a thorough explanation of the information briefly presented in this subsection, the reader is referred to [5].

A complete electrical representation of a PEM electrolyser includes the modelling of: (i) the stack, (ii) the power conversion system (i.e., transformer and electronic converters), and (iii) the balance of plant (i.e., auxiliary equipment such as feedwater and circulation pumps). The electrolysis stack can be reduced to an equivalent electrical circuit composed by the open-circuit DC voltage of the stack and a resistor in series that represents the different internal electrical losses. The electrolyser is coupled to the external grid via an AC–DC rectifier, followed by a DC–DC converter that controls the input power by modulating the electrical current. The described electrolyser system is exemplified in Fig. 12.1.

Even though several electrical models of small electrolysers exist in the literature, practical models of large electrolysers (>1 MW) are not yet available. Existing models typically focus on specific components of small electrolyser systems; therefore, limiting their usability for grid studies, and in order to understand the interaction of large electrolysers with the power system, generic models that capture the dynamics of all the PEM stack, power conversion system, and the balance of plant with sufficient detail are required.

In the present work, the electrolyser model has been based on and validated against the field measurements retrieved from a pilot 1 MW electrolyser installed in the north

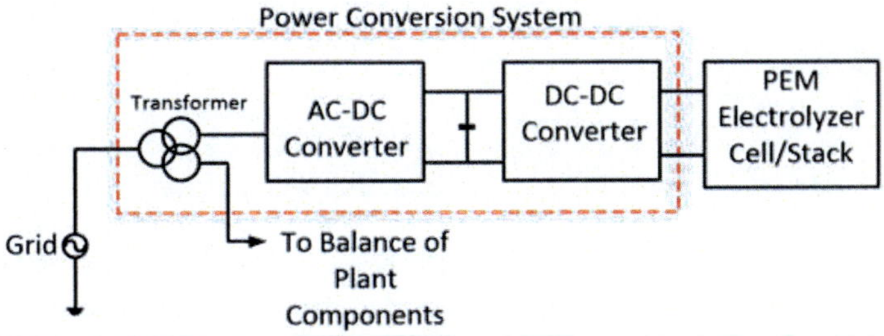

Fig. 12.1 High-level component overview of electrolyser system

of the Netherlands (at Veendam-Zuidwending). The model has been designed to be a simple, yet relatively accurate representation of the electrolyser grid side dynamics, especially with regard to the response of the active power consumption. For frequency studies, the accuracy of the ramping dynamics of the electrolyser constitutes the most relevant aspect of the modelling process.

Based on the field measurements, it is possible to estimate the ramp rate of the pilot electrolyser unit. As later shown in Fig. 12.9, it can be observed that the electrolyser shows a linear response to active power setpoint changes and has a ramp rate of approximately 0.5 MW/s (i.e., 0.5 p.u./s). Assuming that large electrolyser facilities consist of several small electrolysers connected in parallel, then a hypothetical 100 MW electrolyser plant consisting of 100 units of 1 MW could reach a maximum ramp rate of 50 MW/s. To put these data into perspective, the performance of a 40 kW PEM electrolyser studied in literature is also mentioned [6]. This unit shows a nonlinear behavior with limited dependency on the size of the setpoint change. Ramping up or down is generally completed within 0.2 s. Consequently, a capacity change of 50% within 0.2 s gives a ramp rate 0.1 MW/s (2.5 p.u./s). Extrapolating this ramp rate to a hypothetical 100 MW electrolyser plant, the resulting maximum total ramp rate would be 250 MW/s. While it cannot be assumed that the response time will not increase significantly for electrolyser capacities in the MW range, it still gives a rough indication of the range of ramp rates to be considered until more field measurements become available.

12.3 Model Implementation in PowerFactory

12.3.1 Fuel Cell

In PowerFactory, the fuel cell is represented as a static generator that is controlled as a current source via the external signals I_d and I_q. The model for the fuel cell includes the frequency control that allows the fuel cell to participate in the frequency containment reserve (FCR), the fuel cell dynamic model, and the converter that connects the fuel cell to the grid. Figure 12.2 shows a block diagram of the PEM fuel cell model in DIgSILENT PowerFactory which will be implemented using DIgSILENT Simulation Language (DSL).

The frequency control sets the output power P of the PEM fuel cell based on the frequency deviation. The block diagram of the frequency control is shown in Fig. 12.3.

First, the model measures the frequency deviation then applies a dead band filter of 10 MHz to suppress noise signals and minor deviations in frequency. The signal then goes through a droop control block that determines the power change as follows:

$$\Delta P = -P_{\text{bid}} \frac{(f - f_{\text{nominal}})}{\text{Full Bid Frequency Deviation}} \tag{12.10}$$

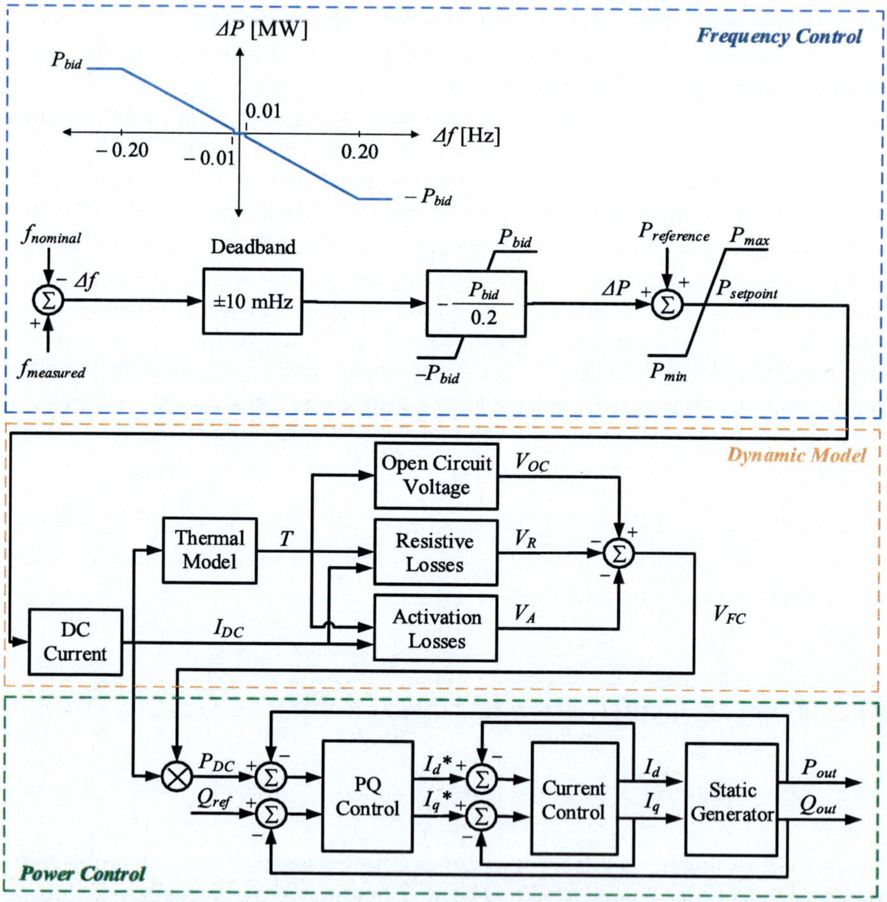

Fig. 12.2 Block diagram of PEM fuel cell model in PowerFactory

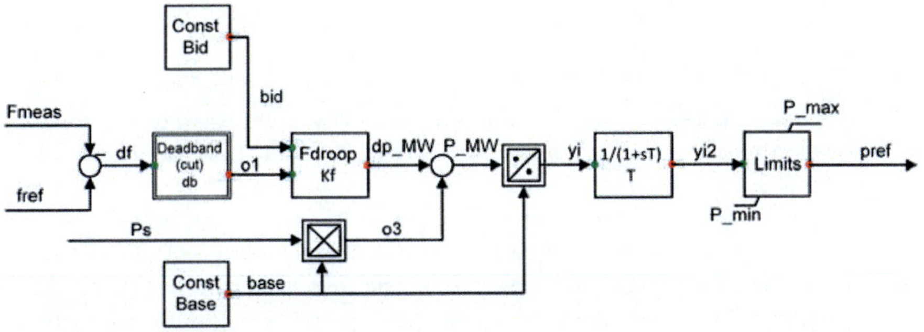

Fig. 12.3 Frequency control block diagram

The output power ΔP is dependent on the FCR bid value (P_{bid}) and the frequency deviation between the measured frequency (f) and the reference frequency (f_{nominal}). The full bid frequency deviation is when the full bid value is activated, it is a constant decided by the system operator. A limiting function is also added to ensure that ΔP does not exceed P_{bid}. The reference power point is then added to ΔP to give the power set point for the fuel cell which is converted to per units and limits are applied to ensure the fuel cell operates in the allowable limits 20–100% of the rated power.

The fuel cell dynamic model is the implementation of Eqs. (12.1)–(12.9) in DSL. Figure 12.4 shows the block diagram for the dynamic model.

First, the "Current Calc" block converts the power set point to watts then to amperes using a look-up table of empirical data from [7].

The current in amperes is used to determine the instantaneous temperature using the "Thermo Model" block. The block determines the initial and final asymptotic temperature using look-up tables and empirical data from [3]. The change in temperature in this block is exponential, which will be reflected in the undershoot and overshoot of the stack voltage.

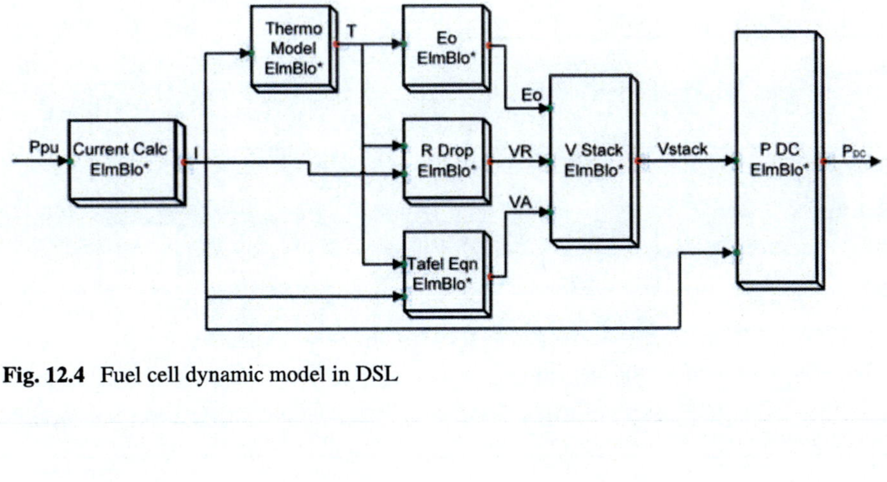

Fig. 12.4 Fuel cell dynamic model in DSL

The current and temperature are used to calculate the voltage "Vstack" through adding the output of the blocks "Eo1", "R Drop", and "Tafel Eqn" which are calculated using Eqs. (12.2)–(12.4). The voltage and current are multiplied to calculate the instantaneous DC power "P_{DC}" in watts which is converted again to per units.

The power control for the fuel cell is done through two feedback loops: an outer slower power control loop and an inner faster current control loop. The block diagram for control of active power *P* and reactive power *Q* is shown in Fig. 12.5.

The deviation between the power reference, which is the output of the dynamic model and the measured *P* is measured, and a dead band filter is applied to filter the noise signal. Then, the deviation is fed into a proportional and integral (PI) controller. The reference power is then added in order to give the set point for the current controller *idset*. The control mechanism for *Q* is the same except that it is maintained at a constant value.

The current loop provides the current signals *id_ref* and *iq_ref* to the static generator. The block diagram for the current control is shown in Fig. 12.6.

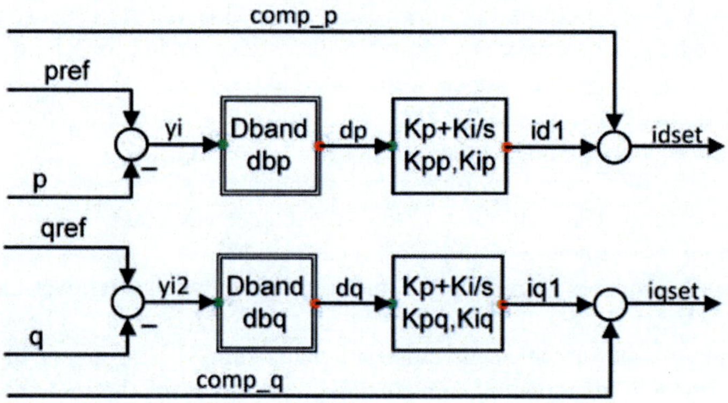

Fig. 12.5 Power control for the fuel cell

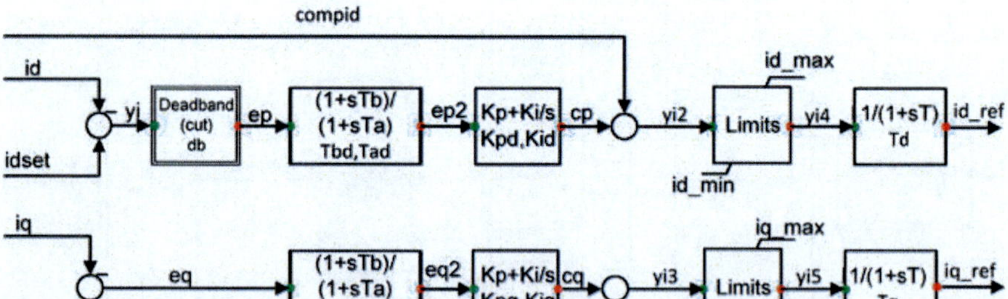

Fig. 12.6 Current control for the fuel cell

To control I_d, the measured value is compared to the reference value from the power control block, then the deviation passes through a dead-band followed by a lead-lag filter and a PI controller. A compensating signal is added, then a limit is applied to ensure the output remains within the operating limits of the fuel cell 20–100% of rated power. The delay of the converter and the power control unit is reflected through a first-order delay function. I_q is control is done in a similar fashion except there is no low-level noise that requires a dead band filter.

Current control initialization:

```
vardef(Kpd)='pu';'id Proportional Gain'
vardef(Kid)='pu';'id Integral Gain'
vardef(Td)='s';'id inverter delay'
vardef(Tbd)='pu';'id Lead-Lag Filter Gain'
vardef(Tad)='pu';'id Lead-Lag Filter Gain'
vardef(Tbq)='pu';'iq Lead-Lag Filter Gain'
vardef(Taq)='pu';'iq Lead-Lag Filter Gain'
vardef(Kpq)='pu';'iq Proportional Gain'
vardef(Kiq)='pu';'iq Integral Gain'
vardef(Tq)='s';'iq inverter delay'
vardef(P_Filter)='pu';'Filter P signal 1on, 0off'
vardef(Q_Filter)='pu';'Filter Q signal 1on, 0off'

inc0(compid)=id_ref
inc0(compiq)=iq_ref
inc(idset)=id
inc(iqset)=iq
inc(xtd)=id_ref
inc(xtq)=iq_ref
inc(xd)=0
inc(xq)=0
inc(xd1)=0
inc(xq1)=0
```

PQ control initialization:

```
vardef(dbp)='pu';'P deadband'
vardef(dbq)='pu';'Q deadband'
vardef(Kpp)='pu';'P Proportional Controller'
vardef(Kip)='pu';'P Integral Controller'
vardef(Kpq)='pu';'Q Proportional Controller'
vardef(Kiq)='pu';'Q Integral Controller'

inc0(qref)=q
inc(comp_p)=idset
inc(comp_q)=iqset
inc(pref)=p
inc(xp)=0
inc(xq)=0
```

Frequency control initialization:

```
vardef(db)='Hz';'Dead Band'
vardef(Kf)='pu';'Frequency Gain(droop)'
vardef(Bid)='MW';'Bid capacity'
vardef(Base)='MW';'FC Plant Base'

inc0(Ps)=Ppu
inc(fref)=Fmeas
inc(Ppu)=p
inc(dp)=0
```

12.3.2 Electrolyser

The electrolyser is modelled in PowerFactory as a general load, plus its associated dynamic model. The balance of plant is modelled by an additional general load, with constant demand depending on the size of the electrolyser.

The control structure of the electrolyser is illustrated in Fig. 12.7. Since all the control blocks involve only constant values, no state variables need to be initialized.

The list of control input variables is displayed in Fig. 12.8.

The dynamics of the electrolyser model are governed by a linear ramp rate block, which is defined with respect to the total electrolyser installed capacity. In this example, for a 100 × 1 MW electrolyser fleet, and given the 0.5 MW/s ramp rate of the 1 MW pilot, the corresponding total ramp rate is 50 MW/s. The rated active power output P_{max} and the technical minimum active power output P_{min} of the electrolyser fleet can be configured with the power limiter. Also, it is critical to verify that the reference demand P_{ref} coincides with the active power set point *plini* as defined in the load flow tab of the load element; otherwise, the dynamic model will not perform properly.

The response of the dynamic model for different positive and negative active power step changes is compared with the measurements retrieved from the pilot

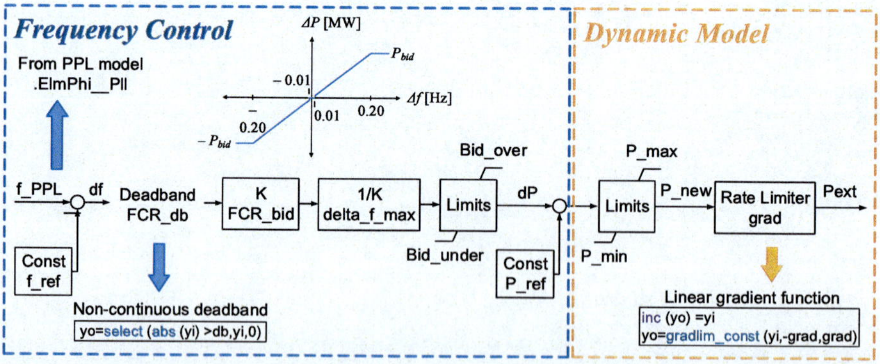

Fig. 12.7 Implementation of the dynamic model and frequency control of the electrolyser

	Parameter
▶f_ref Nominal frequency [Hz]	60,
FCR_db FCR deadband [Hz]	0,01
FCR_bid FCR bid size [MW]	70,
delta_f_max Maximum FCR frequency deviation [Hz]	0,2
P_ref Load flow active power setpoint plini [MW]	100,
grad Electrolyser ramp rate [MW/s]	50,
Bid_under (-) Bid size limit for under-frequency [MW]	-70,
P_min Electrolyser technical minimum active power [MW]	30,
Bid_over (+) Bid size limit for over-frequency [MW]	70,
P_max Electrolyser active power rating [MW]	100,

Fig. 12.8 List of input electrolyser control variables. Example of under-frequency simulation

1 MW electrolyser in Fig. 12.9. It can be observed that due to the inherently linear behaviour of the pilot electrolyser, the implemented model is able to follow the field measurements with acceptable accuracy. The gradient ramp rate block constitutes the most straightforward, yet accurate representation of the electrolyser dynamics at the grid side.

The frequency control of the electrolysers has been designed according to the current European FCR framework (e.g., frequency deadband, maximum frequency deviation, etc.).

Since the electrolyser is an electric load, an inverse droop characteristic has been implemented (i.e., decreased consumption during under-frequency events and increased consumption during over-frequency events). Changing the sign of the summation point for input *delta_P* from positive to negative results in the traditional droop characteristic for generation technologies.

It can be observed that the design is slightly different from the frequency control shown for the fuel cell, as it includes additional control blocks that enable flexibility to the definition of the droop characteristic. In this case, the hypothetical bid awarded to the electrolyser in the FCR market FCR_{bid} is explicitly used instead of transforming it into a droop value. Moreover, bid limits have been included for both over-frequency (positive) and under-frequency (negative) as to enable asymmetric droop characteristics. For instance, if the over- or under-frequency limit is set to zero, the electrolysers will not react to those events.

12.4 Simulation Results

12.4.1 Test System

Figure 12.10 shows the test system implemented in PowerFactory to illustrate the performance of the fuel cells and the electrolysers to support frequency stability. The

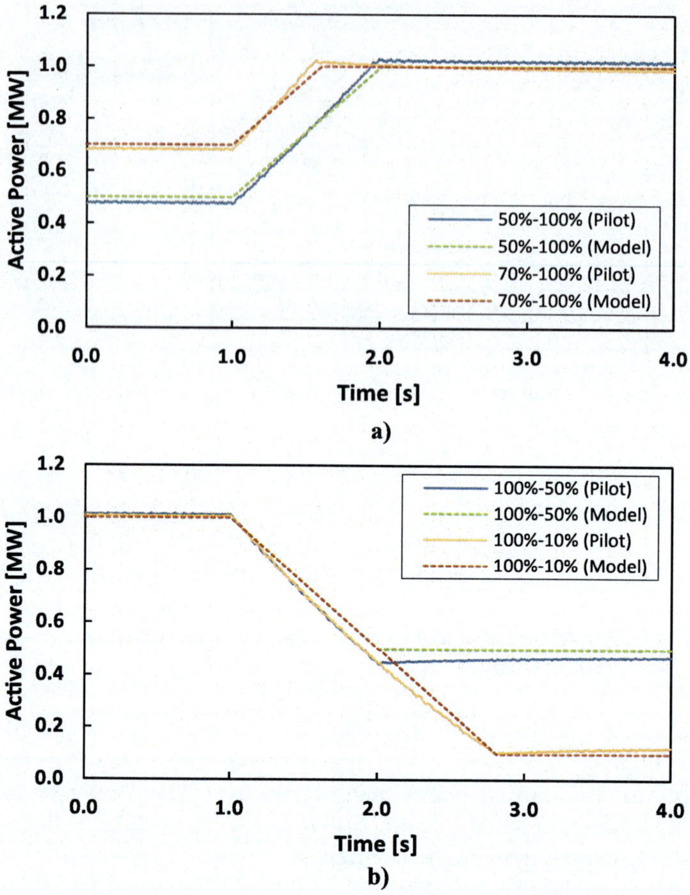

Fig. 12.9 1 MW electrolyser model validation. Active power ramp-up (**a**) and ramp-down (**b**)

test system is a three-area network composed of synchronous generators, wind parks (WP1 and WP2), electrical loads, fuel cells (FC), and electrolysers (P2G).

12.4.2 Study Cases

Two study cases have been considered to analyze the performance of fuel cells and electrolysers to support frequency stability. The first study case is a sudden loss of generation. To perform this test, the total generation of the system was decreased by 90 MW. Figure 12.11 shows the response of the system frequency when the participation of fuel cells or electrolysers in FCR is varied from 0 to 50% by replacing the FCR support of some generators by the fuel cells or electrolysers.

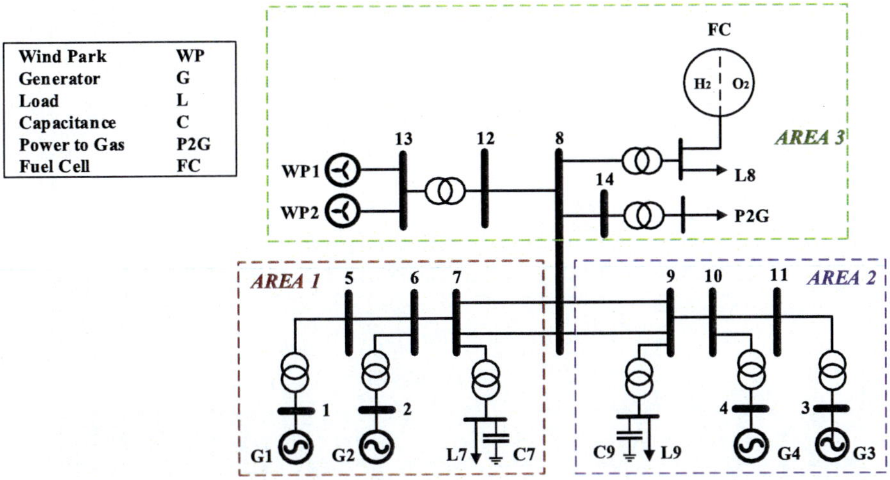

Fig. 12.10 Three-area six-generator test system

The second study case is a loss of load. In this case, the total system load is decreased by 90 MW. Once more, Fig. 12.12 shows the response of the system frequency when the participation of fuel cells or electrolysers in FCR is varied from 0 to 50% by replacing the FCR support of some generators by the fuel cells or electrolysers.

It can be observed from Figs. 12.11 and 12.12 that substituting frequency support from synchronous generators by fuel cells or electrolysers leads to improvements on the frequency nadir of the system, and therefore has a positive impact on the frequency stability of the system. This is due mainly to the fact that fuel cells and electrolysers respond much faster to frequency deviations in comparison with synchronous generators.

12.5 Conclusion

In this chapter, a generic modelling approach of PEM hydrogen technologies for power system stability has been presented. The theoretical background of the models and the implementation in the software have been described for both fuel cells and electrolysers. Moreover, the performed case study, based on a three area test system, proves that the participation of the fuel cells and electrolyser in frequency balancing services can improve power system stability in comparison with conventional synchronous generators, especially in terms of the maximum frequency deviations.

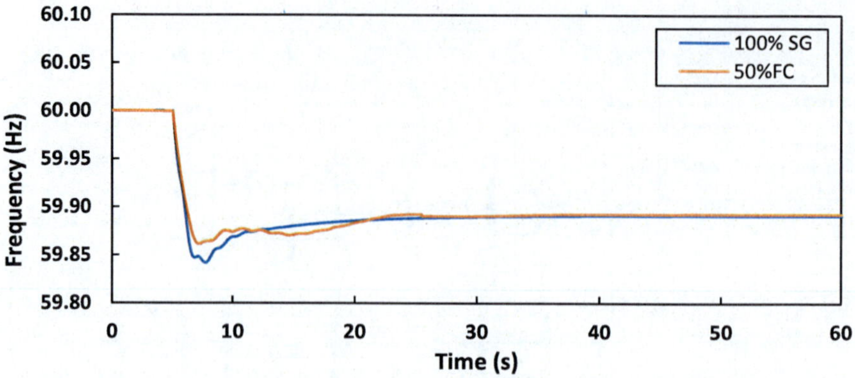

a) **Fuel cell**

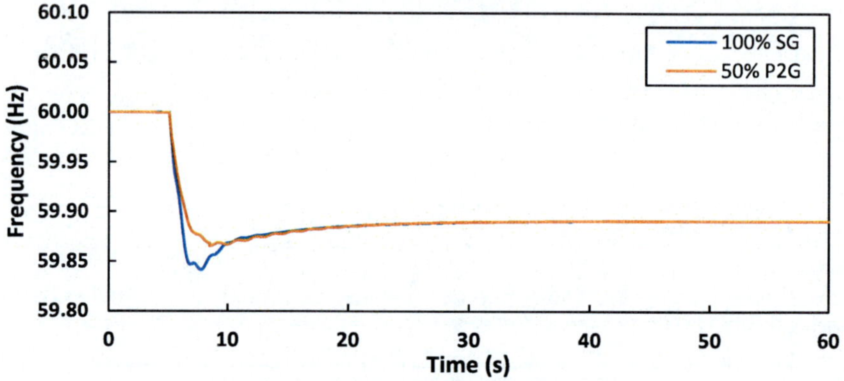

b) **Electrolyser**

Fig. 12.11 Frequency response for 0 and 50% shares of fuel cell or electrolyser in FCR capacity (loss of generation)

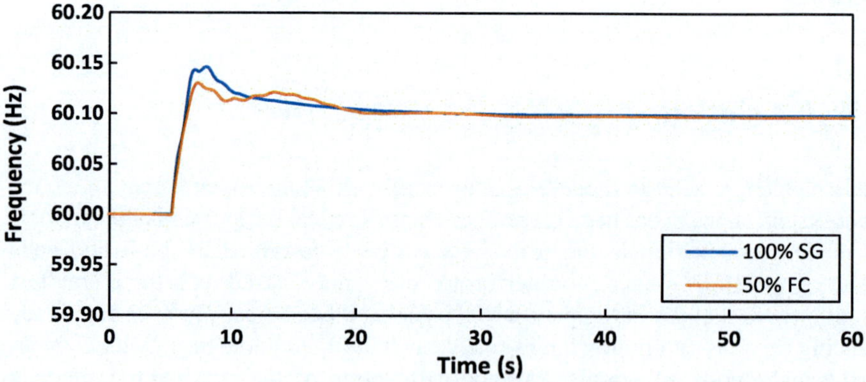

a) **Fuel Cell**

Fig. 12.12 Frequency response for 0 and 50% shares of the fuel cell or electrolyser in FCR capacity (loss of load)

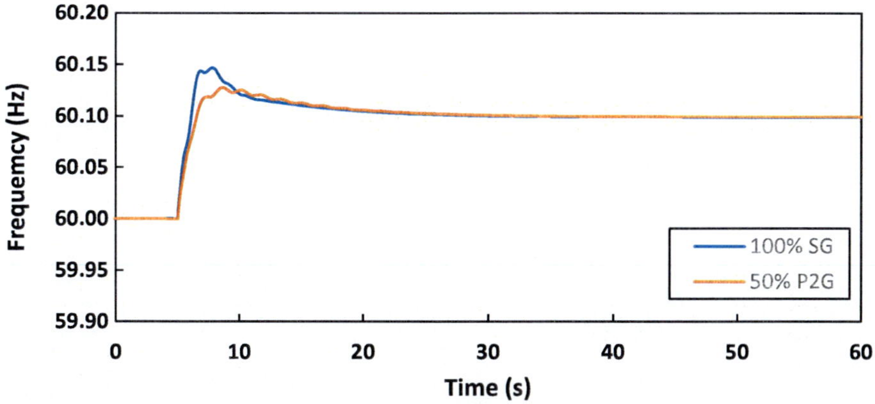

b) Electrolyser

Fig. 12.12 (continued)

References

1. V. Suárez, P. Ayivor, J. Torres, M.A.M.M. van der Meijden, Demand side response in multi-energy sustainable systems to support power system stability. Presented at 16th wind integration workshop, Berlin, Germany (2017)
2. F. Alshehri, V. García, J.L. Rueda, A. Perilla, M.A.M.M. van der Meijden, Modelling and evaluation of PEM hydrogen technologies for frequency ancillary services in future multi-energy sustainable power systems. Elsevier Heliyon **5**(4) (2019)
3. H.-i Kim, C.Y. Cho, J.H. Nam, D. Shin, T.-Y. Chung, A simple dynamic model for polymer electrolyte membrane fuel cell (PEMFC) power modules: parameter estimation and model prediction. Int. J. Hydrogen Energy **35**(8), 3656–3663 (2010)
4. R. O'hayre, S.-W. Cha, F.B. Prinz, W. Colella, *Fuel cell fundamentals* (Wiley, 2016)
5. Stability analysis of an international electricity system connected to regional and local sustainable gas systems. *TSO 2020 Activity 2 Final Report* (2019)
6. J. Eichmann, K. Harrison, M. Peters, Novel electrolyser applications: providing more than just hydrogen, *Technical Report NREL/TP-5400-61758* (National Renewable Energy Laboratory, Denver CO, USA, 2014)
7. *Nexa^{TM} Power Module User's Manual* (Ballard Power Systems Inc., 2003)

Chapter 13
PST-17 Benchmark Power System DSL-Based Model with 90% Power Electronic Interfaced Generation with Black-Start and Grid Restoration Capabilities in DIgSILENT PowerFactory

Leonel Noris, José Luis Rueda Torres, E. Rakhshani, and Abdul W. Korai

Abstract In this chapter, an integral approach for Grid-Forming and Black-Start capability of a large-scale interconnected Power System model is developed. This chapter introduces a model for Electro-Magnetic Transients (EMT) simulations, where a three-area power system is presented, containing different devices interfaced to the transmission network via voltage-source converters (VSC); seventeen wind power plants (WPP), seven battery-energy storage systems (BESS), and two HVDC transmission links. Of the total energy produced in this model, 90% is generated by the WPP and the other 10% by conventional generation units (CGU). The control systems that regulate the WPP and the HVDC stations were upgraded with Grid-Forming capability. Therefore, the power system model is suitable for simulations during both steady-state and transient operational scenarios. If the latter case may derive in a Blackout, it allows simulating Black-Start and Restoration strategies. The proposed grid-forming and black-start capabilities were tested with various EMT simulations reproducing severe short-circuit faults, deriving in a full blackout in one of the areas of the power system. The model also was upgraded with five protection relays with a restoration algorithm that determines the best re-energisation path for the fastest possible restoration strategy. The simulation results demonstrate

---

Electronic supplementary material The online version of this chapter (https://doi.org/10.1007/978-3-030-54124-8_13) contains supplementary material, which is available to authorized users.

L. Noris
DNV GL–Energy. Arnhem, Gelderland, Netherlands

J. L. Rueda Torres (✉) · E. Rakhshani
Department of Electrical Sustainable Energy, Delft University of Technology, Delft, Zuid-Holland, Netherlands
e-mail: j.l.ruedatorres@tudelft.nl

A. W. Korai
DIgSILENT GmbH. Gomaringen, Baden-Württemberg, Germany

F. M. Gonzalez-Longatt and J. L. Rueda Torres (eds.), *Modelling and Simulation of Power Electronic Converter Dominated Power Systems in PowerFactory*, Power Systems,
https://doi.org/10.1007/978-3-030-54124-8_13

that a power system with high penetration of converter-based generation and transmission is completely capable of managing a grid during all circumstances if its control systems are designed to do so, without encountering the problems arising from current injection control.

Keywords DIgSILENT EMT simulation · DIgSILENT simulation language (DSL) programming · Wind turbine control · HVDC control · Grid-forming capability · Renewable generation unit (RGU) · Battery-energy storage systems (BESS) · Soft-energisation method · Critical-time method

13.1 Introduction

To the present day, converter-based generation, which is vital to all renewable-based power generation, is a secondary or complementary type of generation. The grid codes regarding renewables are designed primarily to ensure that they just follow the voltage and frequency reference from the interconnected CGU, thus limiting WPP installations to areas where the electric network is guaranteed to be strong and stable. Besides being limited to following the grid-forming CGU, renewables are designed primarily to contribute in such a way that they do not hinder with the safe and secure operation of the main CGU.

However, it is expected that the share of solar-photovoltaic and offshore/onshore wind power will be of 72% of the global overall installed capacity by 2050 [1], and from these options, offshore WPP is the main ones for large-scale renewable generation development [2]. These projections foreseeing the increment future power generation facilities, based on renewable sources, directly will increase in the use of VSC, which bring new challenges to the stability and the security of supply of the power systems in the future. Future renewable generation units (RGU) are expected to contribute to reactive power regulation and will manage voltage directly in the same manner CGU nowadays do. In circumstances where the share of renewables is less than 40% in modern power systems, the variability associated to renewable power generation is absorbed by CGU, but if this percentage is increased, the ability of the power system to deal with variation and disturbances during operating conditions may be jeopardised, boosting the risk of disruptions and blackouts [3, 4].

To cope with the challenge of the high variability of renewable power generation, a large-scale three-area Benchmark System, ideally used for power system stability studies, was developed in DIgSILENT PowerFactory with the aid of the DSL functionalities subsumed in this software, with the capability of being stable and robust during both steady-state and transient operational scenarios. Even if the latter case study derives into a blackout, this Benchmark System is able to self-heal and preform a power system restoration. The model includes the addition of grid-forming control systems for WPP and HVDC links, BESS storage support and a Protection and Restoration Algorithm. The theoretical background of the control systems involved, an application guide for DIgSILENT PowerFactory and numerical simulations are provided to illustrate the performance of the system during the combined black-start and restoration strategies when operated with high penetration of renewables.

This chapter is organised as follows: In Sect. 13.2, a review on blackouts in power systems is presented briefly. In Sect. 13.3 the PST-16 Benchmark Model is presented. Then, in Sect. 13.4, a review of the classical current control approach for WT is discussed. In Sect. 13.5, the Power System model with Black-Start/Restoration capabilities is presented. In Sect. 13.6, DSL based control implementation in DIgSILENT PowerFactory is explained, and finally, in Sect. 13.7 the performance of the proposed control for Power System modelling with high penetration of Power-Electronic Converters is simulated and analysed for the PST-16 Benchmark System case study with different contingencies. Summary and main conclusions of this chapter are also presented in Sect. 13.8.

13.2 Blackouts in Power Systems

The loss of electric power in the grid is known as shutdown, power outage, or blackout. Shutdowns can keep going for a short period or a long-haul timeframe. There are numerous components that could conceivably help to develop a situation into a power blackout in an electric power system. Deficiencies at power stations, ageing assets, harm to transmission lines, substations or different components of the distribution system, short-circuit faults, and resulting voltages collapsing, the over-burdening of power mains, dispatch-demand unbalance, defective protection and/or redundancy schemes are just some examples of scenarios that could derive into a blackout.

Frequently is the loss of voltage support in the network that causes the inability to recover itself, prompting a voltage breakdown. As the voltage falls, Under-Voltage Relay (UV) programs trip out components, yet, in case that this technique is not done cautiously, such action lessens the capacity of the grid to deal with the crumbling voltage. This could trigger a falling impact in which lines trip in a steady progression until the entire power system passes out. Another potential wellspring of catastrophe is the point at which the system frequency encounters substantial deviations that outperform limit thresholds. Continued operation of power systems is inconceivable if the system frequency is unable to be kept inside these strict cut-off points [3].

There are different types of power outages [5]:

- **Partial system blackout**: Blackout of a fragment of the grid. Partially, with one or more than one portion of the system outaged.
- **Islands**: In case that a part of the network is isolated from the rest because of system splitting or through protection schemes, the detached part needs to play out the restoration and operate independently. This method of activity is named Islanded Operation Mode. One example is when the faulted area of the grid is

Table 13.1 Direct and indirect consequences of blackouts [3]

Direct Impacts of a Blackout
Food spoilage
Manufacturing plant shutdown
Damage to electronic data and loss of computer services
Loss of life-support systems in hospitals, nursing homes and households
Suspension of electrified transport
Traffic congestion and risk of accidents due to the failure of traffic control services (both in motorways and airports)
Indirect Impacts of a Blackout
Short term
Property losses from looting and vandalism
Overtime payments to police, medical and fire personnel
Cancellation of social aids
Medium term
Cost of recovery from looting and vandalism
Lost tax revenues during recovery period
Consequent increase in insurance rate problems
Incarceration of looters
Long term
Litigation costs
Contamination due to sewage disposal
Consequent increased disease

interconnected through HVDC links, which is then secluded from the rest after the fault.

- **Total system blackout**: The network for this situation is under full power outage. All the generation services have stopped working, and there is no supply of outside interconnections to outer networks, and it is not feasible for the all-out system to work again except if it is totally restarted.

Under specific conditions, a de-energised system element can cause current changes in neighbouring portions of the system, prompting a falling failure of a bigger segment of the system. This may extend from a building to a block, to a whole city, or the totality of the power network. In this way, it is changing from one sort of shutdown to an increasingly extreme kind of power outage.

Because of the fundamental requirement for power in the modern world, outages are unwanted because of the social, economic and industrial impacts that sway in the civilisation. Some of the most noteworthy consequences and negative impacts that infer a power outage are indicated in Table 13.1.

13.3 The PST-16 Benchmark Power System

The study case used for the power system restoration analysis is based on the PST-16 Benchmark System (Fig. 13.1) [6], and it depicts a large-scale interconnected network with different voltage levels widespread in three strongly meshed areas. Two out of the three areas are connected by long-distance HVDC systems and the third one by a regular HVAC transmission line. The three areas are *Area A*, *Area B* and *Area C*. The main power flow direction is from areas *A* and *B* to *C*; in the latter is where most of the loads are located. *Area A* is the main exporter. The total installed capacity is around 17 GW. The elements comprising the model case are [3]:

- **Seventeen wind power plants (WPP)**. Designed to represent simplified wind farms composed of ninety Permanent-Magnet Synchronous Generator (PMSG) Type-4 Wind Turbines (WT), each rated at 10 MW. Ready to deliver 900 MW*

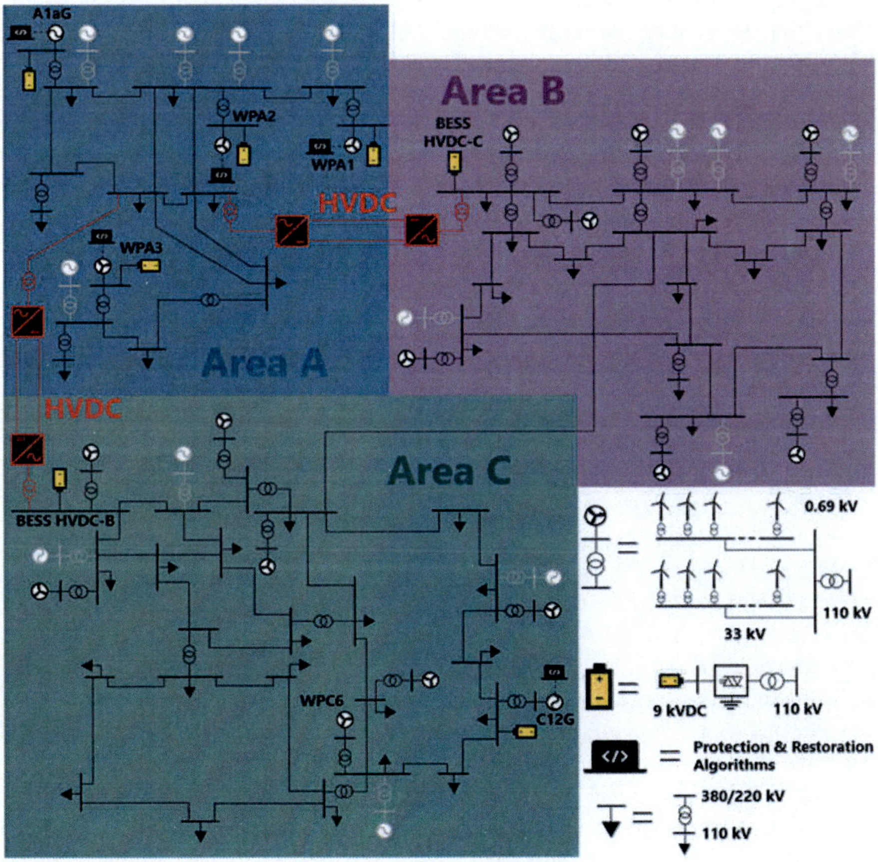

Fig. 13.1 PST-16 benchmark system case designed in DPF [3]

of active power in their neighbouring terminals. Each one includes an External Station Controller that forces the WPP to dispatch specific values of reactive power in Mvar for ensuring a balanced Load Flow.

- **Sixteen synchronous machines (CGU)**. Five of them are hydro, four nuclear and the other seven are a coal-type energy source. All of them have their respective governor, automatic voltage regulator (AVR) and power system stabiliser (PSS). Most of them have approximately four-to-five parallel units connected to them, while every single unit is rated at 220 MVA (hydro), 259 MVA (nuclear) and 247 MVA (thermal). Saturation parameters are enabled (Table 13.5a).
- **Two VSC-HVDC Transmission Links**. *Area A* is connected with the rest of the grid with DC transmission. The sending-end rectifiers operate in *P-Q* mode, while the receiving-end inverters do so in *VDC-Q* mode. Depending on the operation mode, their neighbouring terminals have specific *VDC*, *P*, or *Q* setpoint values for ensuring a smooth load flow. The converter stations are rated at 350 kVAC, 760 kVDC and 2000 MVA. The DC cables are rated at ± 380 kV and 4.08 kA. Their lengths are between 200 and 220 km.
- **Seven battery-energy storage systems (BESS)**. All are composed by a nine kVDC battery connected to a monopolar controlled VSC. Their power rating is 45 MVA with an AC voltage of 4 kV. Two of them are installed as emergency plants next to two synchronous generators in order to provide them with excitation voltage and become them black-start units. Other three are disposed to do the same duty for three WPP located in *Area A*. The remaining two provide frequency support to the sending-end HVDC converter stations.
- **Eighty-one transformers**. Their power ratings vary from 130 to 2760 MVA. Transformer magnetising parameters are enabled for an accurate EMT transient analysis (Table 13.5b). The ones interconnected between the simplified WPP and the grid are step-up type. In real-physical facilities, two transformers are used to step-up from 0.69 to 33 kV, and then from it another one is used to transmission level transformation. For simplicity reasons, seventeen equivalent transformers at 1000 MVA-0.69/110 kV with a reactance of 6.65% of nominal ratings were modelled for the connection of the WPP [8, 9].
- **Two different operational scenarios**. The sixteen CGU and seventeen RGU are not set to all work at the same time. An operational scenario of 100% CGU (hydro, nuclear and thermal) can be activated, as well as an operational scenario of high penetration of RGU (90% wind and 10% CGU). The most relevant scenario for this research is, evidently, the latter one. In the operational scenario running with 90% converter penetration, the only two CGU active that compose the remaining 10% of generation are CGU *A1aG* and *C12G*, sketched in black in Fig. 13.1, while the other CGU are depicted in grey.
- **Fifty-eight load centres**. Their classifications vary from agricultural, domestic, commercial and industrial applications. Their power ratings vary from 20 to 800 MW and from -50 to 100 Mvar. The total load demand is 16 GW.
- **Eight shunt C-type high-pass filters**. Rated at 380 kV. They are employed to supply reactive power and avoid resonances in parallel, as well as to filter

harmonics maintaining zero losses in the fundamental frequency and not depicted in Fig. 13.1.

- **Fifty-one transmission lines.** Voltage ratings vary from 100 to 380 kV. Distances vary from 0.1 to 220 km, while their current ratings go from 3320 to 4080 A.
- **A hundred and sixty-eight busbars.** Voltage ratings vary from 0.55 to 110, 220 and 380 kV.

The PST-16 Benchmark System model case was intended for the appraisal of stability and strength issues identified with low-frequency oscillations (LFO) as it has a vast size, with very-long transmission lines interconnecting the three areas [6]. Further, it was planned based on standard parameters and viewpoints found in the European power network. The impact of the converter-interfaced units on the dynamic performance of the power system depends on their share of total renewable generation, on their location, network topology and the sort of seriousness of the disturbances to be considered. On this rationale, this Benchmark is a phenomenal canvas to assess a conjunct Black-Start/Restoration strategy, thanks to its high multi-faceted nature, and the huge/comprehensive amount of power and control components engaged in combination with the extensive size of the power grid. Given this entangled scenario, if the Grid-Forming Control Systems encourage effectively the aggregate Black-Start and Restoration plans, certainly will perform satisfactorily in other power system models [3].

13.4 Review on Grid-Following Control Systems

When it is expressed that renewable energy integration will be required to partake in active and reactive power management, it implies that the power-electronic devices interfacing the renewables should accompany them with improved control highlights. For this reason, the self-commutated VSC converters are appropriate, on the grounds that they can keep up a consistent polarity of DC voltage; thus, power reversal is reached instead by reversing the direction of the current. Further, the utilisation of MMC converters, framed by several VSC connected in series, provides a higher definition in the signals [3, 10].

In classical power systems, the centrifugal governor that controls the speed of a CGU is responsible for such action. Conversely, on account of renewables, like WPP (Fig. 13.2a) and Photovoltaic (PV) Systems, the controllers of the VSC are the equivalent of the governor systems for CGU. Contemporary control systems for RGU are Grid-Following (or Grid-Feeding). This scheme follows the voltage/frequency reference from the Grid-Forming system attached to traditional CGU [7].

Present-day electrical grids that incorporate a mix of RGU do not have more than 40% sustainable penetration, consequently, being quite steady. Besides, in Power System Modelling, usually the machine-side converter and the WT generator are neglected because for the grid dynamics these elements are assumed to be stable and providing constant and abundant wind conditions [7]. This permits power system

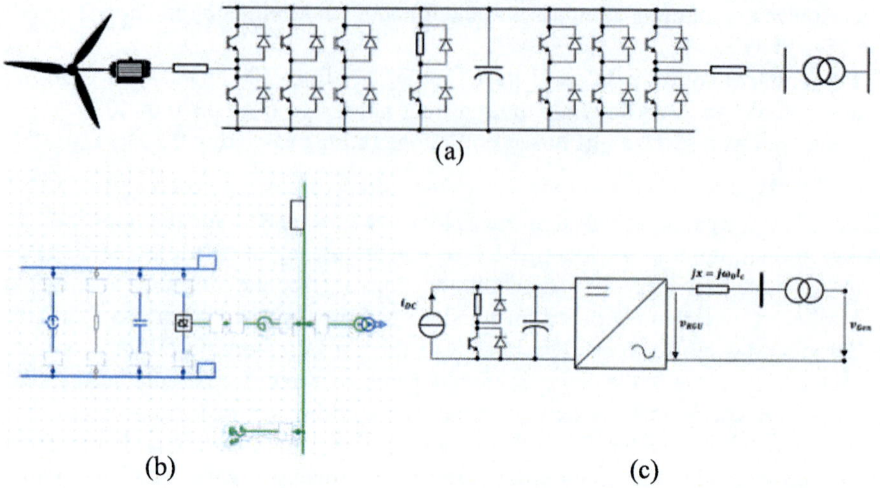

Fig. 13.2 Wind energy conversion model [3]. (a) structure with a back-to-back WPP (b) current-source WPP model in DIgSILENT PowerFactory (c) thevenin equivalent circuit of Fig. 1b

engineers to simplify RGU models in the role of a DC current source, in software like DIgSILENT PowerFactory, as it is shown in Fig. 13.2b [4]. This dimension of detail for most power system studies is normally valid because the frequency of the VSC converters decouples the rotor/generator from the grid. v_{Gen} in Eq. (13.1) is the local measured voltage read at the node where the scaled-RGU model is connected. If using the *d–q* reference frame,

$$v_{\mathrm{RGU}} = v_{\mathrm{Gen}} - jix_{\mathrm{c}} \tag{13.1}$$

Where v_{RGU} is the voltage coming out of the inverter, as shown in Fig. 13.2c. By utilising the real and imaginary currents in an analogical fashion as active and reactive currents (i_{P} and i_{Q}), they can be rewritten as $i_{\mathrm{P}} = i_{\mathrm{d}}$ and $i_{\mathrm{Q}} = -i_{\mathrm{q}}$.

Consequently, Eq. (13.1) can be stripped down as

$$v_{\mathrm{RGU,d}} = v_{\mathrm{Gen}} + i_{\mathrm{q}}x_{\mathrm{c}};\ v_{\mathrm{RGU,q}} = -i_{\mathrm{d}}x_{\mathrm{c}} \tag{13.2}$$

The control system regulating the WPP model in Fig. 13.2b, presented in Fig. 13.3a, is based on the classical current-injection method [4, 7], where the reference parameter that sets the terminal voltages $v_{GSC,d-q}$ to be related to the network connection is given by the *d–q* current components, which in turn are the outcome of a previous reactive current setpoint calculation, where a steady-state reference is superimposed on a dynamic reference (Fig. 13.3b).

As long as the v_{GSC} signal is known, this feedforward scheme works satisfactorily, correcting any error detected when calculating the changing Δ voltage. However, the use of feedforward terms may cause disturbances due to unpredictable parameters,

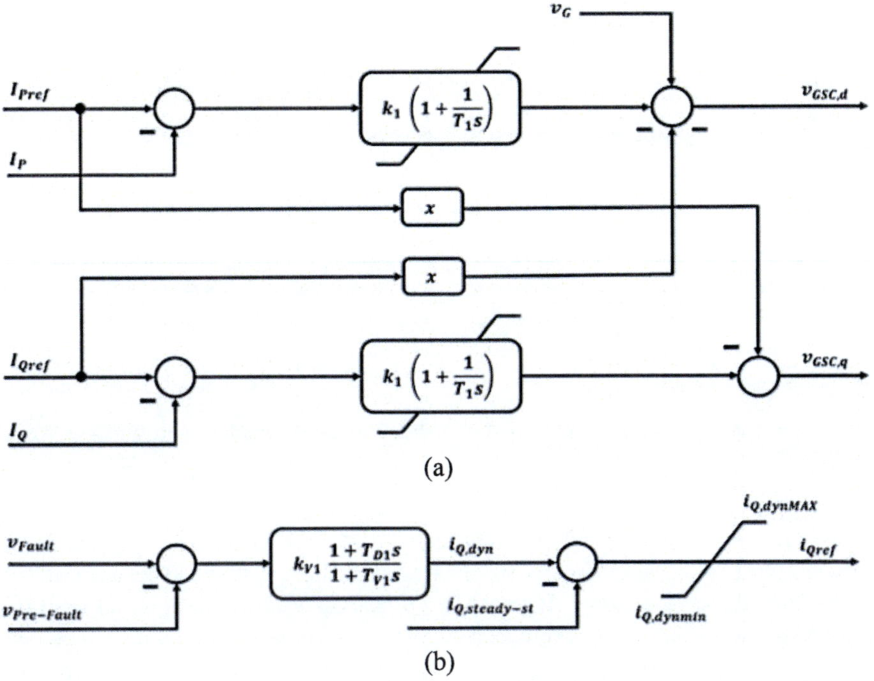

Fig. 13.3 Classical current control method with current setpoint reference [4, 7, 8]. (a) classical current control method (b) reactive current setpoint calculation

measurements, frequency, etc., which could be augmented when used large amounts of power-electronic devices, known for possessing a fast behaviour.

Due to their complexity, RGU is prone to multiple resonance frequencies of wide ranges, and with the upcoming increment of future large-scale RGU installations, transient phenomena are always a risk [4]. A potential consequence of such risk may be the blocking of the VSC stations, bringing a sudden change in the operating condition to an involuntary islanded mode, with very low-to-zero local load. During an Island, there would not be any frequency reference to follow. Without it, the accurate voltage correction required to be on duty in real-time is jeopardised, because Grid-Following systems assume that the current reference can be injected to the network at any time, as it cannot detect the sudden loss of load neither can act against it. This scheme is not equipped with the required voltage-frequency control capabilities, which, in contrast, is one of the main characteristics that, on the other hand, Grid-Forming systems do have [9].

The effect of a massive fault followed by Islanding can be seen in Fig. 13.4. The curves represent the performance of the seventeen WPP in the PST 16 Benchmark System after a massive fault followed by Islanding in *Area A* (presented in Fig. 13.1), in terms of their DC link voltage. During this simulation, all the WPP have classical grid-following current control systems. In *Area A* there are only three

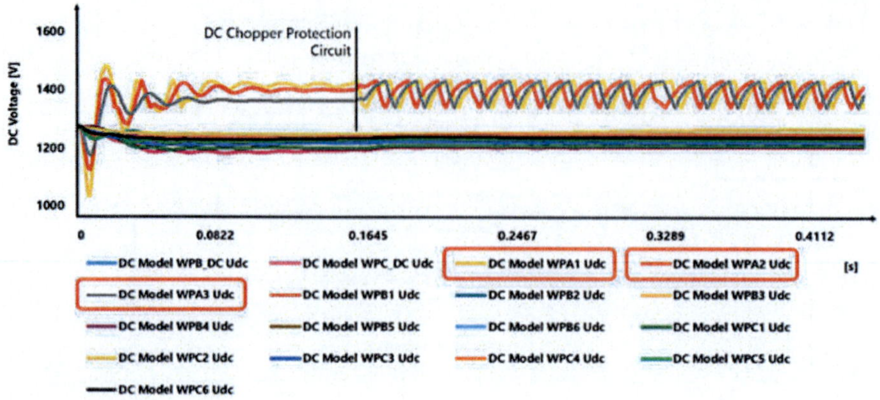

Fig. 13.4 DC voltage of the grid-following WF in the PST-16 benchmark system after a blackout in zone A [3]

WPP, where their behaviour stands out compared to the rest. The changing Δ voltage correction error is not large enough for the overall control system to sense and mitigate it. Thus, there is a sudden increase in voltage, caused by the integral term of the proportional-integral (PI) controller [4, 9]. The current reference i_{Pref} does not change despite the sudden fault, and the consequence is a big mismatch between the current reference and the actual current. After the blackout, the actual current i_{P} drops by nearly zero, but because the current reference value did not change, the integral term of the PI winds up, increasing the error and forcing the VSC converter to inject current regardless the fact that a loss-of-load scenario emerged [4, 9]. Meanwhile, surrounding electrical equipment like transformers, compensators, overhead lines, etc., reach their saturation threshold several times, presenting resonance frequencies [4]. During this scenario, the *d*-component of the current controller output reaches very rapidly the maximum permissible modulation level.

The chopper/capacitor default structure in WPP, as seen in Fig. 13.2, prevents the voltage from the further increase. This maximum modulation level and the high DC voltage would lead to the rise of the fundamental frequency voltage component, potentially exceeding the saturation boundary of the transformers [4].

13.5 Power Systems with Grid-Forming RGU/HVDC Presenting Black-Start and Restoration Capabilities

As explained in the previous section, the current injection control method, applied to grid-following RGU, cannot physically work in a network connected to the grid with a large share of inverters and/or little-to-zero load on its own. Accordingly, a new power-electronic converter control scheme is required, which does not presuppose the current injection. Instead, it mimics the operation of classical control systems that regulate CGU, which presents a Grid-Forming control characteristic [4]. Such a control system is capable of [9]:

- Regulate directly the voltage and reactive power directly of the RGU (instead of by means of current control).
- Operate the network during both grid-connected mode and islanded mode.
- Stabilise the converter dominated network frequency tightly around the grid nominal frequency.
- Limit the voltage rise following load shedding.

Still, on large-scale power systems scenarios with high penetration of power electronic converters, the grid-forming control systems for RGU alone are not enough to guarantee Black-Start and Restoration capabilities. Generation and load have to be perfectly matched in an unstable island. For that, the addition of storage units to be coupled with RGU can significantly improve the response of the grid-forming RGU during complicated situations. Besides, system topology and resource mix will govern any restoration process. Another vital factor is the way the re-energisation method of the disconnected passive elements of the grid is carried out. A careless implementation can lead to a negative impact, increasing the risk of a blackout relapse. Therefore, the remaining characteristics of the Benchmark System need to acquire are:

- Support for energy storage.
- Intelligent algorithms that re-energise and reconnect outaged elements of the grid without further damaging it.
- Tuning of CGU units for converter-interfaced systems environment.

13.5.1 RGU/HVDC Grid-Forming Control

Figure 13.5 illustrates two loops. Each has as output the terminal voltage of the grid-side converter in terms of the *d–q* reference frame, respectively. This scheme has been taken from the previous feedforward current injection control method shown in Fig. 13.3. From there, the integral term was removed, the proportional element was moved towards the output terminal of the controller, and a washout filter that serves as a virtual resistance for current damping was added. As long as the d–q components of the inverter voltage v_{VSC} are correctly known, the feedforward term works satisfactorily. Besides, the scheme contains a current limiter control, which forbids excessive current injection and prevents to surpass pre-set maximum allowable currents. The converter current is limited through the voltage thresholds $v_{\mathrm{GSCd,MAX}}$ and $v_{\mathrm{GSCq,MAX}}$. Figure 13.6 unveils the origin of the signals $v_{\mathrm{GSCd,MAX}}$ and $v_{\mathrm{GSCq,MAX}}$ along with a phasor/circle diagram that explains the method for current limitation in the converter. For limiting the real current, if

$$\left(\left|i_{\mathrm{d}} + j i_{\mathrm{q}}\right| - i_{\mathrm{MAX0\_ref}}\right) > 0 \rightarrow i_{\mathrm{MAX\_ref}} = i_{\mathrm{MAX0\_ref}} - k_{\mathrm{red}}\left(\left|i_{\mathrm{d}} + j i_{\mathrm{q}}\right| - i_{\mathrm{MAX0\_ref}}\right) \tag{13.3}$$

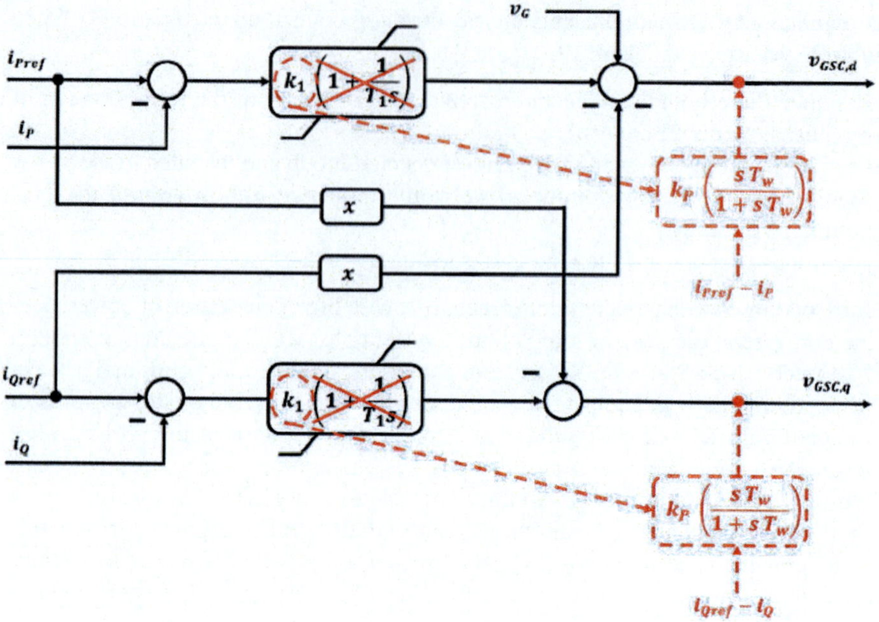

Fig. 13.5 Proposed grid-forming control method [4, 7, 8]

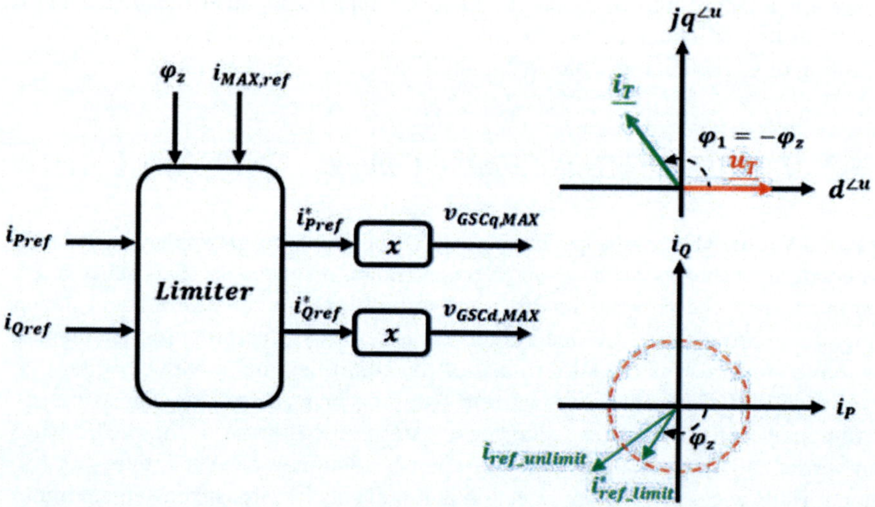

Fig. 13.6 Converter current limitation control [4, 7, 8]

Otherwise

$$i_{\text{MAX\_ref}} = i_{\text{MAX0\_ref}} \tag{13.4}$$

For limiting the current reference, in case is not bounded

$$v_{\text{GSCd,MAX}} = v_{\text{GSCq,MAX}} = x i_{\text{MAX,ref}} \tag{13.5}$$

Otherwise

$$v_{\text{GSCd,MAX}} = x i^{*}_{\text{Q,ref}};\ v_{\text{GSCq,MAX}} = x i^{*}_{\text{P,ref}} \tag{13.6}$$

When the real converter current exceeds the pre-set limit, a new maximum current value is computed. The new computed maximum current is used in the limiter control block (Fig. 13.6) with the grid angle impedance seen by the VSC. The current reference limits of the *d–q* VSC control voltages are limited based on the impedance of the grid and the new computed current value. Consequently, the converter adjusts the reference currents based on the impedance seen by it in real-time [4, 7]. The Grid-Forming control scheme is named as Voltage/Power/Frequency Control, and each branch is in charge of regulating the reactive power/voltage and active power/frequency combinations. The outputs of each control loop are the inverter voltage signals v_{GSCd} and v_{GSCq} [4, 7].

Figure 13.7 depicts the AC voltage/reactive power control loop. It works on a hierarchical process. An upper-level slow-acting controller tracks setpoint changes dictated by the system needs. The reactive power control has a droop characteristic for the definition of the reactive power reference q_{ref}, followed by a PI for reactive power control. This part is designed to stabilise the system when relatively lethargic transients take place, acting within a time frame of 5–30 s. With this scheme, there is no significant control action during network short circuit because the response time is large enough to avoid it yet is still small enough to avert unnecessary tap movement in on-load tap-changing (OLTC) transformers [4, 9]. Further, in case heavy voltage dips

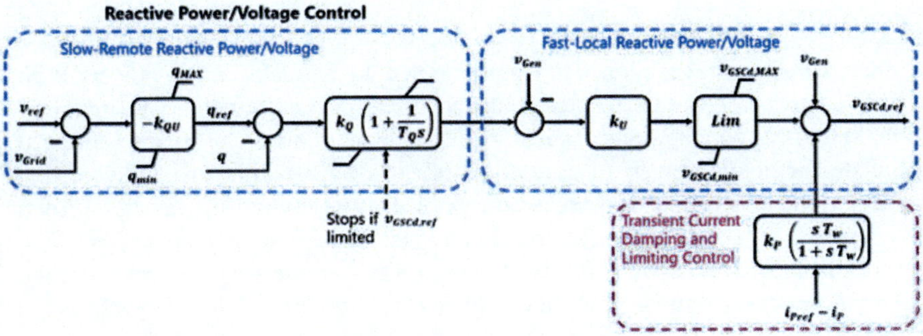

Fig. 13.7 Voltage/reactive power control loop [4, 7, 8]

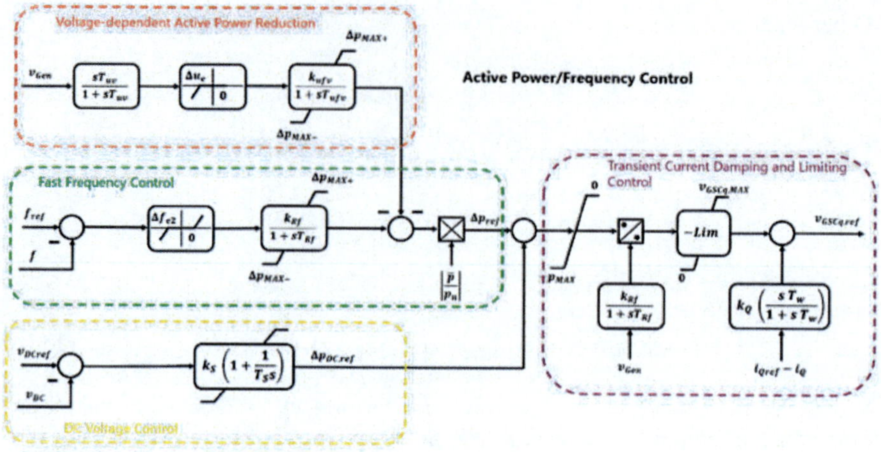

Fig. 13.8 Active power/frequency control loop [4, 7, 8]

are detected, the controller changes its operation mode, moving to the downstream part of this control loop. The fast-local regulator was designed to cope with sudden events demanding a faster response. It presents a proportional behaviour in steady-state, which can be increased with a lead-lag block to acquire a dynamically varying gain. There is no deadband in order to ensure continuous voltage control. This can be accompanied by a stabilising feedback signal, which is common practice in CGU AVR [4, 9].

Figure 13.8 depicts the active power–frequency control loop. The output of the DC voltage control provides a reference for the active power–frequency control. The output of the DC voltage is the active power injected to the network. Considering Eqs. (13.1) and (13.2), the active

power can be controlled by using the q-component of the converter voltage, which can also be used to limit excessive currents:

$$p = v_{\mathrm{Gen}} i_{\mathrm{P}} = -v_{\mathrm{Gen}} \frac{v_{\mathrm{DC,q}}}{x_{\mathrm{c}}} \tag{13.7}$$

Hence, direct voltage control is implemented to regulate active power. With Eq. (13.7) can be seen that reactive current injection requirement is no required in this theoretical concept, and instead, direct voltage control can be implemented to regulate the active power. In case of an Island or a blackout or any event that derives in transients or faults, there is no risk of integrator wind-up, because direct, reactive power/voltage/frequency control can make the dispatched current to adjust itself for the objective of grid stability in response to the changing converter voltage [8]. Therefore, the scheme is more like conventional voltage control in synchronous generators.

This channel also has additional branches that are active in case of contingencies [8]. A voltage-dependent active power reduction (VDAPR) loop performs power injection based on the real-time sensed terminal voltage of the grid-side converter. So, in case of a fault, the VDAPR loop prevents the converter from injecting current to the grid, then reducing the power setpoint reference of the VSC instantly [3]. Similarly, a fast frequency control loop mitigates both increasing/decreasing frequency deviations, the latter case, though, is controlled with further effort with a gradient limiter control block [4].

This developed control system can be implemented into the embedded controllers of the VSC that interface both the WPP and the HVDC links with the transmission network. They can smoothly transit between successive steady-state operating points and between steady-state and dynamic operating modes in contingency situations. The equations that exemplify the controllers for the VSC are adequate for EMT simulations involving small-time steps as integration to represent the high-switching frequency levels [10].

13.5.2 BESS-Storage Support

The addition of seven BESS units presented several advantages. It adjusts the power injected from the WPP to match with the load demand and thus regulate the system frequency. Besides, BESS can compensate for the slow response of the CGU governor systems very fast, in terms of seconds to a few minutes [11, 12]. Moreover, the results of the simulations concluded that BESS units were crucial because they provide the excitation voltage that all the generation units need to black-start [3].

Large battery-storage units are being increasingly used in Flexible AC Transmission Systems (FACTS) applications to improve the voltage, frequency, oscillatory and/or transient stability of the system, enhancing so the reliability of power supply [11]. Battery systems aid to not only provide start-up power but convert it, to allow a reset generator to achieve synchronisation, so, they can be a very useful tool to secure supply and demand balancing when there is a shortage of generation resources [12]. They can as well limit the steep output power of WPP by absorbing the excessive power injection themselves instead of the grid, thus avoiding imbalances in the network [3]. Storage systems also possess Fault-Ride Through Capability for ensuring that they will keep connected to the network during heavy voltage dips caused by heavy faults of the likes of loss of main generation resources or involuntary Islanded Operation. This means BESS can also be actively used for restoring a network after a blackout. Coupling a WPP with storage should be able to allow the wind power plant to act as a Black-Start Resource [11, 12]. The BESS can smooth the plant output, provided that the wind plant is never loaded near 100% [3].

As can be seen in Fig. 13.9, BESS is also interfaced via VSC to the transmission network. Yet, the control loops that compose the storage models are simpler compared to the ones defining the WPP/HVDC applications, because the main functions of the BESS are to act against frequency deviations and provide excitation voltage.

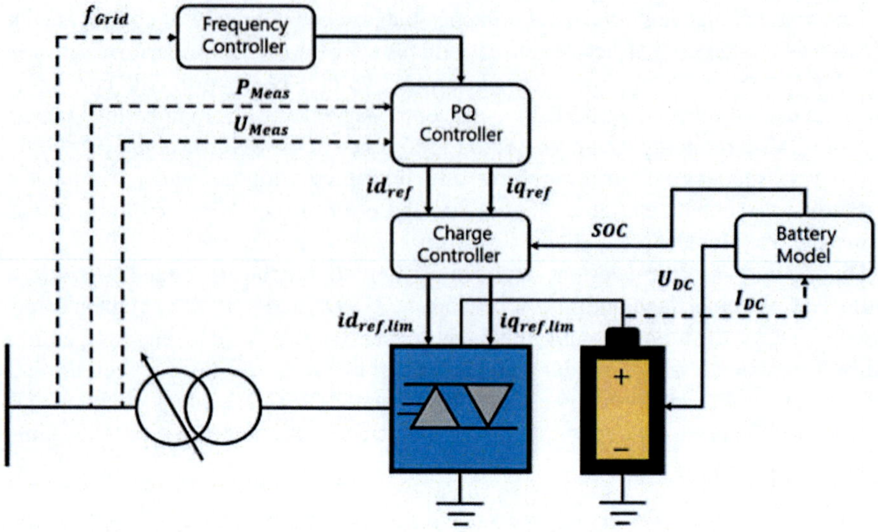

Fig. 13.9 BESS-storage scheme [3]

Therefore, a priority to control the behaviour of the VSC at a fundamental frequency level is enough, and so are the equations that represent them [10].

The BESS model has a simplified electrochemical battery model, a charge controller, a reactive power/voltage controller, and a frequency deviation control loop. The voltage level was modified and stepped-up from 0.9 to 9 kV by connecting more cells in series, according to the electrochemical battery parameters [11]. The frequency deviation controller also presents changes, an additional selector block to prevent the battery from discharge during unnecessary circumstances was added, setting a threshold of ±0.5 Hz. The VSC-based controller administrates the current dispatch with respect of its *d*-*q* components, which could be seen as controlling the BESS in terms of active and reactive power [10, 11]. With the active power output, it is possible to control the frequency of the grid. Similarly, with the reactive power, the voltage can be controlled. The reactive power output of the BESS is not bound to the battery capacity, so the AC voltage could be controlled continuously.

For the success of the operation of the BESS, some boundary conditions have to be imposed [10]:

- The storage systems only can consume active power if their battery is not fully loaded (state of charge (SOC) < 1).
- The storage systems can always supply active power unless its battery is fully depleted (SOC > 1).
- The battery should be automatically recharged if its SOC has reached a certain lower limit. With this characteristic, the BESS has always the ability to control the active power in both ways.

- The total output of active and reactive power must not be greater than the complex rated power. Therefore, a priority is needed for active or reactive power.

13.5.3 Protection and Restoration Algorithm

Any large-scale power system to have restoration capabilities must count with protection systems with the extra capability to reconnect and re-energise elements that were previously shut down due to a failure or a prior protection scheme action as quickly, reliably and safely as possible while minimising the stress on network components [13].

For that, a DSL Protection & Restoration Algorithm was designed and incorporated in some of the elements that have a major role in the restoration simulation. In case there is a fault that could bring the outcome of a blackout, the protection systems detect faults via measurement instruments found in the DIgSILENT PowerFactory default metering stations. When identified, the logic acts with a set of various time delays. Given the size and seriousness of the faults, the model can be arranged to play out a controlled isolation scheme, shutting down parts of the grid in order to preserve the integrity of most of it. For example, if a massive fault was located in *Area A*, in Fig. 13.1, one of the primary prompt activities is the obstructing of the HVDC converter stations and opening the contacts of their AC-side circuit breakers to keep away from the fault widespread into the rest of the grid. The protection of all the power elements in that area was also protected by the same protection scheme, disengaging all generation units, loads, transmission lines and transformers. Once a power outage has occurred, the restoration plans must begin.

There are three main stages that compose this method [3, 5]:

1. **Preparation**. This phase can be summarised as an optimisation problem concerning pathfinding. It helps to calculate a realistic time and avoid overly optimistic/pessimistic estimations of the probability of completion. The error can be eliminated by averaging many simulated trials of the near-critical paths. It is based on a system that limits the blackout span by treating the term of every movement as a variable, and for every action, it is chosen three conceivable time lengths, an optimistic time x, a most-likely time y and a pessimistic time z [5]. Assuming a β distribution for considering the probability of the possible duration of the activities, and with a standard deviation σ that calculates the $\pm$bias of such probability, the expected time for each activity is calculated with Formulas (13.8) and (13.9) [5, 14].

$$t_c = \frac{(x + 4y + x)}{6} \tag{13.8}$$

$$\sigma = \frac{(y - x)}{6} \tag{13.9}$$

The expected total length of the critical path is the accumulated activities from (13.8) and (13.9), computed with Formulas (13.10) and (13.11)

$$T_c = \sum t_c \tag{13.10}$$

$$\sigma_c = \left(\sum \sigma^2\right)^{\frac{1}{2}} \tag{13.11}$$

2. **Black-Start**. In the second stage, only necessary elements are reconnected to be picked up by the first generators restored and synchronised. The priority at this stage is the reintegration of some generation units and network backbone path. This is generally accomplished by cranking up elements in slow increments that can be well met with the inertia or frequency response of the restored units [14]. Lower-voltage transmission lines are energised first as very high-voltage lines need a ton of reactive power, which is a scarce/critical resource. Without it, transmission lines cannot be picked up, because when they are started to be served, they are open-ended and act as large capacitors due to the Ferranti Effect, which make voltage management hard [3, 14]. At this stage, the network is very weak. Lack of enough inertia and electro-magnetic issues related to reconnection of heavy elements/generation units cause transient voltages and deep frequency variations which may delay the whole process. Further, the protection relays that are reconnected and can always trip again. Therefore, it is very important to follow the structured strategy calculated in the previous stage [14].
3. **Restoration**. The third stage minds the re-energisation of the unserved loads. It is common that the integration of the loads is carried one by one. This makes the task very time-consuming, but otherwise reconnecting several elements at once (or restoring them in large increments) is risky because could trigger severe transients and abnormal frequency that might make the system re-incite in an outage. The common practice is that feeders equipped with frequency relays are cranked up at this phase when the system frequency tends back to stability again [3, 14]. As this stage evolves later, however, the grid is more robust as now more generation units and meshed transmission paths are engaged in the process [14]. The scheduling of the load cranking and the permissible load size to be picked up will depend on the response of the rate capabilities of the generators that were previously getting online. If the whole process has reached this point, then it is allowed to restore loads at higher increments [14].

For the last two stages, an approach known as Soft-Energisation Method is applied in the model [13]. Some of the benefits of this method are enlisted in Table 13.2. This method connects first overhead lines, cables, transformers and loads in that hierarchical order in a dead-busbar configuration, via their substations and to a power plant without activated voltage excitation [13]. Once some of the critical paths have been identified, the black-start generators start increasing their excitation voltage slowly in their terminals, from 0 kV up to their nominal values. This controlled approach of voltage ramp-up averts transformer saturation, and as a result, all the

Table 13.2 Benefits and considerations of the soft-energisation method [13]

Benefits of soft-energisation method	Considertions of soft-energisation method
Simultaneous and gradual energisation	Under-voltage relays must be blocked
Transient oscillations are avoided	Dead-busbar configuration must me allowed with circuit breakers
Stress on network components reduced	BESS-storage units required
Applicable to any grid regardless to topology, resource mix, etc.,	Review of generator and network protection settings
Reliable, safe and fully controlled	Special operation and training needed

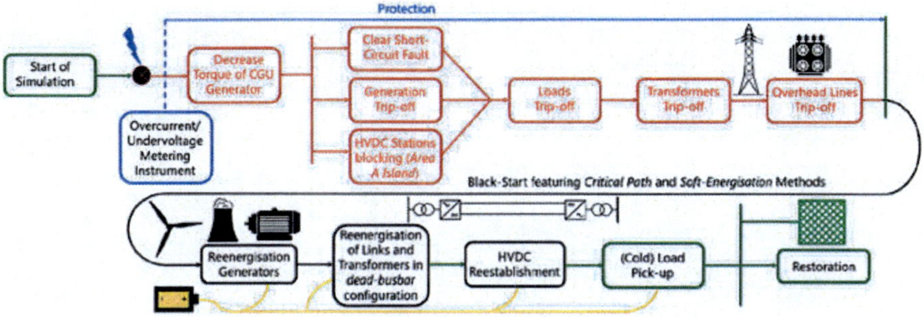

Fig. 13.10 Flowchart describing the protection and restoration scheme [3]

elements in the outaged grid are energised gradually and simultaneously without any transient oscillation, lessen any stress on the components in the network and substation assets [13].

By the time the selected critical paths are reconnected, the BESS-storage units provide excitation voltage to the large-scale power plants, therefore black-starting them gradually. An illustrative Flowchart describing Measurement, Protection and Restoration DSL Model is displayed in Fig. 13.10. The Protection & Restoration Algorithm is shown in DSL-code Eq. 13.2, in Sect. 13.6.1.

13.6 DSL Implementation in DIgSILENT PowerFactory

The complete implementation in DIgSILENT PowerFactory is composed of the following elements:

- Seventeen WPP simplified models based on PMSG WT with Protection & Restoration Algorithms.
- Two VSC-HVDC Transmission Systems with Protection & Restoration Algorithms.
- Seven BESS.

- Two Synchronous Generators, one hydro and one coal CGU with Protection & Restoration Algorithms.

13.6.1 Wind Power Plants (WPP)

The model of the seventeen WPP as implemented in the PST-17 Benchmark system is unveiled in Fig. 13.11, with the composite model that contains the following slots (Fig. 13.11a), from right to left and from top to bottom: (1) Wind Turbine Generator Farm (ElmGenstat), (2) Modulation Limitation (ElmDsl*), (3) Protection

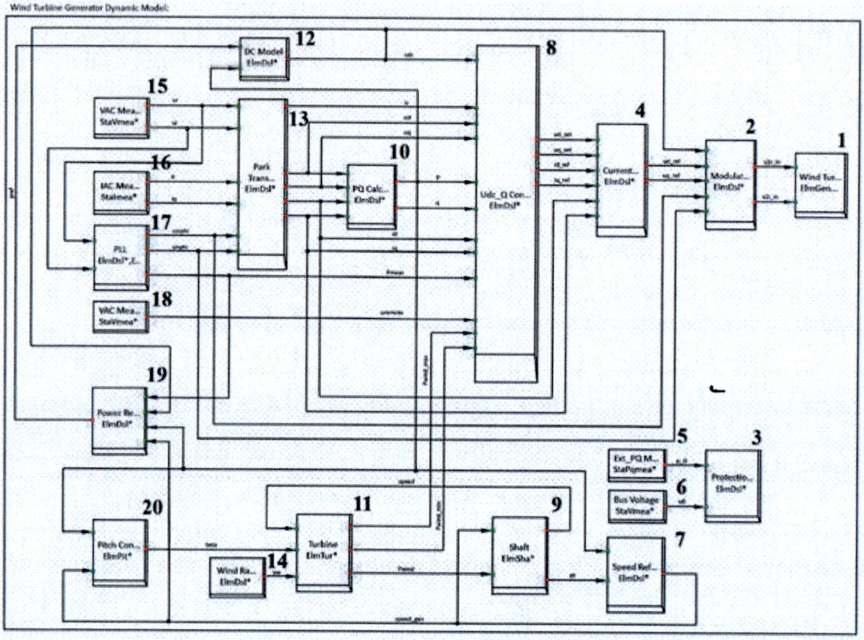

(a)

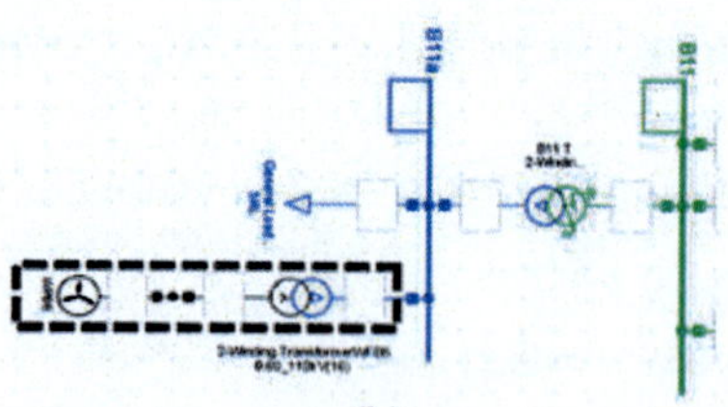

(b)

Fig. 13.11 WPP model in DIgSILENT PowerFactory [3]. (a) composite frame structure of the WPP in DIgSILENT PowerFactory (b) static generator (ElmGenstat*) depicted as a scaled WPP

Relay (ElmDsl*), (4) Current Damping Terms (ElmDsl*), (5) External_PQ Measurement (StaPqmea*), (6) Bus Voltage (StaVmea*), (7) Speed Reference (ElmDsl*), (8) Udc_Q Controller (ElmDsl*), (9) Shaft (ElmSha*), (10) PQ-Calculation (ElmDsl*), (11) Turbine Aerodynamics (ElmTur*), (12) DC Model (ElmDsl*), (13) Park Transformation (ElmDsl*), (14) Wind Ramp (ElmDsl*), (15) VAC Measurement (StaVmea*), (16) IAC Measurement (StaImea*), (17) PLL (ElmDsl*, ElmPhi*); (18) VAC Measurement Remote (StaVmea*), (19) Power Reference (ElmDsl*) and (20) Pitch Control (ElmPit*).

All the WPP contain an External Station Controller that forces the WPP to dispatch specific values of reactive power for ensuring a balanced load flow, according to the network characteristics. Each WPP also comes with a two-quadrant reactive power Capability Curve that defines their capacity limits at 1.1 per unit (p. u.), as seen in Table 13.3a.

They all have a matrix wind/power array that permits to perform dynamic simulation under different wind conditions. It is assumed that the WPP perceive wind speeds in the range of 12–17 m/s, and the speed ranges can be modified as well (Table 13.4).

Correct initialisation of a model in a power-system simulation-tool avoids fictitious electrical transients and makes it possible to evaluate correctly the real dynamic performance of the system. Therefore, the DSL-code equations of the most relevant dynamic block of these composite models for this design are presented below [7]:

DSL-code Eq. 13.1: *Modulation Limitation.*

```
inc(ud_ref)=(ua*cos(phi)+ub*sin(phi))/(Udc*sqrt(3)/(2*sqrt(2)*UacN))
inc(uq_ref)=(-ua*sin(phi)+ub*cos(phi))/(Udc*sqrt(3)/(2*sqrt(2)*UacN))
inc(xua)=ua
inc(xub)=ub
inc(uaa)=ua
inc(ubb)=ub

vardef(Ucmax)='p.u.';'Maximum Modulation Index of Converter'
vardef(UacN)='V';'Nominal AC Voltage of Converter'
```

Table 13.3 Capability curves [3]. (a) two-quadrant capability curve for all the WPP (b) four-quadrant capability curve for all the HVDC stations

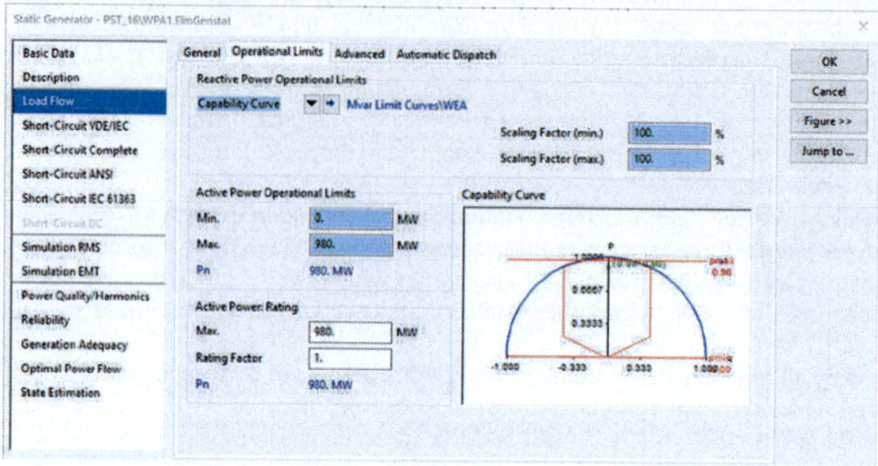

(a)

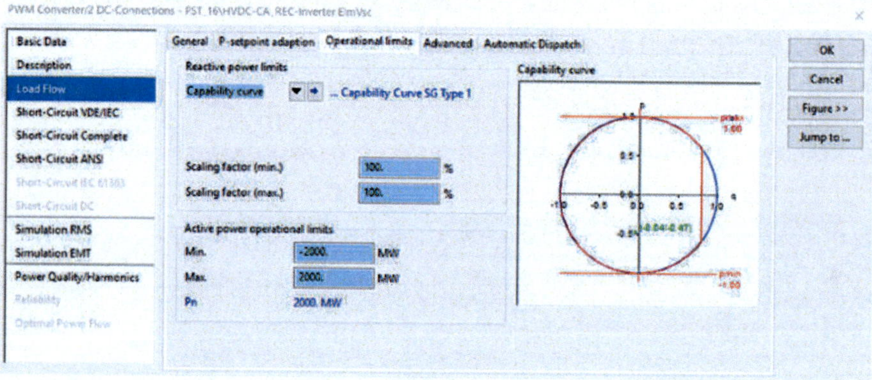

(b)

DSL-code Eq. 13.2: *Protection & Restoration Relay.*

Table 13.4 66*49 wind power matrix array [3]

Common Model - PST_16\PMSG_Wind Farm_WPA1\Turbine WPA1.ElmDsl

Basic Data | Description

General | Advanced 1 | Advanced 2 | Advanced 3

OK | Cancel | Events

Two Dimensional Characteristics:

	Axis	i1	i2	i3	i4	i5
Size	66.	49.	0.	0.	0.	0.
Axis	0.	0.8	1.2	1.6	2.	2.4
1	-2.	0.0022211	0.0039615	0.00723615	0.02064635	0.04816785
2	-1.5	0.00248425	0.00446405	0.0083182	0.0233966	0.05227565
3	-1.	0.00274645	0.00496185	0.0094601	0.02620765	0.0563407
4	-0.5	0.0030077	0.0054549	0.01063525	0.0289978	0.06027845
5	0.	0.00326515	0.0059413	0.0118826	0.03175185	0.0641763
6	0.5	0.0035226	0.00642865	0.01317745	0.03459235	0.0679155
7	1.	0.0037791	0.0069122	0.0145369	0.03737205	0.0714837
8	1.5	0.0040299	0.00738815	0.0159638	0.0401356	0.07489515
9	2.	0.00428165	0.00786885	0.01745055	0.0427975	0.0781432
10	2.5	0.00453245	0.0083524	0.0189373	0.04548695	0.0812782
11	3.	0.0047785	0.00882645	0.0205333	0.04813365	0.0843752
12	3.5	0.00502455	0.00931	0.02216255	0.0506977	0.08727175
13	4.	0.00526965	0.00980305	0.0237861	0.05314775	0.09013695
14	4.5	0.0055119	0.01029515	0.025422	0.0555427	0.09292425
15	5.	0.00575225	0.0108357	0.02714625	0.0578797	0.0956422
16	5.5	0.0059907	0.01137815	0.02884295	0.0601008	0.09817585
17	6.	[illegible]	[illegible]	[illegible]	[illegible]	[illegible]

```
vardef(umin)='p.u.';'Bus Voltage Threshold'
vardef(pmax)='p.u.';'Permissible Max Active Power'
vardef(pmin)='p.u.';'Permissible Min Active Power'
vardef(TSCC)='s';'Short-Circuit Clearing Delay'
vardef(TMtd)='s';'Generator Torque Decrease Delay'
vardef(TSwtG)='s';'Switch-off Generator Delay'
vardef(TIslA)='s';'Area A Isolation Delay'
vardef(TSwtLA)='s';'Trip A Loads Delay'
vardef(TSwtEA)='s';'Trip A Elements Delay'
vardef(Trec1)='s';'Time of Restoration Stage (1)'
vardef(Trec2)='s';'Time of Restoration Stage (2)'
vardef(Trec3)='s';'Time of Restoration Stage (3)'
vardef(Trec4)='s';'Time of Restoration Stage (4)'
vardef(Trec5)='s';'Time of Restoration Stage (5)'
vardef(Trec6)='s';'Time of Restoration Stage (6)'
vardef(Trec7)='s';'Time of Restoration Stage (7)'
vardef(Trec8)='s';'Time of Restoration Stage (8)'
vardef(Trec9)='s';'Time of Restoration Stage (9)'
vardef(Trec10)='s';'Time of Restoration Stage (10)'
vardef(Trec11)='s';'Time of Restoration Stage (11)'
vardef(Trec12)='s';'Time of Restoration Stage (12)'
vardef(Trec13)='s';'Time of Restoration Stage (13)'
vardef(Trec14)='s';'Time of Restoration Stage (14)'
vardef(Trec15)='s';'Time of Restoration Stage (15)'
vardef(Trec16)='s';'Time of Restoration Stage (16)'
vardef(Trec17)='s';'Time of Restoration Stage (17)'
vardef(Trec18)='s';'Time of Restoration Stage (18)'
vardef(Trec19)='s';'Time of Restoration Stage (19)'
vardef(Trec20)='s';'Time of Restoration Stage (20)'
vardef(Trec21)='s';'Time of Restoration Stage (21)'
vardef(Trec22)='s';'Time of Restoration Stage (22)'
vardef(Trec23)='s';'Time of Restoration Stage (23)'
vardef(Trec24)='s';'Time of Restoration Stage (24)'

inc(ptmin)=0
ptmin=select(time()>9,1,0)

inc(surgp)=0
surgp=p>pmax.or.p<pmin

inc(SCC)=0
SCC=picdro(u<umin,TSCC,0)  !Short-Circuit Clearing

surgisl=time()>90.56,surgp
!90.56 = Protection Action Time + TIsLA (0.15) + 0.01

surgld=time()>90.61,surgp
!90.61 = Protection Action Time + TSwtLA (0.2) + 0.01

surgelm=time()>90.71,surgp
!90.71 = Protection Action Time + TSwtEA (0.3) + 0.01

recstt=time()>135.01,surgp
!135.01 = Protection Action Time + Trec2 (75) + 0.01

recsyg=time()>165.01,surgp
!165.01 = Protection Action Time + Trec2 (75) + 0.01

recwp3=time()>330.01,surgp
!330.01 = Protection Action Time + Trec13 (240) + 0.01

rechvdc=time()>345.01,surgp
!345.01= Protection Action Time + Trec14 (255) + 0.01

pul=time()>435.01,surgp
!435.01 = Protection Action Time + Trec20 (345) + 0.01

recend=time()>495.01,surgp
!495.01 = Protection Action Time + Trec24 (405) + 0.01


!FAULT STAGE (1)
!------------------------------------------------------------------------------
!EVENTS 1.1: CLEAR SHORT CIRCUIT FAULT

!Clear Short-Circuit Fault in Bus A1 (Area A)
event( 1, SCC, 'name=A1_SC_Clear target=A1')
!Clear Short-Circuit Fault in Bus A2 (Area A)
event( 1, SCC, 'name=A2_SC_Clear target=A2')
!Clear Short-Circuit Fault in Bus A7a (Area A)
event( 1, SCC, 'name=A7a_SC_Clear target=A7a')
output(surgp,'SHORT-CIRCUIT FAULT in Area A. Four faults detected in this Zone. Risk of MASSIVE BLACKOUT')

!EVENTS 1.2: LOSS OF GENERATION A
```

```
!Decrease Minimum Mech. Power A1aG Hydro SyM (Area A)
event(1, surgp, 'name=Gov_A1aG_PtminDwn dtime=TMtd')
!Decrease Mech. Torque A1aG Hydro SyM (Area A)
event(1, surgp, 'name=MTorq0_A1aG dtime=TMtd')
!Disconnect A1aG Hydro SyM (Area A)
event(1, surgp, 'name=SyM_A1aG_Out dtime=TSwtG')
!Decrease Mech. Torque PMSG WPA1 (Area A)
event(1, surgp, 'name=MTorq0_WPA1 dtime=TMtd')
!Disconnect PMSG WPA1 (Area A)
event(1, surgp, 'name=SyM_WPA1_Out dtime=TSwtG')
!Decrease Mech. Torque PMSG WPA2 (Area A)
event(1, surgp, 'name=MTorq0_WPA2 dtime=TMtd')
!Disconnect PMSG WPA2 (Area A)
event(1, surgp, 'name=SyM_WPA2_Out dtime=TSwtG')
!Decrease Mech. Torque PMSG WPA3 (Area A)
event(1, surgp, 'name=MTorq0_WPA3 dtime=TMtd')
!Disconnect PMSG WPA3 (Area A)
event(1, surgp, 'name=SyM_WPA3_Out dtime=TSwtG')
output(SCC,'Remedial Protection Scheme Started: Short-Circuit Clearded. AREA-A SHEDDING in order to preserve the integrity of the grid. A
BLACK-START and NETWORK RESTORATION will be executed at the proper time.')

!EVENTS 1.3: AREA A ISLAND
!Disconnect HVDC AB Circuit Breaker
event(1,surgp,'name=HVDC_AB_Out dtime=TIslA')
!Disconnect HVDC BA Circuit Breaker
event(1,surgp,'name=HVDC_BA_Out dtime=TIslA')
!Disconnect HVDC AC Circuit Breaker
event(1,surgp,'name=HVDC_AC_Out dtime=TIslA')
!Disconnect HVDC CA Circuit Breaker
event(1,surgp,'name=HVDC_CA_Out dtime=TIslA')
output(surgisl,'Area A is now ISOLATED from the rest of the Network')

!EVENTS 1.4: LOSS OF LOADS A
!Disconnect General Load (Area A)
event(1,surgp,'name=GL_Out dtime=TSwtLA')
!Disconnect General Load 1 (Area A)
event(1,surgp,'name=GL_Out(1) dtime=TSwtLA')
!Disconnect General Load 2 (Area A)
event(1,surgp,'name=GL_Out(2) dtime=TSwtLA')
!Disconnect General Load 3 (Area A)
event(1,surgp,'name=GL_Out(3) dtime=TSwtLA')
!Disconnect General Load 4 (Area A)
event(1,surgp,'name=GL_Out(4) dtime=TSwtLA')
!Disconnect General Load 5 (Area A)

event(1,surgp,'name=GL_Out(5) dtime=TSwtLA')
!Disconnect General Load 6 (Area A)
event(1,surgp,'name=GL_Out(6) dtime=TSwtLA')
!Disconnect General Load 7 (Area A)
event(1,surgp,'name=GL_Out(7) dtime=TSwtLA')
!Disconnect General Load 8 (Area A)
event(1,surgp,'name=GL_Out(8) dtime=TSwtLA')
output(surgld,'Disconnection of Loads in Area A')

!EVENTS 1.5: LOSS OF DEVICES A
!Disconnect M_A1a_2T Trafo (Area A)
event(1,surgp,'name=M_A1a_2T_Out dtime=TSwtEA')
!Disconnect 110_kV_A_TF1 Trafo (Area A)
event(1,surgp,'name=A_TF1_Out dtime=TSwtEA')
!Disconnect 110_kV_A_TF2 Trafo (Area A)
event(1,surgp,'name=A_TF2_Out dtime=TSwtEA')
!Disconnect 110_kV_A_TF3 Trafo (Area A)
event(1,surgp,'name=A_TF3_Out dtime=TSwtEA')
!Disconnect 110_kV_A_TF4 Trafo (Area A)
event(1,surgp,'name=A_TF4_Out dtime=TSwtEA')
!Disconnect 110_kV_A_TF5 Trafo (Area A)
event(1,surgp,'name=A_TF5_Out dtime=TSwtEA')
!Disconnect 110_kV_A_TF6 Trafo (Area A)
event(1,surgp,'name=A_TF6_Out dtime=TSwtEA')
!Disconnect A1-A2 Line (Area A)
event(1,surgp,'name=A1-A2_Out dtime=TTSwtEA')
!Disconnect A1-A4 Line (Area A)
event(1,surgp,'name=A1-A4_Out dtime=TSwtEA')
!Disconnect A2-A3 Line (Area A)
event(1,surgp,'name=A2-A3_Out dtime=TSwtEA')
!Disconnect A2-A5a Line (Area A)
event(1,surgp,'name=A2-A5a_Out dtime=TSwtEA')
!Disconnect A2-A5b Line (Area A)
event(1,surgp,'name=A2-A5b_Out dtime=TSwtEA')
!Disconnect A4-A5a Line (Area A)
event(1,surgp,'name=A4-A5a_Out dtime=TSwtEA')
!Disconnect A4T Trafo (Area A)
event(1,surgp,'name=A4T_Out dtime=TSwtEA')
!Disconnect A5a-A5b Line (Area A)
event(1,surgp,'name=A5a-A5b_Out dtime=TSwtEA')
!Disconnect A5a-A7a Line (Area A)
event(1,surgp,'name=A5a-A7a_Out dtime=TSwtEA')
```

```
!Disconnect A5b-A7a Line (Area A)
event(1,surgp,'name=A5b-A7a_Out dtime=TSwtEA')
!Disconnect A6T Trafo (Area A)
event(1,surgp,'name=A6T_Out dtime=TSwtEA')
!Disconnect A6-A7 Line (Area A)
event(1,surgp,'name=A6-A7_Out dtime=TSwtEA')
!Disconnect A7T Trafo (Area A)
event(1,surgp,'name=A7T_Out dtime=TSwtEA')
output(surgelm,'Disconnection of all elements in Area A. Now Area A is FULLY OUTAGED.')

!RESTORATION STAGE (2)
!------------------------------------------------------------------------------
!EVENTS 2.1
!Reconnect A1aG Hydro SyM (Area A)
event(1,surgp, 'name=SyM_A1aG_Rec dtime=Trec1')

!EVENTS 2.2
!Reconnect M_A1a_2T Trafo (Area A)
event(1,surgp,'name=M_A1a_2T_In dtime=Trec2')
output(recsyg,'Hydro Synchronous Generator A1aG Reconnected.')
output(recstt,'NETWORK RESTORATION STAGE BEGINS.')

!EVENTS 2.3
!Reconnect A1-A2 Line (Area A)
event(1,surgp,'name=A1-A2_In dtime=Trec3')
!Reconnect 110_kV_A_TF2 Trafo (Area A)
event(1,surgp,'name=A_TF2_In dtime=Trec3')

!EVENTS 2.4
!Reconnect PMSG WPA1 (Area A)
event(1,surgp, 'name=SyM_WPA1_Rec dtime=Trec4')

!EVENTS 2.5
!Reconnect A2-A3 Line (Area A)

event(1,surgp,'name=A2-A3_In dtime=Trec5')
!Reconnect 110_kV_A_TF3 Trafo (Area A)
event(1,surgp,'name=A_TF3_In dtime=Trec5')

!EVENTS 2.6
!Reconnect PMSG WPA2 (Area A)
event(1,surgp, 'name=SyM_WPA2_Rec dtime=Trec6')

!EVENTS 2.7
!Reconnect A1-A4 Line (Area A)
event(1,surgp,'name=A1-A4_In dtime=Trec7')
!Reconnect A2-A5a Line (Area A)
event(1,surgp,'name=A2-A5a_In dtime=Trec7')
!Reconnect A2-A5b Line (Area A)
event(1,surgp,'name=A2-A5b_In dtime=Trec7')

!EVENTS 2.8
!Reconnect A4-A5a Line (Area A)
event(1,surgp,'name=A4-A5a_In dtime=Trec8')
!Reconnect A5a-A5b Line (Area A)
event(1,surgp,'name=A5a-A5b_In dtime=Trec8')

!EVENTS 2.9
!Reconnect A5a-A7a Line (Area A)
event(1,surgp,'name=A5a-A7a_In dtime=Trec9')
!Reconnect A5b-A7a Line (Area A)
event(1,surgp,'name=A5b-A7a_In dtime=Trec9')

!EVENTS 2.10
!Reconnect A7T Trafo (Area A)
event(1,surgp,'name=A7T_In dtime=Trec10')

!EVENTS 2.11
!Reconnect A6-A7 Line (Area A)
event(1,surgp,'name=A6-A7_In dtime=Trec11')

!EVENTS 2.12
!Reconnect A6T Trafo (Area A)
event(1,surgp,'name=A6T_In dtime=Trec12')

!EVENTS 2.13
!Reconnect PMSG WPA3 (Area A)
event(1,surgp, 'name=SyM_WPA3_Rec dtime=Trec13')
output(recwp3,'Wind Farms of Zone A Reconnected.')

!EVENTS 2.14
!Reconnect HVDC AB Circuit Breaker
event(1,surgp,'name=HVDC_AB_In dtime=Trec14')
!Reconnect HVDC BA Circuit Breaker
event(1,surgp,'name=HVDC_BA_In dtime=Trec14')
!Reconnect HVDC AC Circuit Breaker
event(1,surgp,'name=HVDC_AC_In dtime=Trec14')
```

```
!Reconnect HVDC CA Circuit Breaker
event(1,surgp,'name=HVDC_CA_In dtime=Trec14')
output(rechvdc,'HVDC Transmission lines are now Reconnected.')

!EVENTS 2.15
!Increase Mech. Torque A1aG Hydro SyM (Area A)
event(1, surgp, 'name=MTorqUp_A1aG dtime=Trec15')
!Increase Minimum Mech. Power A1aG Hydro SyM (Area A)
event(1, surgp, 'name=Gov_A1aG_PtminUp1 dtime=Trec15')
!Increase Mech. Torque PMSG WPA1 (Area A)
event(1, surgp, 'name=MTorqUp_WPA1 dtime=Trec15')
!Increase Mech. Torque PMSG WPA2 (Area A)
event(1, surgp, 'name=MTorqUp_WPA2 dtime=Trec15')
!Increase Mech. Torque PMSG WPA3 (Area A)
event(1, surgp, 'name=MTorqUp_WPA3 dtime=Trec15')

!EVENTS 2.16
!Reconnect 110_kV_A_TF1 Trafo (Area A)
event(1,surgp,'name=A_TF1_In dtime=Trec16')

!EVENTS 2.17
!Reconnect 110_kV_A_TF4 Trafo (Area A)
event(1,surgp,'name=A_TF4_In dtime=Trec17')
!Reconnect 110_kV_A_TF5 Trafo (Area A)

event(1,surgp,'name=A_TF5_In dtime=Trec17')

!EVENTS 2.18
!Reconnect A4T Trafo (Area A)
event(1,surgp,'name=A4T_In dtime=Trec18')

!EVENTS 2.19
!Reconnect 110_kV_A_TF6 Trafo (Area A)
event(1,surgp,'name=A_TF6_In dtime=Trec19')

!EVENTS 2.20
!Connect Schalter-Ereignis Load (Area A)
event(1,surgp,'name=Schalter-Ereignis dtime=Trec20')
output(pul,'Initialisation of Load Pick-Up Process.')

!EVENTS 2.21
!Reconnect General Load 1 (Area A)
event(1,surgp,'name=GL_In(1) dtime=Trec21')
!Reconnect General Load 4 (Area A)
event(1,surgp,'name=GL_In(4) dtime=Trec21')
!Increase Minimum Mech. Power A1aG Hydro SyM (Area A)
event(1, surgp, 'name=Gov_A1aG_PtminUp2 dtime=Trec21')

!EVENTS 2.22
!Reconnect General Load (Area A)
event(1,surgp,'name=GL_In dtime=Trec22')
!Reconnect General Load 3 (Area A)
event(1,surgp,'name=GL_In(3) dtime=Trec22')
!Reconnect General Load 8 (Area A)
event(1,surgp,'name=GL_In(8) dtime=Trec22')

!EVENTS 2.23
!Reconnect General Load 2 (Area A)
event(1,surgp,'name=GL_In(2) dtime=Trec23')
!Reconnect General Load 7 (Area A)
event(1,surgp,'name=GL_In(7) dtime=Trec23')

!EVENTS 2.24
!Reconnect General Load 5 (Area A)
event(1,surgp,'name=GL_In(5) dtime=Trec24')
!Reconnect General Load 6 (Area A)
event(1,surgp,'name=GL_In(6) dtime=Trec24')
output(recend,'NETWORK RESTORATION IN ZONE A COMPLETED.')
```

DSL-code Eq. 13.3: *Current Damping Terms.*

```
incfix(id_ref)=id
incfix(iq_ref)=iq
inc(ud_set)=(ud_ref+ud_washout)
inc(xdw)=dId
inc(xqw)=dIq
inc(uq_set)=(uq_ref+uq_washout)

vardef(Tw)='s';'Washout Filter Time Constant'
vardef(Kw)='-';'Washout Filter Propotional Constant'
vardef(ud_min)='p.u.';'Minimum Modulation Index d-Channel'
vardef(uq_lim)='p.u.';'Modulation Index Limitation of q- Channel'
vardef(ud_max)='p.u.';'Maximum Modulation Index d-Channel'
```

DSL-code Eq. 13.4: *Speed Reference DSL-Code Equations.*

```
xp.=(pt-(-1*p))/Td
speed_gen=lim(xp,0,1.2)
inc(xp)=1
inc(speed_gen) =xp
inc(p)=-1
inc(pt)=0

vardef(Td)='p.u.';'Damping Torque Coefficient Based on Power'
```

DSL-code Eq. 13.5: *Udc_Q Controller DSL-Code Equation.*

```
incfix(uref)=-q/Kqu+uremote
inc(xdc)= p/u
incfix(udc_ref)=udc/(UdcN*1e3)
inc(xuVDAPR)=0
inc(xud)=u
inc(x)=u
inc(xf)=p_fdapr
incfix(f0)=Fmeas
inc(xp_avg)=p
inc(xvar_remote) = 1+ u*(1-Ku)/Ku - ud_set/Ku

vardef(Kq_remote)='-';'Remote Var Control P. Constant'
vardef(Tq_remote)='s';'Remote Var Control Integral Time Constant'
vardef(Kdc)='-';'DC Link Control P. Constant'
vardef(Tdc)='s';'DC Link Control Integral Time Constant'
vardef(Kqu)='-';'Static Gain for the Reactive Power Control'
vardef(Ku)='-';'Fast Voltage Control Gain'
vardef(T_rate_limit_freq)='s';'Frequency Control Delay'
vardef(l)='p.u.';'Converter Reactance'
vardef(Tud)='s';'Delay for the AC Voltage Measurement'
vardef(T)='s';'Delay for the VDAPR'
vardef(imax)='p.u.';'Maximum Converter Current'
vardef(db_voltage_VDACR)='p.u.';'Deadband for the VDAPR'
vardef(KuVDAPR)='-';'Gain for the VDAPR'
vardef(TuVDAPR)='s';'Delay for the VDAPR'
vardef(Tp_aver)='s';'Delay for the Measurement of the Active Power'
vardef(qmax)='p.u.';'Maximum Reactive Power in Steady State'
vardef(k_ir)='-'; 'Gain for the Current Limitation'
vardef(uset_min)='p.u.';'Lower Limit of Remote Var PI Control'
vardef(uset_max)='p.u.';'Upper Limit of Remote Var PI Control'
vardef(id_max)='p.u.';'Maximum Active Current of Converter'
vardef(id_min)='p.u.';'Minimum Active Current of Converter'
vardef(UdcN)='kV';'Nominal DC Voltage'
```

DSL-code Eq. 13.6: *Shaft DSL-code equations.*

```
inc(dphi12)=pt*Pgbase/Ptbase/speed_gen/Ktg/Ktg
inc(xH)=speed_gen
inc(Pwind)=pt*Pgbase/Ptbase
inc(speed_gen)=speed_tur
inc(speed_tur)=1
inc(pt)=0

vardef(H)='s';'Wind Turbine Inertia Constant'
vardef(Ktg)='p.u./rad';'Shaft Stiffness'
vardef(Dtg)='p.u.';'Shaft Torsional Damping'
vardef(Pgbase)='MW';'Nominal Power'  !Pmon
vardef(Ptbase)='MW';'Mechanical Power' !Pmech
vardef(ombase)='rad/s';'Nominal Speed'
```

DSL-code Eq. 13.7: *PQ-Calculation.*

```
inc(p)=ud*id+uq*iq
inc(q)=uq*id-ud*iq

p=ud*id+uq*iq
q=uq*id-ud*iq
```

DSL-code Eq. 13.8: *Turbine-Aerodynamics.*

```
inc(Pwind)=rho/2*pi()*sqr(R)*Cp*pow(vw,3)/1E6
inc(Cp)=0.4
inc(Pwind_max)=lim(Pwind,0.1,1.5)
inc(Pwind_min)=0.1
inc(vw)=12.3
inc(speed_tur)=1

vardef(R)='m';': Rotor Blade Radius'
vardef(rho)='kg/m^3';': Air Density'
vardef(vw)='m/s';'Wind Speed'
```

DSL-code Eq. 13.9: *DC Model.*

```
inc(Udc)=xDC
inc(xDC) = UdcN*1000

vardef(Cdc)='F';'Capacitor on DC Busbar'
vardef(UdcN)='kV';'Nominal DC Voltage'
vardef(Sbase)='MVA';'Base Mega-Volt-Amperes of Converter'
vardef(Pnom)='MW';'Rated Active Power of the Converter'
vardef(Udcmax)='kV';'Maximum DC Voltage'
vardef(uDC_Chon)='p.u.';'Chopper Activation Voltage Level'
vardef(uDC_Choff)='p.u.';'Chopper Deactivation Voltage Level'
```

DSL-code Eq. 13.10: *Wind Ramp.*

```
hold=select(time()<Tstart,0,1)
xw.=1/Tw*vw0*hold
vw=limstate(xw,vw0,vwmax)
inc(xw)=12.3
inc(vw)=12.3

vardef(vw0)='m/s';': Initial Air Velocity'
vardef(vwmax)='m/s';': Air Gust Increase'
vardef(Tw)='s';': Wind Gust Duration'
vardef(Tstart)='s';': Wind Gust Event Time'
```

DSL-code Eq. 13.11: *Power Reference.*

```
inc(xp_act)=-p
inc(p)=-1
inc(xp) =0
inc(idc)=0
inc(pref)=0.9

inc(xpdc)=pref
inc(speed_gen)=1
inc(Udc)=1

vardef(tp_act)='s';'Filter Time Constant for Active Power'
vardef(Pnom)='MW';'Nominal Power'
vardef(Pmech)='MW';'Mechanical Power'
vardef(Kp)='-';'Voltage/Power Gain'
vardef(Tp)='s';'DC Voltage Time Constant'
vardef(Tpp)='s';'Active power Time Constant'
vardef(p_switch_nor)='-';'Power Disconnection Switch'
vardef(Pmax)='p.u.';'Maximum AC Power Limit'
vardef(Pmin)='p.u.';'Minimum AC Power Limit'
vardef(Pdc_max)='p.u.';'Maximum DC Power Limit'
vardef(Pdc_min)='p.u.';'Minimum DC Power Limit'
vardef(NomVoltage)='V';'Nominal DC Voltage'
vardef(Tstartd)='s';': Wind Gust Event Time with Delay'
```

DSL-code Eq. 12: *Pitch Angle Controller.*

```
inc(xB)=Bmin;
inc(xFil_Beta)=Bmin;
inc(xservo)=Bmin;
inc(xfilter)=xservo;
inc(beta)=0
inc(wref_wt)=speed_gen
inc(p)=-1
inc(speed_gen)=1
inc0(pref)=p*NomPower

vardef(tFil_Beta)='s';'Filter Time Constant for Pitch Angle'
vardef(Kp_pitch)='deg/p.u.';'Proportional Gain of Pitch Controller'
vardef(Ti_pitch)='s';'Time Constant of Pitch Controller'
vardef(Tservo)='s';'Time Constant of Pitch Actuator'
vardef(Bmin)='deg';'Lower Pitch Angle Limit'
vardef(Bmax)='deg';'Upper Pitch Angle Limit'
vardef(Bratemin)='deg/s';'Minimum Rate Limit for Pitch Angle'
vardef(Bratemax)='deg/s';'Maximum Rate Limit for Pitch Angle'
vardef(Load_Ctrl)='p.u.';'Proportional Load Controller'
vardef(Tfilter)='p.u.';'Measurement Delay'
vardef(NomPower)='MVA';'Nominal Power of Generator'
```

13.6.2 *Tuning of CGU Control Systems*

In either of the two existent operational scenarios, there will be at least two active CGU. If it is desired to simulate a conjunct Black-Start/Restoration strategy under a scenario with no renewable integration, the control systems attached to each CGU should respond satisfactorily, since classical CGU control systems are already grid-forming type. The model for all the CGU is shown in Fig. 13.12. Such control schemes are the standard IEEE models for the AVR-Exciter and Governor (EMAC1T and IEEEG1, respectively) [4], and can be found as available templates in the DIgSILENT PowerFactory library, therefore, their default equation settings do not need to be modified. Yet, two of the sixteen governors were added a secondary frequency

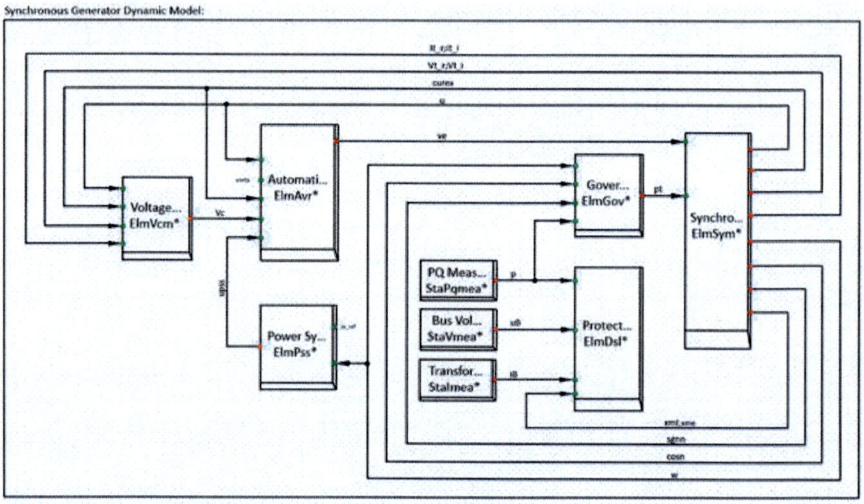

(a)

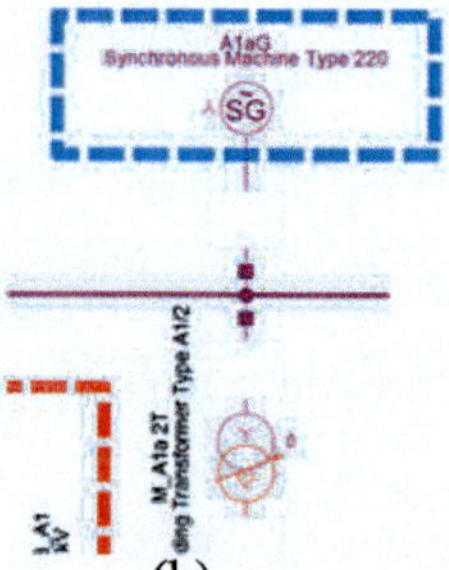

(b)

Fig. 13.12 CGU model in DIgSILENT PowerFactory [3]. (a) composite frame structure of the CGU in DIgSILENT PowerFactory (b) CGU graphic model in DIgSILENT PowerFactory

Table 13.5 Saturation/magnetising parameters [3]. (a) synchronous generator saturation parameters (b) transformer saturation parameters

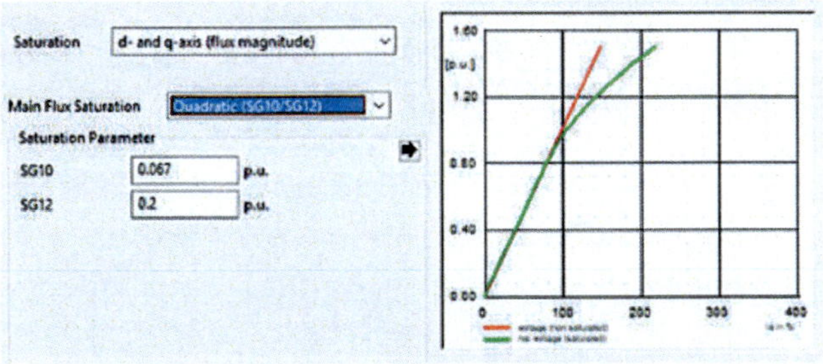

(a)

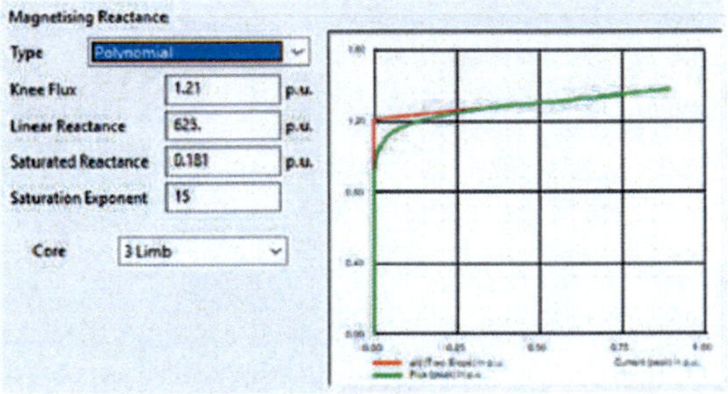

(b)

controller with a power measurement unit, and some of their Common Model parameters were tuned to improve their frequency responses for coping with the high-speed response of the converter-dominated scenarios, included in the WPP, BESS and HVDC systems. PSS can also be enabled to improve generator rotor oscillation damping [4]. Generator saturation parameters are enabled for an accurate EMT transient analysis (Table 13.5a).

13.6.3 VSC-HVDC Transmission Systems

Figure 13.13 shows the model of the HVDC transmission stations based on MMC converters. The design of this complex system was based on the CIGRÉ Technical Report No. 604 [10]. In turn, the MMC-HVDC model was provided by DIgSILENT GmbH [15]. The two HVDC Systems installed in the PST-17 Benchmark System Network are therefore replicas based entirely on the former reference. Each VSC contents a voltage-dependent reactive power Capability Curve. Because VSC HVDC allows power to flow bidirectionally, these capability curves are four-quadrant type (Table 13.3b) [3]. Depending on the operation mode, their neighbouring terminals have specific *VDC*, *P*, or *Q* setpoint values for ensuring a smooth load flow. Figure 13.13a shows the control scheme of the MMC, with loops for both slow and

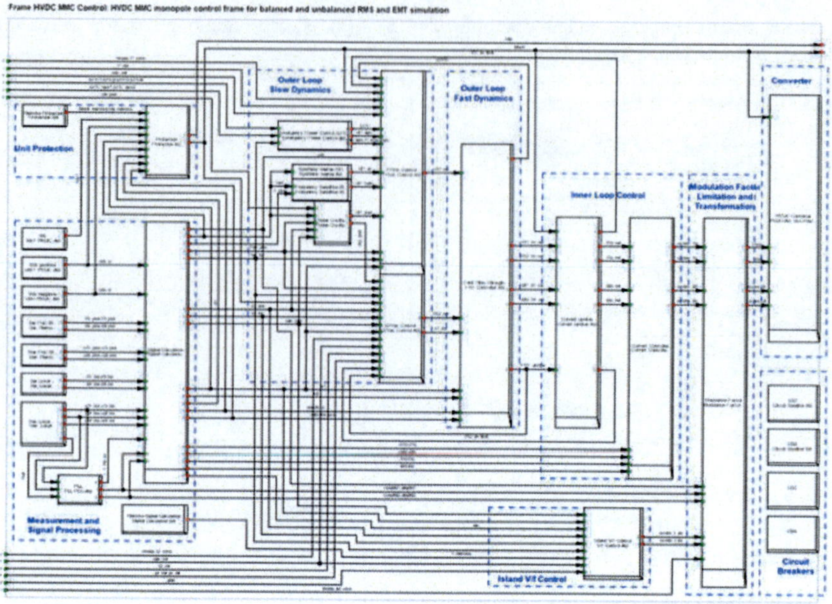

(a)

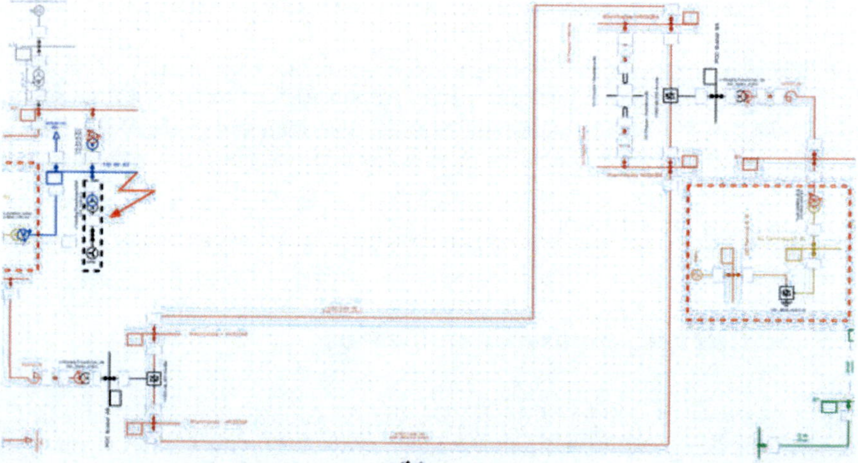

(b)

Fig. 13.13 HVDC station models in DIgSILENT PowerFactory [15]. (a) composite frame structure of the HVDC stations in DIgSILENT PowerFactory (b) MMC-HVDC transmission system that interconnects *Area A* with *Area B* fast dynamics. Figure 13.13b shows the model of the HVDC transmission station based on MMC converters connecting *Area A* with *Area B*. There is another identical DC link between *Area A* and *Area C*.

13.6.4 Battery-Energy Storage Systems (BESS)

Figure 13.14 shows the model of the BESS-Storage units. The design of the control system was based on the BESS Application Example Technical Report [11]. Thus, the seven BESS installed in the PST-17 Benchmark System Network are therefore replicas based entirely on the former reference. Yet, some modifications were applied to this reference, which took place in the following elements [3]:

- **Frequency Controller Model Definition**. A selector block was added before the output signal of this model definition to prevent the storage units to work under unnecessary situations that don't require emergency support, in terms of ±frequency.
- **Battery Common Model**. Additional battery cells were connected in series to increase the battery voltages from 0.9 to 9 kV.
- **BESS Composite Frame**. Two additional wired signals were added to the original BESS Composite Frame; the sine and cosine of the angle reference of the BESS PWM converter (signals $sinphi$ and $cosphi$).

13.7 Performance Evaluation in DIgSILENT PowerFactory

13.7.1 Load Flow Analysis

The scaled WPP models shown in Fig. 13.11b are placed in the PST 16 Benchmark System. It can be seen in Fig. 13.15 all the seventeen full-converter WPP surrounded with an orange-reddish colour glow, while most of the CGU are surrounded by grey colour, meaning that they are inactive for this scenario. Orange-reddish colour denotes an element of the network reaching their nominal value, while green colour denotes a neat 1 p.u. voltage level [3]. It is admissible in power system analysis to measure the generation units loading close to their nominal values, though this is not the same case for transformers and busbars. Therefore, for this case study the network holds a healthy load flow, with acceptable values in the generators and all passive elements absorbing adequate quantities of active and reactive power [3].

In the settings for the Load Flow Analysis, it was selected a positive-sequence type. Automatic tap adjustments for transformers were disabled, but reactive power limits and the automatic model adaptation for convergences were enabled. The output window in DIgSILENT PowerFactory did not throw any error or warning messages while the Newton–Raphson converged relatively fast with nine iterations. The WPP

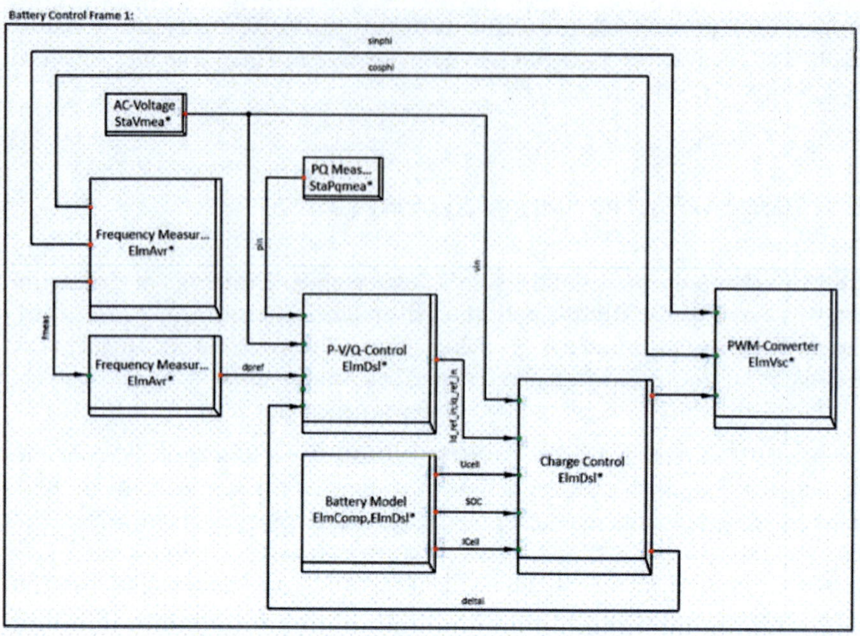

(a)

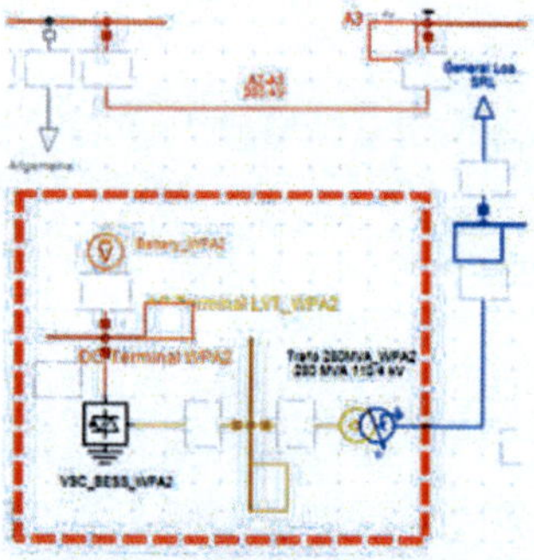

(b)

Fig. 13.14 BESS models in DIgSILENT PowerFactory [3]. (a) composite frame structure of the BESS in DIgSILENT PowerFactory (b) graphical structure of the BESS in DIgSILENT PowerFactory

needed to have the assistance of the DIgSILENT PowerFactory External Station Controller, added in each of the WPP. With them, an established reactive power setpoint was placed in a cubicle inside the busbar, each WPP was connected [3].

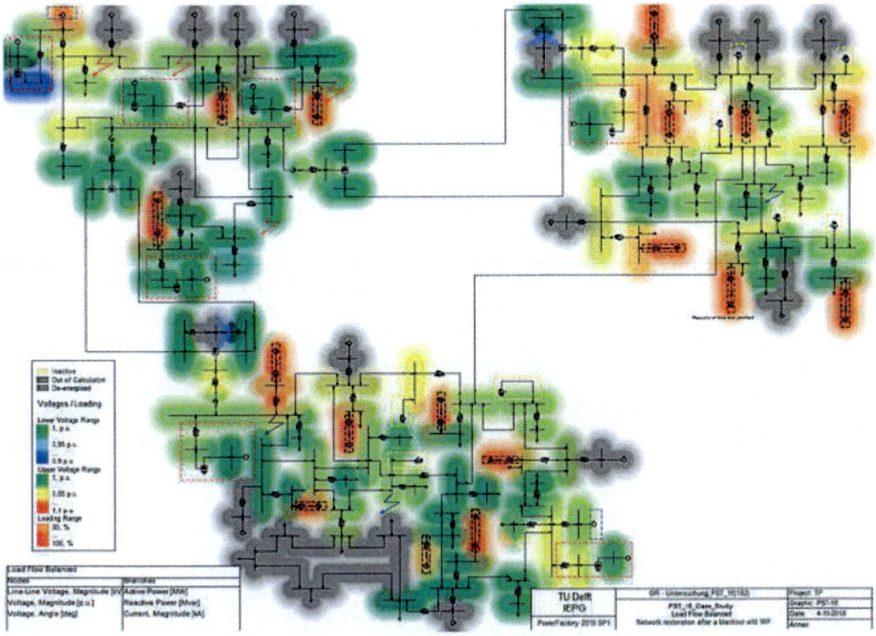

Fig. 13.15 Load flow analysis applied to the 90% wind share PST-17 benchmark system network [3].

13.7.2 EMT Simulation

A 17-min EMT simulation was performed for the assessment of the modified PST-17 Benchmark System. The settings and the initial conditions of this simulation are shown in Table 13.6. In general, these options were selected in order to improve the speed of the overall test yet without sacrificing too much definition in the plots.

The Automatic step size adaptation allows increasing the step size in moments when the program detects the simulation does not require high detailed curves, e. g., during long periods of steady operation. This setting allows to go back to a small-time step if EMT transients take place; an obvious consequence is that then the simulation slows down. An acceptable maximum time step size for the software in order to perform an EMT simulation without errors is 100 microseconds; any number slightly larger than that brings non-convergence errors. The solution of the linear equations works well with both direct method or iterative method. Several simulations were performed with both options, and no noticeable effect was distinguished, however, the second option was selected because it is recommended to use it if the case study is large and complex [3]. The total simulation duration time (absolute) was defined after several executed experiments but also considering the relatively large size of the PST 17 Benchmark System Network. The simulation started with the network operating in steady-state conditions during the first ninety seconds (to illustrate proper

Table 13.6 EMT simulation settings [3]

Active elements	
Conventional Generation Units (CGU)	2
Wind Power Plants (WPP)	17
VSC-HVDC Links	2
Battery-Energy Storing Systems (BESS)	7
Basic settings	
Verify initial conditions	Active
Automatic step size adaptation	Active
Solution of linear equations	Iterative method
Step size	
EMT Small time step	100 μ seconds
Maximum step size	200 s
Solver options	
Solve dynamic model equations at initialisation	Active
Enable partial initialisation in case of deadlock	Active
Fast convergence check	Active
Fast computation of outputs	Active
Maximum number of iterations	200
Integration	
Maximum error for dynamic model equations	1%
Damping factor	99
Iteration	
Maximum error for bus equations	1 MVA
Maximum error for model network equations	0.10%
Maximum number of iterations	25
Iteration limit to recompute Jacobian matrix	5
Simulation times	
Time between initialisation and blackout faults	1.5 min (approximately)
Time to restoration system to begin	1 min (after fault)
Total simulation duration time (absolute)	17 min

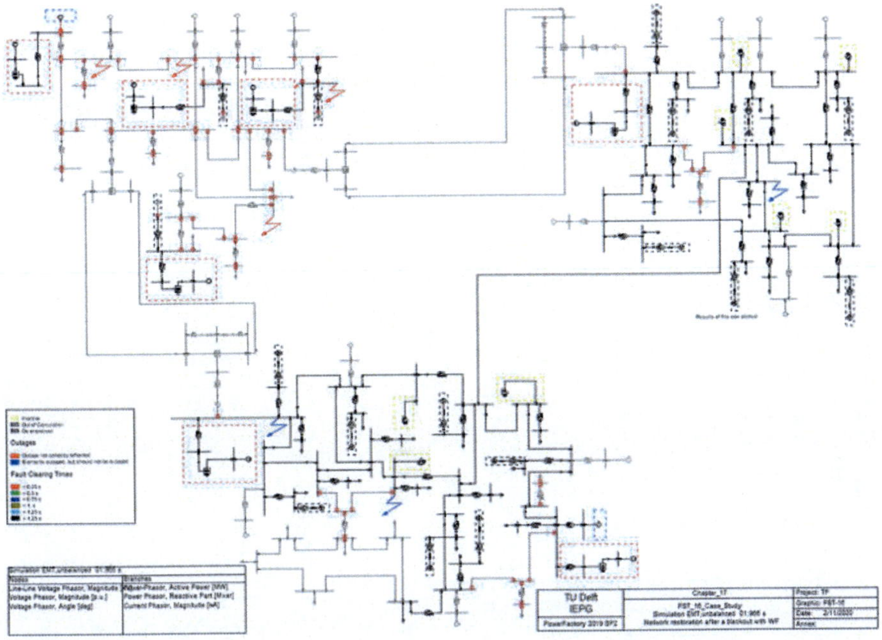

Fig. 13.16 EMT Simulation of the 90% converter penetration after a blackout [3]

initialisation of the model). Then, to simulate an extreme case, four short-circuit faults occurred in *Area A*, in the busbars under the red-arrow highlights, as denoted in Fig. 13.16. Three of these faults occurred in busbars where three diverse power stations are connected to the grid.

Thus, the protection system acted by turning off these generators; this caused a substantial-frequency plunge, which constrained the protection scheme to disengage the rest of the generators in *Area A*. To keep the heavy disturbance from spread out to the next two areas, the HVDC links were blocked and separated as well. *Area A* turned into an island from that instant. As there was a loss of generation, all components and loads must be tripped off too. The red dots in the network displayed in Fig. 13.16 represent circuit breakers that have been tripped off, and all the elements sketched in grey colour are no-longer energised and dead. Therefore, there is a Blackout status in these areas.

This test was intended to isolate *Area A* from the rest of the grid, as it was the area where the faults ensued. However, the results showed that some generators and transformers within areas *B* and *C* got dangerously overloaded, therefore, some sections of these areas had to be shut down as well as otherwise, these currents could increase to values of 2 p.u [3]. When *Area A* was totally shut down, the Black-Start and Restoration plans began [3].

Table 13.7 displays the events that have been programmed in case a burdensome disturbance derives into a blackout. The events shown in grey colour took

Table 13.7 Protection and restoration relay DSL-code equations [3]

Simulation Events/Fault - Study Cases\PST_16_Case_Study\Simulation Events/Fault :

Close

Name	Time	Object	Out of Service	Object modified
110_kV_B_TF5_Out	0.6403167	110_kV_B_TF5	☐	06/10/2018 07:38:56 p.
GL_Out(20)	0.6403167	General Load(20)	☐	06/10/2018 07:38:56 p.
C3-C5_Out	0.6403167	C3-C5	☐	06/10/2018 07:38:56 p.
C5-C6_Out	0.6403167	C5-C6	☐	06/10/2018 07:38:56 p.
C5T_Out	0.6403167	C5 T	☐	06/10/2018 07:38:57 p.
GL_Out(41)	0.6403167	General Load(41)	☐	06/10/2018 07:38:57 p.
GL_Out(34)	0.6403167	General Load(34)	☐	06/10/2018 07:38:57 p.
110_kV_C_TF13_Out	0.6403167	110_kV_C_TF13	☐	06/10/2018 07:38:57 p.
C12-C13_Out	0.6403167	C12-C13	☐	06/10/2018 07:38:57 p.
C13-C14_Out	0.6403167	C13-C14	☐	06/10/2018 07:38:57 p.
110_kV_C_TF8_Out	0.6403167	110_kV_C_TF8	☐	06/10/2018 07:38:57 p.
SyM_A1aG_Rec	60.00845	A1aG	☐	06/10/2018 07:35:14 p.
M_A1a_2T_In	75.00845	M_A1a 2T	☐	06/10/2018 07:35:08 p.
A1-A2_In	90.00845	A1-A2	☐	06/10/2018 07:35:08 p.
A_TF2_In	90.00845	110_kV_A_TF2	☐	06/10/2018 07:35:08 p.
SyM_WPA1_Rec	105.0084	WPA1	☐	06/10/2018 07:35:04 p.
A2-A3_In	120.0084	A2-A3	☐	06/10/2018 07:35:08 p.
A_TF3_In	120.0084	110_kV_A_TF3	☐	06/10/2018 07:35:08 p.
SyM_WPA2_Rec	135.0084	WPA2	☐	06/10/2018 07:35:08 p.
A1-A4_In	150.0084	A1-A4	☐	06/10/2018 07:35:08 p.
A2-A5a_In	150.0084	A2-A5a	☐	06/10/2018 07:35:08 p.
A2-A5b_In	150.0084	A2-A5b	☐	06/10/2018 07:35:08 p.

Ln 9 | 123 object(s) of 123 | 1 object(s) selected

place already, and all of them correspond to activities exclusively related to protection. The DSL-code equations behind these events are shown in the Protection & Restoration Relay DSL-code Eq. 13.2 in Sect. 13.6.1. This algorithm was designed to command their protection events just in case if a prior fault took place; otherwise, no action is ordered so no further events shall appear in Table 13.7. In this manner, when a predefined fault has arisen, this protection system is triggered to protect the network, shedding as many elements as necessary in ordered to irreversibly blow [3]. The events written in black are yet to occur and are based on the Protection & Restoration Algorithm, as explained in Sect. 13.5.3. The algorithm is also responsible for the reconnection of the lost elements and the restart of the Black-Start-capable generators.

Figure 13.17 shows the execution results of the two CGU active amid the operational scenario with 90% WPP share. The only active synchronous machine in *Area A*, sketched in black in Figs. 13.1 and 13.16, was shut down. Notwithstanding, after a few minutes of the area blackout, it was effectively black started, with the aid of the storage unit associated in the same busbar. Its governor system succeeded in keeping the frequency of this component near 50 Hz after the power outage and amid the first stages of reconnection. During the ensuing re-incorporation of the other power plants, its frequency went up because there was still no load brought back on the grid. The over-frequency deviation was reduced as soon as the load pick-up stage was reached,

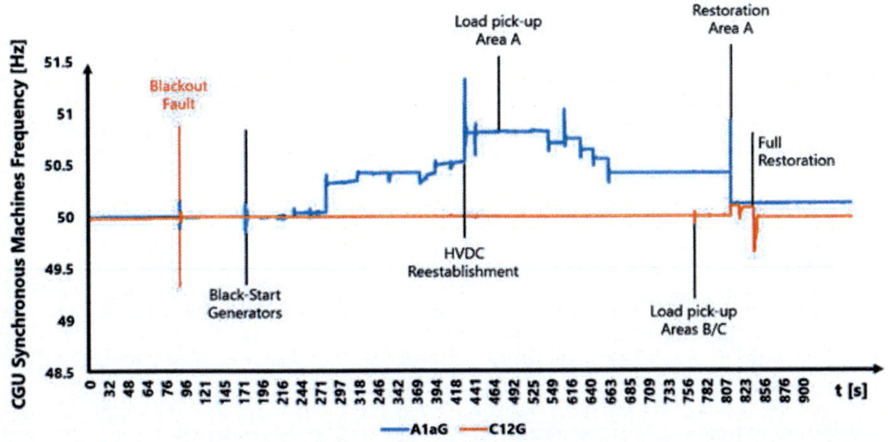

Fig. 13.17 Frequency plots of the two CGU units for the EMT simulation [3]

as dictated by the reaction of the protection & restoration relays. Step-by-step, the frequency came closer to its initial value before the faults and blackout.

When the reconnection of the HVDC links occurred, at around seven minutes, the frequency of the CGU in *Area A* (blue plot in Fig. 13.17) went up, and the active power dispatched by the WPP in *Area A* went down, simultaneously. This is because the network is working as one big system. By the time the HVDC reconnection occurred, most transmission lines were at that point re-energised, but the transformers and the loads were yet to be re-energised. The last part of the restoration process is the re-energisation of loads; by the time this milestone has reached, the frequencies of the two CGU were converging to their initial values before the faults. Similarly, the WPP were getting back their initial active power setpoint as more loads were brought back online.

Figure 13.18 draws the execution results of the WPP before, amid and after the faults followed by the power outage in terms of active power conveyed. Sixteen of the seventeen WPP are rated at 900 MW, and the remaining one (*WPC6* in Figs. 13.1 and 13.18) at 500 MW. The WPP situated in *Area A* are *WPA1*, *WPA2* and *WPA3* in Fig. 13.18. After the blackout, these three WPP were delivering 0 MW, but the remaining WPP located in the other areas kept working. By the time the load pick-up process started, each of the *Area A* WPP had their active power output slowly increasing at the same time their respective frequencies were converging to 50 Hz. The grid-forming extended capability prevailed in the restart and brought back to normal operation these three previously shut down WPP located in *Area A*, accomplishing re-incorporation to the network regardless of the troubling development they encountered during the entire restoration process.

Figure 13.19 demonstrates the busbar AC voltage magnitude for some of the disconnected loads in *Area A*. This plot indicates unmistakably the advantages of using the grid-forming control systems in a power system with 90% converter-interfaced generation. Without the grid-forming controllers, the load pick-up ruled

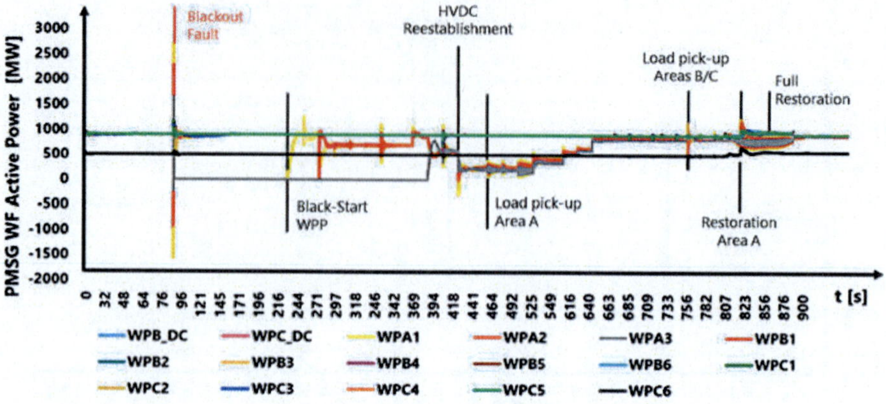

Fig. 13.18 Active power plots of the seventeen WPP for the EMT simulation [3]

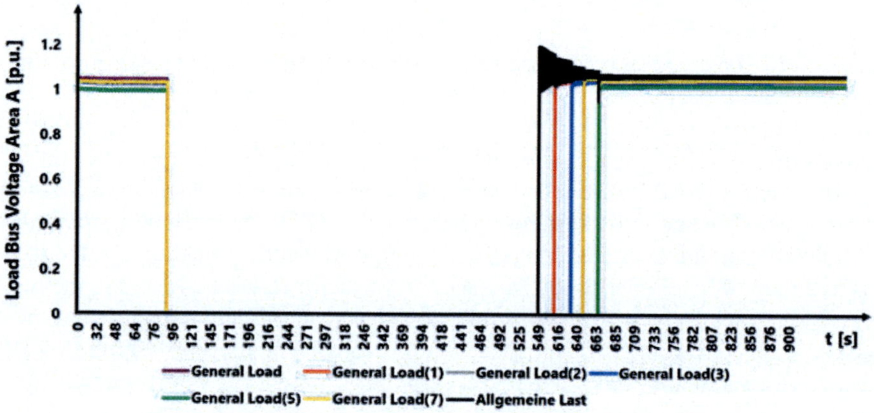

Fig. 13.19 AC voltage of the busbars connected with the loads of *Area A* in p. u. [3]

by the WPP could not have been conceivable, since the present renewable energy-based generators cannot re-energise loads. The frequency and active power plots previously shown coincide in their dynamic behaviour every time there is a common event. This can be reflexed as well in Fig. 13.19, where it can be noticed that as soon as any load represented in this plot started to be served, their respective AC bus voltages come back to 1 p.u. By the end of the simulation, all loads were served by their nominal values, as it was before the faults.

Figures 13.20 and 13.21 demonstrate the performance results of the seven BESS during the whole simulation in terms of their state of charge (SOC) and their active power conveyed. The correct operation of these storage devices is relied upon to be noted just when there are contingency situations were auxiliary voltage and power is required, but otherwise, these components ought not to work. Indeed, in Figs. 13.20 and 13.21, during steady-state operation, the active power dispatched is zero, and

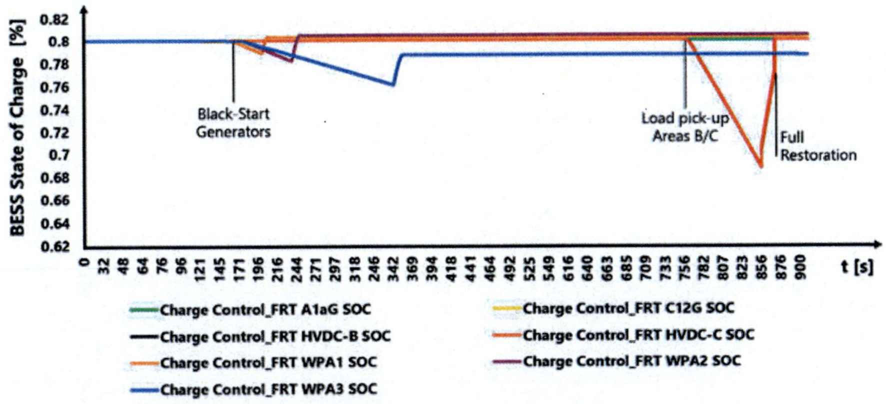

Fig. 13.20 State of charge of the seven BESS for the EMT simulation [3]

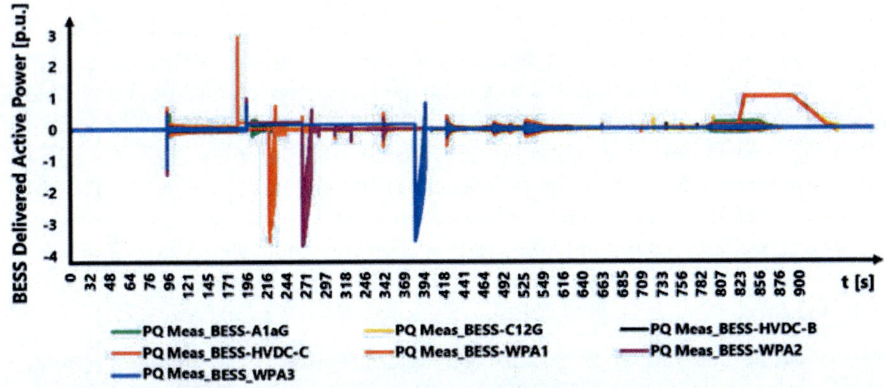

Fig. 13.21 Active power of all the BESS-storage units before, during and after the outage and restoration events [3]

the SOC stays unaltered. Thus, the BESS reacted quickly when the system required them to provide ancillary services when circumstances of emergency arose. At that point, the SOC of most of the batteries went down, which implies that they dispatched active power for various reasons, including giving excitation voltage to extensive-size plants, extra-supply of power to anticipate CGU overloads, and support active power balance to mitigate the frequency deviations that took place during the disturbances. In Fig. 13.21, the highlighted peaks allude to the fact that most transmission lines were re-energised during the first stage of the restoration strategy. As being reconnected in a dead-busbar arrangement, they were going about as huge capacitors. But in turn, the soft-energisation strategy kept the loads and electrical components from unsafe overloading and voltage oscillations while transformers were shielded from saturation.

Figures 13.17, 13.18, 13.19, 13.20 and 13.21 likewise show a strange behaviour in the last phase of the restoration. This was on the grounds that, in defiance of the *Area A* being isolated to avoid transients to spread out, the unsettling disturbances proliferated to the other two areas. Therefore, a fifth protection & restoration relay had to be included in the model, placed in the generator *C12G*, as shown in Fig. 13.1. The fifth DSL-code-based relay led a load-shedding strategy for easing the overloading disturbances resulted by the previous generation shutdown in *Area A*. The steep variations in all the plots shown in Figures 13.17, 13.18, 13.19, 13.20 and 13.21, during the last minutes of the simulation, describe the additional load pick-up development of the loads that previously had to be tripped off in *Area B* and *Area C* to preserve the integrity of the rest of the system. The tripped-off elements in *Area B* and *Area C* are indicated in Fig. 13.16.

13.8 Conclusions

In this chapter, a Benchmark for dynamic simulations in DIgSILENT PowerFactory was presented. The model presents a three-area power system with a 90% renewable generation share, and it also includes BESS and HVDC transmission. The control implementation was designed to provide this Benchmark to be strong enough against heavy disturbances that even may derive in a blackout; in this case, the Benchmark being capable of performing a Black-Start/Restoration scheme. Performing a conjunct Black-Start/Restoration in a power system with high penetration of power-electronic converters can only be possible if such a grid is equipped with these elements [3]:

- Grid-forming control in the power-electronic converters that interface renewable power plants and HVDC links. With such asset, WPP and HVDC converter stations now can regulate electrical power grids alone, featuring direct voltage/frequency control.
- BESS-storage devices that can smooth the variable output power of renewables and provide voltage excitation and frequency balance.
- A strategy for ensuring a safe Black-Start and Restoration without the saturation of transformers, neither endangering other electric equipment and loads.

References

1. Renewables, Power and Energy Use Forecast to 2050—Energy Transition Outlook 2017. DNV GL, Tech. Rep. April 2018
2. R.W. Garvine, W. Kempton, Assessing the wind field over the continental shelf as a resource of electric power. J. Mar. Res. **66**, 751–773 (2008)

3. L. Noris, Modelling and Assessment of Restoration in Electrical Power Systems with High Penetration of Power-Electronic Converters. Master's Thesis Report, Technische Universiteit Delft, 2018
4. I. Erlich, A. Korai, T. Neumann, M.K. Zadeh, S. Vogt, C. Buchhagen, C. Rauscher, A. Menze and J. Jung, New control of wind turbines ensuring stable and secure operation following Islanding of wind farms. IEEE Trans. Energy Conv. **32** (2017)
5. M.M. Adibi, D.P. Milanicz, Estimating restoration duration. IEEE Trans. Power Syst. **14** (4) (1999)
6. S.P. Teeuwsen, I. Erlich, M.A. El-Sharkawi, U. Bachmann, Generic algorithm and decision tree based oscillatory stability Assessment, 2005 IEEE Russia Power Tech. (2005)
7. J.L. Rueda, F. Gonzalez-Longatt, Advanced Smart Grid Functionalities based on PowerFactory (Springer)
8. I. Erlich, A.W. Korai, Description Modelling and Simulation of a Benchmark System for Converter Dominated Grids (Part I) (Universität Duisburg-Essen, 2018)
9. A.W. Korai, Dynamic Performance of Electrical Power Systems with High Penetration of Power Electronic Converters: Analysis and New Control Methods for Mitigation of Instability Threats and Restoration, Doktors der Ingenieurwissenschaften (Dr.-Ing.) genehmigte Dissertation, Universität Duisburg-Essen (2019)
10. 'Guide for the Development of Models for HVDC Converters in a HVDC Grid (604). CIGRE, Tech. Rep. (2014)
11. Application Example—Battery Energy Storing Systems BESS. DIgSILENT GmbH. Tech. Rep. (2018)
12. F. Fernandez, Models of Battery Storage Systems for Power System Analysis: International Seminar on "Energy Storage Options for Renewable Energy Integration", DIgSILENT GmbH, Seminar (2018)
13. H. Pustjens, 'Reducing the risks of network restoration, DNV GL, Tech. Rep. (2016)
14. Power system restoration dynamics (Issues, Techniques, Planning, Training & Special Considerations). IEEE, Tech. Rep. (2014)
15. DIgSILENT PowerFactory Seminar HVDC and FACTS, DIgSILENT GmbH. Tech. Rep. (2018)

Chapter 14
A Generic RMS-Based Wind Turbine Model for the Simulation of Large Power Systems

Abdul W. Korai, E. Rakhshani, M. Ebrahim Adabi, José Luis Rueda Torres, and Mart A. M. M. van der Meijden

Abstract The main objective in this chapter is to develop and present a generic model for wind turbine (WT) which can be used for both DFIG- and FSCG-based WT for large-scale multi-machine power system dynamic studies. The presented model is developed for RMS simulation on PowerFactory, and it can be used as a replacement for both DFIG- and FSCG-based WTs without making any changes in the generic model itself. The generic RMS model is appropriate for the stability studies of large grids where the detailed dynamics, i.e., control action in the range of milliseconds, of the power electronic converter-based controllers do not play an important role.

Keywords DIgSILENT RMS simulation · DSL programming · Wind turbine control · Wind power integration · Large-scale power system

14.1 Introduction

Application of suitable software, like Digsilent, with a generic WT model for analyzing the large-scale power systems with low-inertia, will be necessary. Considering simulation time and computation burden in multi-machine interconnected power systems with a high share of power electronics, application of generic WT model suitable for multi-machine timescale dynamics simulation is the motivation of presented details in this chapter.

Electronic supplementary material The online version of this chapter (https://doi.org/10.1007/978-3-030-54124-8_14) contains supplementary material, which is available to authorized users.

A. W. Korai · E. Rakhshani · M. E. Adabi · J. L. Rueda Torres (✉) · M. A. M. M. van der Meijden
Department of Electrical Sustainable Energy, Delft University of Technology, Delft, The Netherlands
e-mail: j.l.ruedatorres@tudelft.nl

M. A. M. M. van der Meijden
TenneT TSO B.V, Arnhem, The Netherlands

F. M. Gonzalez-Longatt and J. L. Rueda Torres (eds.), *Modelling and Simulation of Power Electronic Converter Dominated Power Systems in PowerFactory*, Power Systems,
https://doi.org/10.1007/978-3-030-54124-8_14

This chapter will provide a detailed explanation for developing a generic model for WT in DIgSILENT PowerFactory. The focus of such models is the behavior of the synchronous machines considering a large generation from power electronic converter-based generation units. As in such studies, the timescale of simulations can be very large; hence, a proper model is needed which can represent sufficiently the dynamics of the controllers of the WTs while on the other hand being as simple as possible thus making the power system faster in the simulations. The generic RMS model should be able to match the active and reactive power generation of the detailed models for the large grid studies. The simulation effort can be significantly reduced if the WT model can be converted into a simple first-order system, i.e., eliminating any proportional-integral controllers and complex transfer function, maintaining the appropriate accuracy of the results.

Certain studies cannot be performed by this model in which the dynamics and sizing of the DC capacitor play an important role, i.e., inertia emulation and DC voltage dynamics in specific scenarios, i.e., blocking of the power electronic converter, connection, and disconnection of loads and operation of the chopper.

This chapter is organized as follows: In Sect. 14.2, the control structure and the way of implementation and initialization of DSL blocks in DIgSILENT PowerFactory are presented and discussed. While in Sect. 14.3, the test system with validation of the proposed generic model is presented, and the results are compared. Summary and the main conclusions of this chapter are also presented in Sect. 14.4.

14.2 Wind Turbines Generic Model

Figure 14.1 shows the control structure of the generic WT model. The time constant T_{d} is optimized with MVMO [1] to match the results of both DFIG- and FSCG-based WTs, which were verified with the manufacturer models.

The optimization task is given by Eq. (14.1), which minimizes the difference between active and reactive power between the detailed and generic model by modifying the PT1 time constant T_{d}. The obtained and range values of T_{d} are given by [0.015, 0, 0.5] sec.

$$\min F = \sum_{i=1}^{n} \Delta P_{\mathrm{WT}_i}(t) + \Delta Q_{\mathrm{WT}_i}(t) \tag{14.1}$$

As shown in Fig. 14.1, the DC model has been eliminated, and the PI control part of the current control is removed. The control is the first-order system with the controller action done by the feedforward terms.

In addition, it is to be noted that the PLL is not used in the model; instead, the voltage angle is calculated directly from the measured voltage. The output voltages of the controller are limited according to the maximum voltage of the power electronic

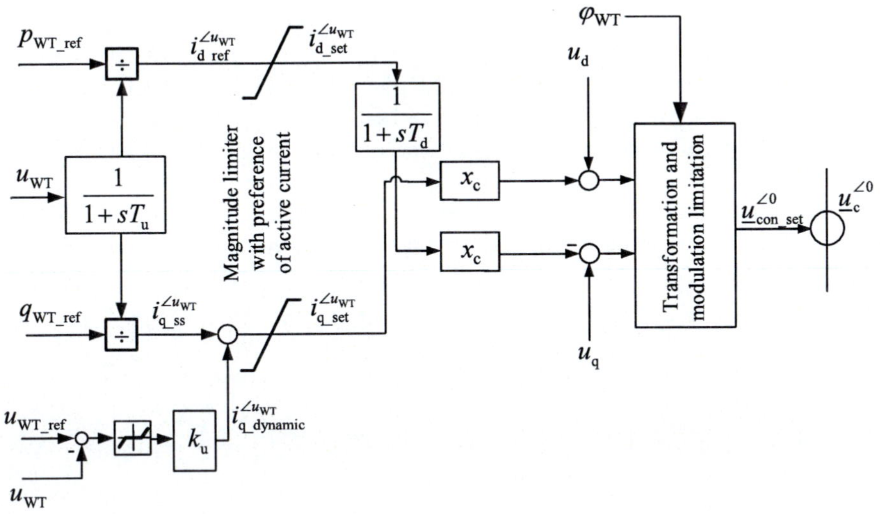

Fig. 14.1 WT generic model

converter. The outer loop of the controller generates the active and reactive voltage references. The fast voltage controller of the WT is also part of the model.

Figure 14.2 shows the model of the wind turbine control, which was built as a composite model and contains the following components: (a) converter used in Windfarm model by the static generator of PowerFactory (ElmGenstat); (b) WT controller as explained before (ElmDsl); (c) Over frequency control block (ElmDsl),

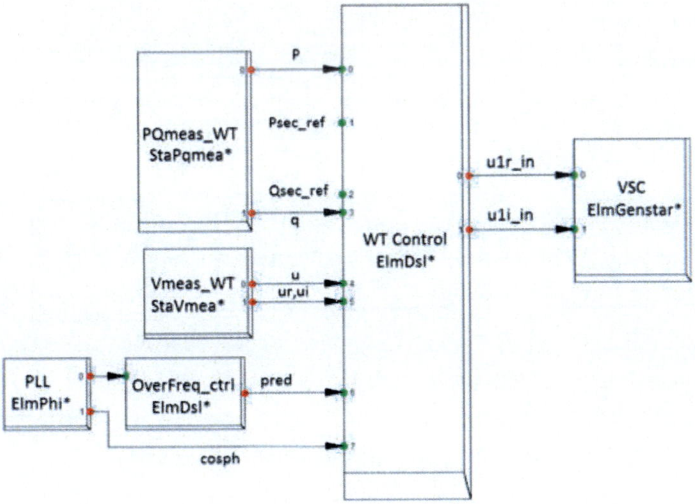

Fig. 14.2 Structure of wind turbine control frame in PowerFactory

(d) P and Q power measurements (StaPqmea), (e) AC voltage of WT measurement (StaVmea), (f) PLL block for angle measurement (Elmphi). The details of the wind turbine control block will be the same as the one presented in Fig. 14.1, while the frequency control approach is implemented in DSL s shown in Fig. 14.3.

Correct initialization of a model in a power-system simulation-tool avoids fictitious electrical transients and makes it possible to evaluate correctly the real dynamic performance of the system. Therefore, the initialization equations of an important dynamic block of this composite model are presented below:

WT control:

```
vardef(Ku) = '-'; 'Gain of local voltage control of WT'
vardef(LVRT)  ='pu';  'Voltage  below  which  reactive  current  has
priority'
vardef(Tu) = 's'; 'PT1 time constant for the voltage measurement'
vardef(l) = 'pu'; 'Inductance of the converter'
vardef(Td) = 's'; 'Optimized time constant of PT1 (dont change)'
vardef(db) = 'pu'; 'Voltage deadband'
vardef(id_max) ='pu'; 'Maximum active current of the converter'
vardef(irmax) ='pu'; 'Maximum current of the converter'
vardef(y_max_r) ='pu'; 'Maximum voltage of the converter in d-axis'
vardef(y_min_r) ='pu'; 'Minimum voltage of the converter in d-axis'
vardef(y_max_i) ='pu'; 'Maximum voltage of the converter in q-axis'
vardef(y_min_i) ='pu'; 'Minimum voltage of the converter in q-axis'
vardef(Tpick) = 's'; 'LVRT and HVRT detection time delay'
vardef(HVRT)='pu';'Voltage support  (priority to reactive current)'
```

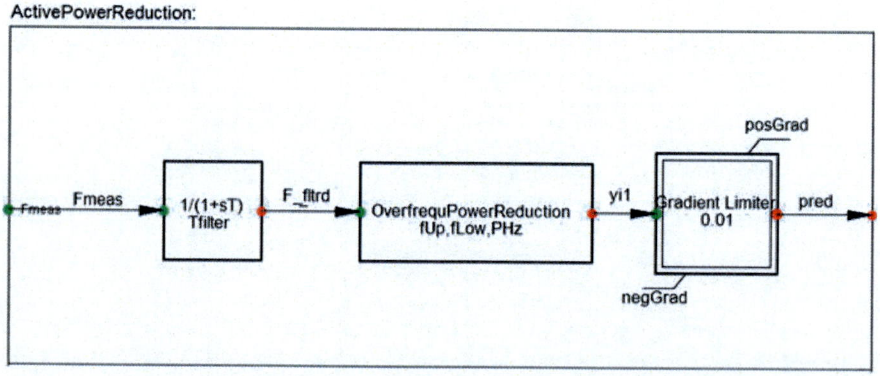

Fig. 14.3 Frequency strategy for active power reduction in PowerFactory

```
vardef(Tdrop) ='s'; 'LVRT and HVRT deacativation time delay'
vardef(T_fault) ='s'; 'Duration of the fault'
vardef(Ramp) ='%'; 'Active power ramp up '
inc(u_ref) = u
inc(xu) = u
inc(Psec_ref) = p
inc(Qsec_ref) = -q
inc0(ip) = p/u
inc0(iQ) = q/u
inc(id) = ip
inc(iq) = -iQ
inc(xdi) = id_set
```

Active power and frequency controller:

```
vardef(fUp)='Hz';'Start of Act. Power Reduction'
vardef(fLow)='Hz';'End of Act. Power Reduction'
vardef(PHz)='%/Hz';'Gradient of Act. Power Reduction'
vardef(Tfilter)='s';'PT1-Filter Time Constant'
vardef(posGrad)='pu/s';'Pos Gradient for Power Change'
vardef(negGrad)='pu/s';'Neg Gradient for Power Change'
inc(pred)=1
inc(x)=Fmeas
```

14.3 Test Grid and Simulation Results

Figure 14.4 shows the test grid used for the validation of the generic model with both the DFIG- and FSCG-based WTs. The grid has been modelled for EMT and RMS studies [2]. A short circuit with a voltage sag of 50% is simulated with a duration of 150 ms. The saturation of the WT transformer in the model is activated for the simulation of the manufacturer models.

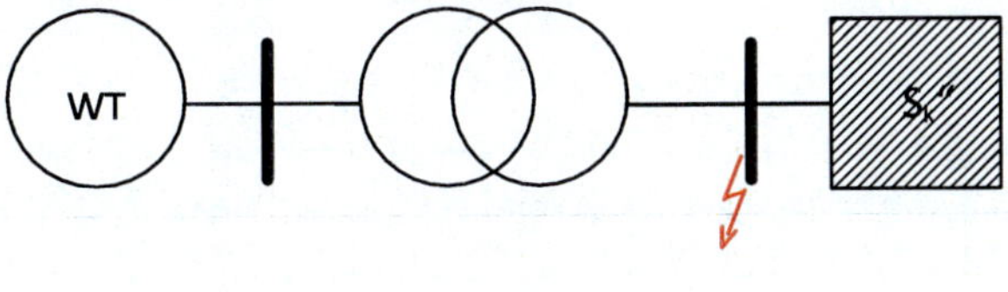

Fig. 14.4 Test grid for the validation of generic model

Figure 14.5 shows the comparison of the detailed WT models with the generic WT model. The generic model matches the detailed models with a good accuracy.

14.4 Conclusions

In this chapter, a generic and simplified wind turbine model is developed, which is suitable for the large-scale stability studies reducing the computation time while retaining sufficient accuracy. Thanks to the proposed model for the stability studies; it is possible to save computation time while at the same time, and it is possible to retain the acceptable accuracy in large grids.

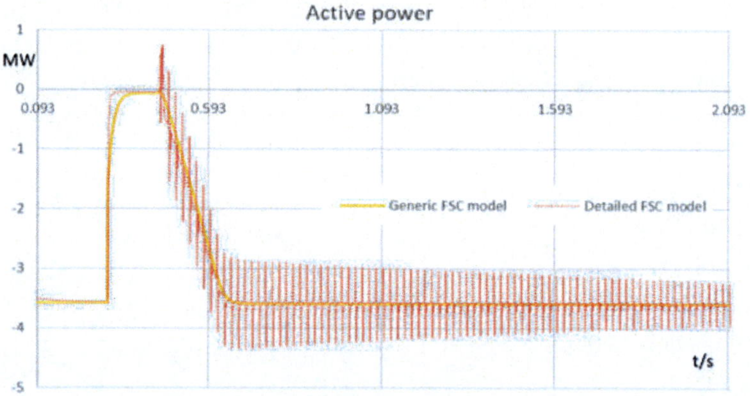

(a)

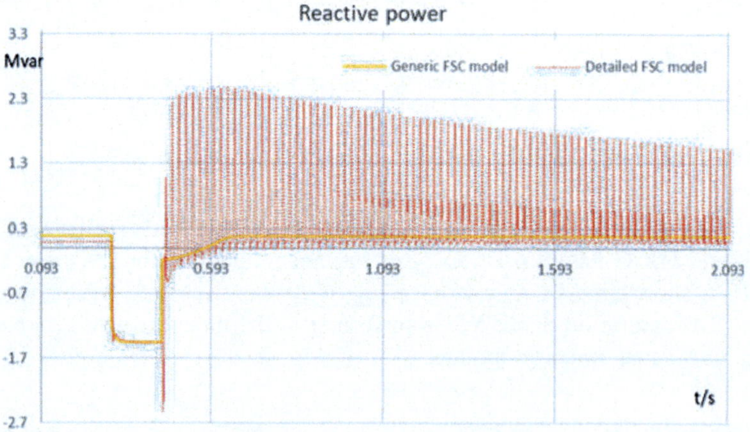

(b)

Fig. 14.5 Comparison of generic model with the detailed models

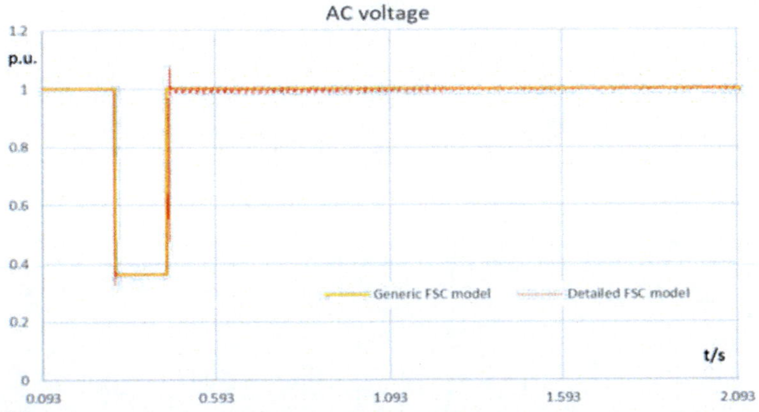

(c)

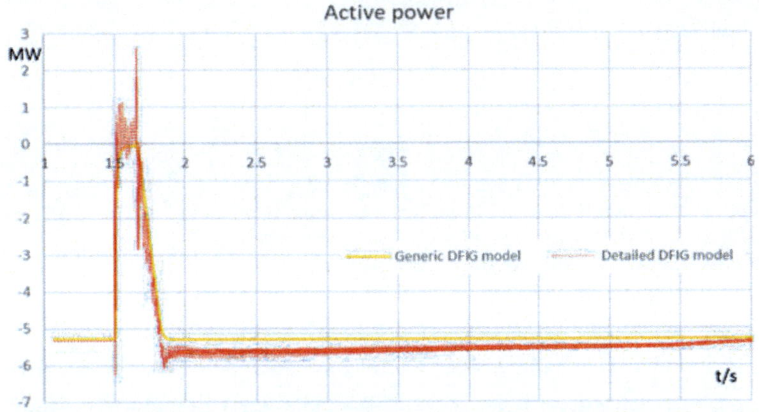

(d)

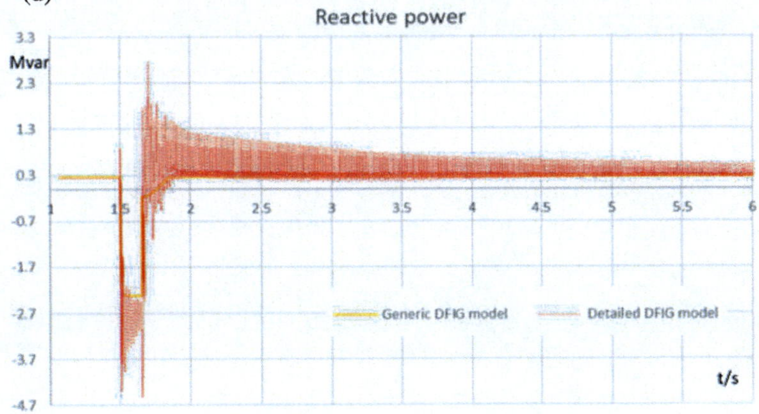

(e)

Fig. 14.5 (continued)

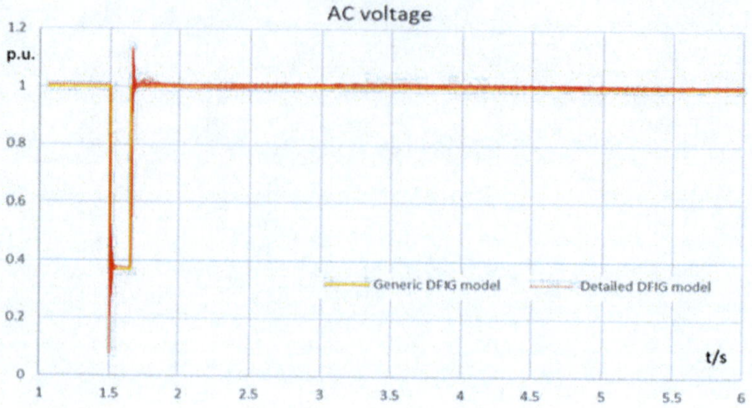

(f)

Fig. 14.5 (continued)

References

1. J. Rueda, I. Erlich, in Testing MVMO on learning-based real-parameter single objective benchmark optimization problems. *IEEE Congress on Evolutionary Computation (CEC)*, Sendai, Japan, 2015
2. PowerFactory 2017, User Manual, DIgSILENT GmbH, 2017

Chapter 15
Applications of PowerFactory for the Study of Basic Notions of Power System Dynamics in Graduate Courses

Ajay Shetgaonkar, Arcadio Perilla, Ilya Tyuryukanov, Digvijay Gusain, Claudio López, and José Luis Rueda Torres

Abstract This chapter presents a general overview of the experience learned with the use of DIgSILENT PowerFactory in the design of theoretical lectures and practical sessions of a power system dynamics course at postgraduate level. This chapter focuses on the experiences acquired in the course that is part of the MSc program in Electrical Engineering of TU Delft, Department of Electrical Sustainable Energy. The discussion provided in this chapter focuses on power systems application with special focus on (i) Steady-state, Dynamic, (ii) Voltage Stability and (iii) rotor angle stability. The main objective of using PowerFactory at MSc level is to expose the postgraduate students to real-life application, however, without lack of generalisation this chapter is dedicated to the is to expose to the application above by using a very well-known two area-four machine test power system (2A4G), it gives students insights and experience with cases closer to actual power systems. Results of this pedagogical experience demonstrate the importance of incorporating appropriate power system simulations tools in the postgraduate level.

Keywords DIgSILENT PowerFactory · Steady-state · Dynamic · Voltage stability · Rotor angle stability

15.1 Introduction

Today, the source of electricity is mutated from conventional energy sources to renewable energy resources. With the proper exploitation of renewables energy sources, electricity will play a dominating role in the future [1].

Electronic supplementary material The online version of this chapter (https://doi.org/10.1007/978-3-030-54124-8_15) contains supplementary material, which is available to authorized users.

A. Shetgaonkar · A. Perilla · I. Tyuryukanov · D. Gusain · C. López · J. L. Rueda Torres (✉)
Faculty of Electrical Engineering, Mathematics and Computer Science,
Technical University of Delft, Building 36, Mekelweg 4, 2628 CD Delft, The Netherlands
e-mail: j.l.ruedatorres@tudelft.nl

F. M. Gonzalez-Longatt and J. L. Rueda Torres (eds.), *Modelling and Simulation of Power Electronic Converter Dominated Power Systems in PowerFactory*, Power Systems,
https://doi.org/10.1007/978-3-030-54124-8_15

In upcoming decades, a revolution will promote the generation and transmission of electrical power, encouraged by the large-scale installation of renewable and distributed power generation. These changes demand environmentally friendly and sustainable components and the liberalisation of the energy markets [1, 2]. Graduates of electrical engineering program are needed to design new components keeping focus towards sustainable materials and efficient conversion processes, as well as to integrate them into a smart and adaptable electricity infrastructure. They should be able to solve problems of power systems under current and future characteristics and challenges. Those challenges include having less inertia in the system (because of introducing smaller machines, reducing stability), reactive power support and temporal variability of renewable energy sources.

The electrical power system is the very complex and nonlinear network, the hand calculation will be complicated and need a significant amount of time. Hence, due to large input handling capacity, with higher accuracy with lower computation time, the power system simulation is needed [3]. One of the leading power system analysis software is PowerFactory. The PowerFactory consists of the full range of function from the standard to the sophisticated one like power flow, state estimation and system modelling [4]. Hence, the power system dynamics course harmonises the theoretical knowledge of the dynamic behaviours of the electrical power system with actual software simulation on the PowerFactory. Hence, broadens knowledge.

15.1.1 Course Overview

The course Power Systems Dynamics is dedicated to the study behaviour of electrical power systems in the quasi dynamic. Knowledge of the dynamic behaviour of electrical power systems and their components is vital to predicting whether the system (or a part thereof) remains in a stable state after a disturbance. The time frame of observation ranges from milliseconds to several minutes, depending on the kind of disturbance and the influence of applied controls. For each time frame, particular models of the network components are derived.

In this course, the model of the synchronous generator, the voltage source converter, and their associated control systems are highlighted due to their foremost influence on the dynamic stability of the power system. In the computer assignments, the theory is put into practice employing software-based simulations in DIgSILENT PowerFactory as well as scripting in Python and Matlab. Also, in this course, the progressive development of the dynamic concept is established, which is applied in the simulated environment, as shown in Fig. 15.1.

Investigated subjects are: modelling of synchronous generator and voltage source converter for RMS simulations, frequency and voltage controls, model-based and measurement-based eigenvalue analysis, wide-area monitoring, damping control, and RMS simulation of interconnected power systems.

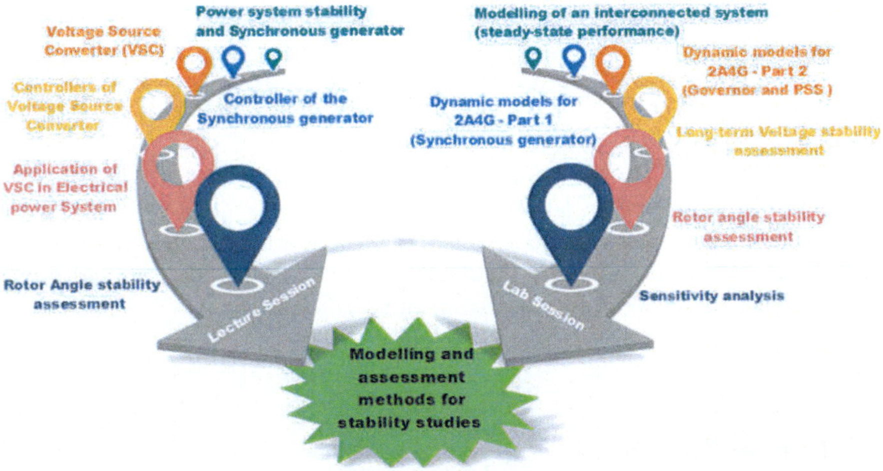

Fig. 15.1 A general overview of the key components of the course "Power Systems Dynamics"

This course is offered as a specialisation course in Quarter 4 for the master student. Typical enrolment range of the student for this course is around 10–20 with the majority from the Electrical power Engineering track. Also, students from sustainable energy technology are enrolled. However, most of the students are a full-time master student, but there are also Ph.D. scholars who wish to revise their knowledge in the power system dynamic.

15.2 Course Structure, Teaching and Evaluation Methods

The prerequisite for this course is two follow up courses Power System Analysis II (ET4107) and Intelligent Electrical Power Grids (SET3065), which are specialisation course offered in quarter 2 and quarter 3 respectively. This course provides a basic understanding of fault computation, power system components, power system stability, modelling and simulation of grids and risk assessment.

This course has different segments: lectures, Lab session, self readings and final examination. The lecture provides insight into the main principles, theories, methods and concepts of Power System dynamics. However, the Lab session gives a practical experience of how to use these concepts and methods in the Software tool. Lectures and Lab sessions are typically handled by the lecturers. In the Lab session, extra interactions are done with students so that deep questions can be discussed. Also, One to one interaction with a student is done in 4 h scheduled Lab session, which is typically conducted by the teaching assistants. The lecturers are also expected to provide two hours per week to handle students' questions and problems. The weekly self readings given to the students allow students to get more familiar with the concepts discussed in the lectures.

The students are expected to attend two hours lecture and four hours of lab session per week over a nine-week period. Students are expected to spend 112 h, out of the 62 h is dedicated to self-study. Couse consists of seven theory lectures and nine lab sessions.

At the end of each laboratory, the session report is to be generated, which contributes to 40% of the final grade. The final exam is three-hours long and contributes 60% of the final grade—the final exam, including a theoretical questions component and a computer-based component. The computer-based part of the test requires familiarity with the models built during the lab assignments. The students are expected to run time-domain simulations and give insight into the observed phenomena.

15.3 Course Contents

The power system dynamic course is carried out, as shown in Table 15.1.

Each new topic is incorporated with the lab session so that student get more insight into the concepts.

The power system dynamic course is classified into Six significant topics which are elaborated as follows.

Table 15.1 Course outline and lab assignments

Week	Lecture session	Lab session
1	Introduction to the power system stability and synchronous generator	Tutorial for PowerFactory
2	Controllers of the synchronous generator	Modelling of an interconnected system
3	The voltage source converter	Dynamic models for 2A4G—Modelling of Governer and AVR
4	Controllers of the voltage source converter	Dynamic models for 2A4G—Modelling of PSS and Sencondary frequency control
5	Applications of voltage source converter in electrical power systems	Long-term voltage stability assessment
6	Rotor angle stability	Rotor angle stability assessment
7	Small-disturbance (small-signal) rotor angle stability (measurement based methods)	Sensitivity analysis—Application VSC for STATCOM

15.3.1 Introduction to Power System Stability and Synchronous Generator

This lecture oversees the general characteristics of the modern power system, power system controls and operating states. Also, Concepts and definition of the power system stability are explained. Importance and classification of power system stability are overseen. Also, one of the objectives of this lecture is to define a line of difference between reliability, security and stability. To understand the power system, a brief introduction about the PS components is given.

The main objective of this session is to introduce the models of the synchronous generator. In this segment, the different order of synchronous models and its parameters is elaborated.

In the Lab session, a hands-on overview of the features that DIgSILENT PowerFactory offers for steady-state analysis (power flow—also known as load flow calculation) and transient stability analysis (RMS simulations) was provided. This was accomplished by building the single-machine infinite bus system in the DIgSILENT PowerFactory. The tutorial was supported with the set of tutorial along with the task to be performed.

Power factory is a computer tool developed by DIgSILENT to aid in the analysis of industrial, utility, and commercial electrical power systems. PowerFactory is used around industry and academia to carry out routine and specific power system studies like load flow calculations, fault analysis, transient stability analysis, contingency analysis, voltage regulator placement and coordinated voltage control of several voltage regulation devices within the power system. These functions are spread out over a wide time scale. Figure 15.2 provides a rough visualisation of the functions that can be performed in PowerFactory and its relative time scale.

PowerFactory also provides support for integration with open source languages such as Python to help in automation of trivial and advanced tasks.

15.3.2 Controllers of the Synchronous Generator

The learning objectives of this lecture is to understand the dynamic behaviour of the uncontrolled synchronous generator, Excitation system and Turbine-governor system. Generally, the generator is supported with additional control units like AVR, Governor and PSS, which provide stable operation of the network. An automatic voltage regulator (AVR) is an electronic device which automatically maintaining generator output terminal voltage to a reference value under varying load and operating temperature. Governor is Mechanical device which is used to adjust the speed of the generator during steady-state or in the transient. A PSS identifies the changing of generator output power, controls the excitation value, and reduces the power swing quickly [3].

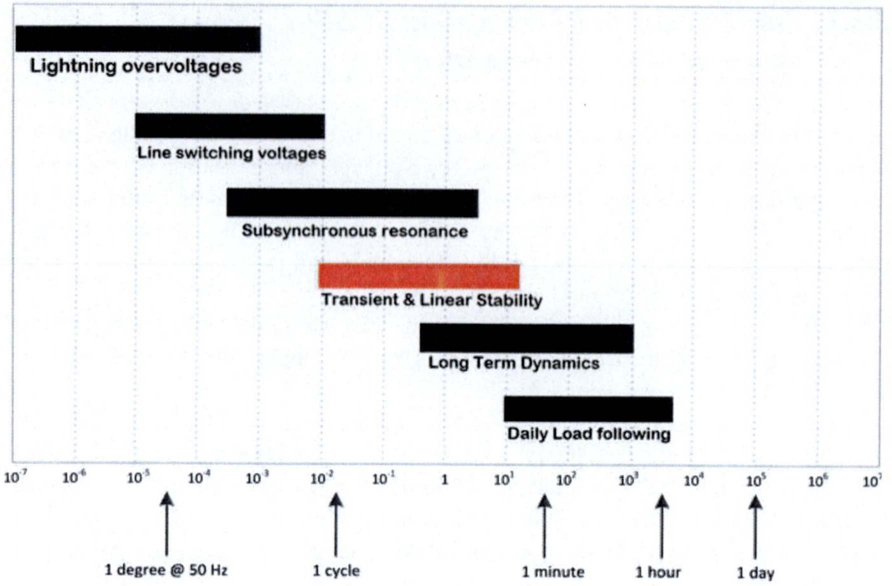

Fig. 15.2 Time scales of power system events. The portion in red depicts the time scale where PowerFactory can perform to study the phenomena of the main focus of this course [5]

To understand this controller, in-depth considerable amount of time was utilised in the lab session. Hence lab session for week 2, 3 and 4 were put together for this purpose. In Week 2, Two area network with four generators was built on power factory. Power flow analysis and the verification of outcomes was performed as shown in Fig. 15.3. Also, the performance of the test network was studied with a various event, for example, disconnection of Tie lines and Increase in load demand.

For the next week, the dynamic analysis was carried out on the same network. In this lab session, additional composite frame and common model of AVR/exciter were built to the 2A4G network. Each generator was assigned with these models

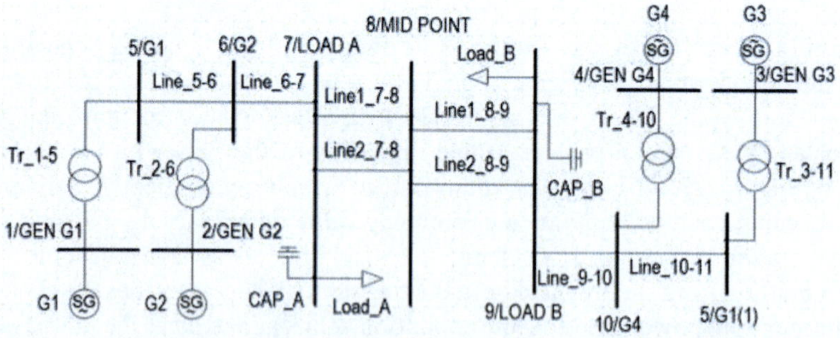

Fig. 15.3 Overview of two areas and four generators

with the help of the composite frame. A task which was performed in this session where load change of 5% at the load B, changes in the Vref and 3 phase short circuit in the transmission line. However, this task was performed for both cases, i.e. with AVR and without AVR.

To conclude the controllers for the synchronous generator, Power system stabiliser (PSS) and Governor model were build to the 2A4G network using the composite model and frame. Along with AVR, these models were assigned to each generator. Using the RMS/EMS Simulation, time-domain analysis of 5% load change in Load B was analysed. Notably, the terminal voltage of each generator and speed were studied. Also, the presence of PSS on the system was observed along with the governor during the disturbance.

Similarly, the effect of change in the droop characteristic was analysed. For example, it was observed that During the primary frequency control, the slope of the droop curve is essential. Stepper the curve higher is the frequency drop. With a reduced slope of the droop curve, generator speed drops to a lower value as compared to the slope of 0.05. Similarly, the drop of terminal voltage is lower and has a faster rise. Figure 15.4 illustrates the Variation of terminal voltage and speed of generator in the presence of PSS alone under a load variation of 15% at Load B.

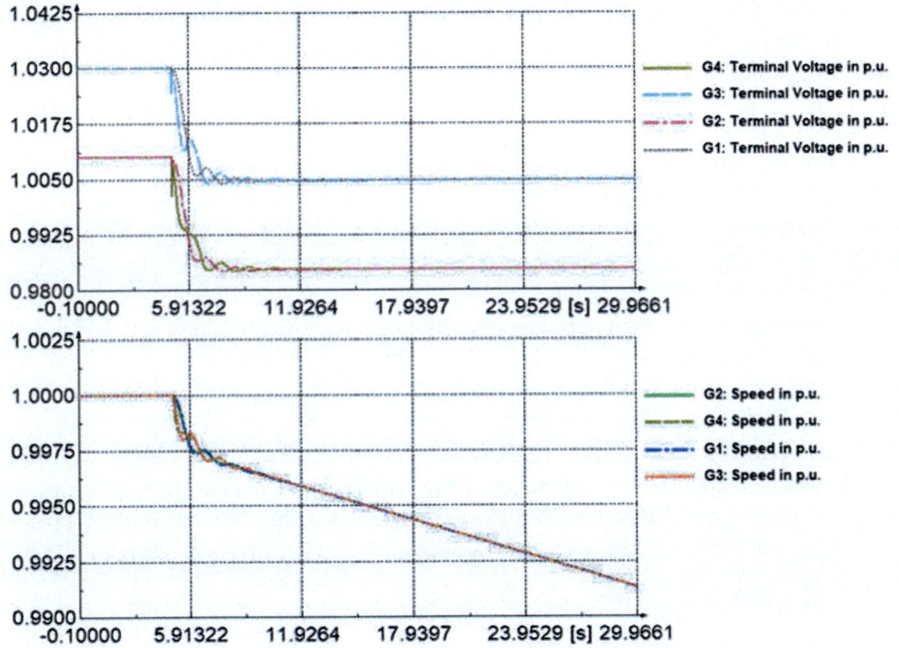

Fig. 15.4 Variation of terminal voltage and speed of generator in the presence of PSS alone under a load variation of 15% at Load B

15.3.3 The Voltage Source Converter—Controllers and Application in Electrical Power System

The lecture in week 3, 4 and 5 are utilised to understand the voltage source converter, its controller and the application in the power grid. HVDC technology is referred to an electrical system which is usually used for transferring the bulk amount of power from one energy system to another one. Hence the Voltage source converter plays a vital role in the technology. In week 3, the concept of the VSC is discussed along with the difference between the VSC and LCC. The importance and working principle of the Multilevel converters is conveyed—the VSC modelling assumption where highlighted. A considerable amount of time is spent on the space–time vector theory (STVT). The fundamental concept of the current controlled VSC is explained. The importance of PLL, inner loop and outer loop are illustrated. Hierarchical control of the VSC system is explained.

With the understanding of the VSC, modelling concepts of VSC is conveyed in week 3 lecture session. The concept of the static generator, which is a component in the PowerFactory is explained, which is used to model the VSC. A software demonstration is given to understand the controllers for the VSC.

The application of the static generator is discussed in week 5. Also, the principle and importance of various FACTS are explained. Concept of voltage stability and how PV curves and PQ curves can be analysed for stability purpose is being illustrated in this lecture.

The week 5 lab session demonstrates the application of the python interface with the power factor via a tutorial. The objective of the lab session is to carry out Voltage stability. In voltage stability analysis was performed on two platforms, one of the python script and another via power factory in-build tool. 2A4G was utilised as a test network for this lab. A python script was build which would commence load flow analysis for various incremental change in the load. All bus voltages and total power data were exported to the CSV file.

In contrast, power factor provides direct analysis which generates the PV curve or nose curve via in build tool. Additional, the effect of the reduced number of the parallel line was analysed on the nose curve. Similarly, Generator capability curve effect was identified in the PV curve.

For example, Fig. 15.5 demonstrates that with including the generator compatibility curve, the voltage remains constant until the collapse because the generator provides reactive power until reactive power demand is within the capability curve. So once it exceeds the region voltage collapses.

15.3.4 Transient and Small-Signal Rotor Angle Stability

In this 2 lecture, the main concept of transient stability, small-signal and damping control are conveyed. The different assessment techniques for transient stability

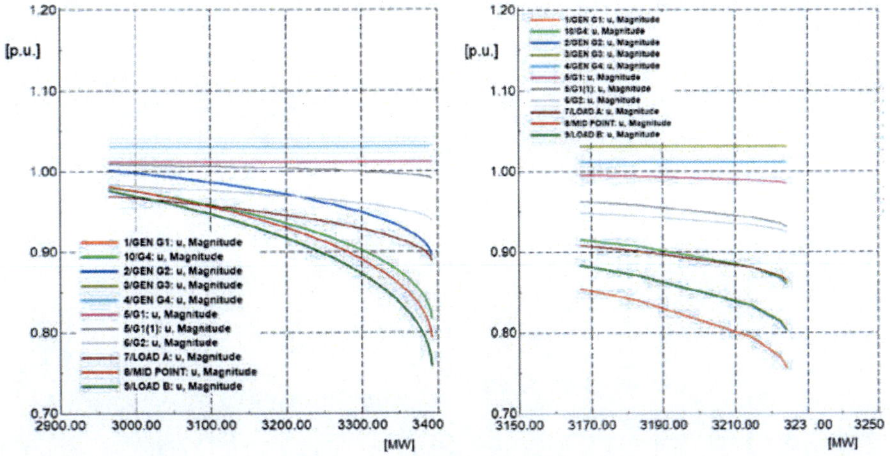

Fig. 15.5 Comparison of the PV curve, constant impedance loads with-out and with generator limits enabled

are discussed like time dome simulation, energy function, equal-area criteria and trajectory sensitivity. Influencing factor for transient stability is analysed.

Different modes in small-signal stability are defined. Importance of these modes in the stability study is illustrated. Analysis tool like model-based and signal recorded method is discussed.

In the week 7 lecture, application of Ponay analysis and spectral analysis is discussed. Also, a brief intro to the WAMs is given.

In the week 6 lab session, an unbalanced modal of 2AG4 was analysed. Firstly, the analysis was carried on the Unregulated modal using the modal analysis from the power factory tool. From the Eigenvalue plot, inter-area and local mode were identified. For example, Fig. 15.6 illustrated that for uncontrolled (unregulated), there are two local modes and one inter-area mode. This mode exists in the conjugate pair in the complex plane. This among these modes the inter-area mode is least damped which lies above 0.025 damping line. This indicates that any disturbance can lead this pole to move into the right side of the S-plane. Hence, by increasing the damping of the system, we can increase the stability of the system, which accomplished by making use of AVR, PSS and Governor.

Also, the phasor plot provides a precise classification of the inter-area and local Electromagnetic mode. The time-domain analysis was also carried out. In this, the active power of all the generators was plotted and compared.

The individual effect of the AVR, PSS and Governor, was analysed by enabling the composite frame and composite modal for a load change of 1% at Load A. For each case the eigenvalue, phasor plot and Participation factor were plotted. Prony and spectrum analysis was performed on the MATLAB. Both analyses were carried out on the recorded signal of generator 4 active power, And the comparison of the modal analysis and prony was performed.

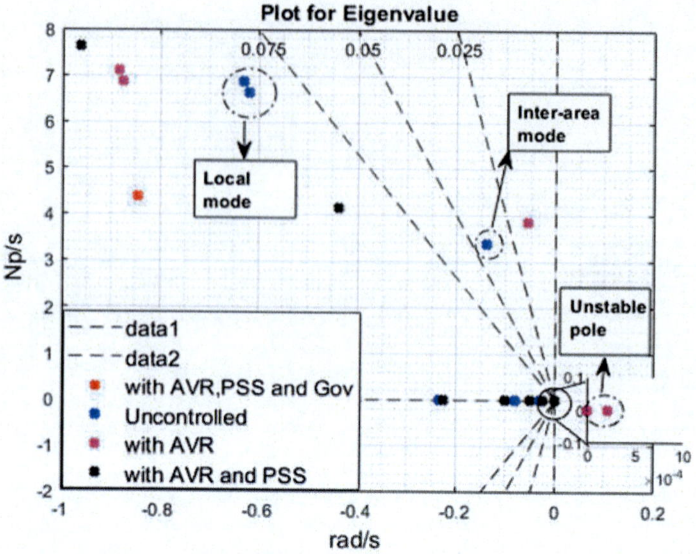

Fig. 15.6 Eigenvalue plot for various scenarios

15.4 Conclusion

In summary, the course covers some current topics and methods in Power system dynamics. In-depth the current state-of-the-art research are discussed. This course provides essential background and training for power engineering students to start research in power system stability. The course bridges the gap between textbook examples and actual system analysis. The methods and tools provided in this course will also serve as a platform for students to solve engineering and industrial applications. Very positive and encouraging feedback has been received from the student.

Appendix 1: Lab Session 1: Powerfactory Tutorial

Introduction

The objective of this lab session is to provide a hands-on overview of the features that DIgSILENT PowerFactory offers for steady-state analysis (power flow—also known as load flow calculation) and transient stability analysis (RMS simulations). To achieve this, you will build the single-machine infinite bus system displayed in Fig. 15.8 (hereafter referred to as SMIB), c.f. Sect. 15.3 With this document, you also receive the solution to do today's assignment that will help you carry out some of the tasks. You are expected to follow the instructions provided in this tutorial, build the system as specified, and compare your results with those from the solution you have received.

In this tutorial you will learn how to:

1. Draw the one-line diagram of a power system.
2. Specify the models associated with each component in the system for steady state (load flow) and transient stability (RMS) simulations.
3. Run a load flow calculation.
4. Run a transient stability (RMS) simulation.
5. Plot and export results.
6. Create short circuit events.
7. Create controls for a synchronous generator in DSL (DIgSILENT Simulation Language).
8. Carry out a modal analysis.

DIgSILENT PowerFactory

PowerFactory features three integration characteristics: functional integration, vertical integration and database integration. Functional integration implies that PowerFactory is a single executable program. Vertical integration means that the system models of power system network can be shared by all of the analysis functions (like load flow, RMS/EMT simulation, short circuit, etc.) in one place. This is shown in Fig. 15.7.

Finally, database integration allows that models, case studies, simulation functions and results can be stored in a single project file. This is done via the Data Manager. Data Manager, as the name suggests, is a place where you can access all the data related to one project. A sample view of the data manager is shown in Figs. 15.20 and 15.21. The Data Manager manages users, global and local libraries, system settings, application configuration, and projects. The folders available inside project are shown in Fig. 15.8.

There are two kinds of libraries available in PowerFactory. The first one is the Global Library (Fig. 15.20), which contains openly available models and templates developed by PowerFactory. These models are those of generators, transformers, lines, exciters, governors, etc. and can be modified and used as needed in projects. The other one is the project library (Fig. 15.21) that contains the models and templates relevant to and used within the project. Two subfolders within project library (from now on referred to as 'library') are the most important to understand: Equipment Type Library and User Defined Models. Their use is explained later as this tutorial progresses.

All of the grid data is stored inside the Network Model folder. The power system elements and their graphical information is present inside this folder. Other folders such as Operation Scenarios stores variations in network model according to scenarios developed by the user. This is one of the more advanced functionalities

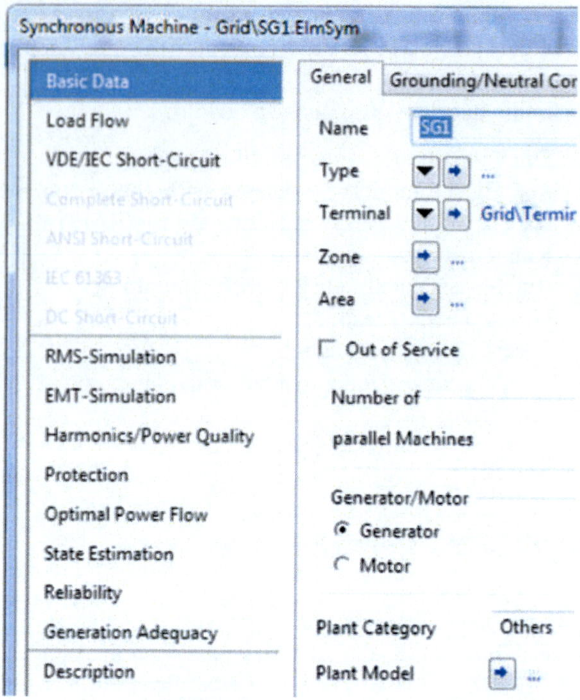

Fig. 15.7 Vertical Integration feature allows user to assign conditions for various functions in a single window

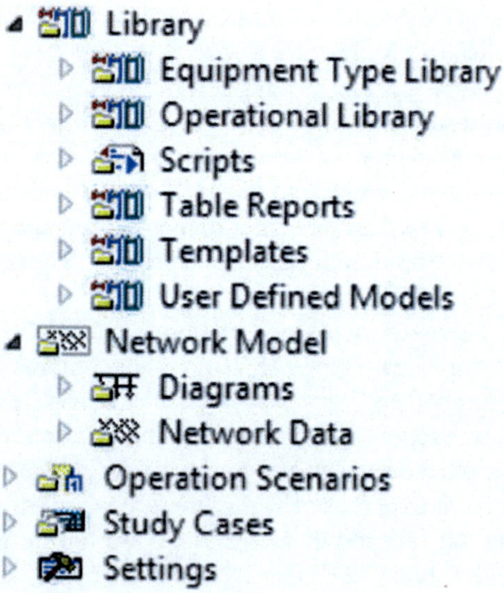

Fig. 15.8 Folders in project file in data manager

present and need not be discussed right now. The Study Cases folder stores data pertaining to the various study cases, such as simulation events, result variables, data export objects, graphical data information etc.

Another important feature that makes DIgSILENT PowerFactory such a great tool is its accessibility to programming languages for specific applications such as DIgSILENT Simulation Language (DSL) for creation of user defined dynamic models, DIgSILENT Programming Language (DPL) for scripting automatic calculation procedures and Python for scripting automatic calculation procedures and interface to other software such as MATLAB®. For this course, we will use DSL and Python extensively and hence these languages are explained in a bit more detail in the next sections.

Dynamic Modelling and DIgSILENT Simulation Language

To completely define the system behaviour during transient analysis, models that affect system stability need to be defined. These models include those of automatic voltage regulators (AVR), power system stabilisers (PSS), turbine governors, etc. DIgSILENT PowerFactory contains various IEEE standard models for such components in its global library. However, there are circumstances when the user may need to define his/her own dynamic models. This can be done in DIgSILENT's own powerful modelling language called DIgSILENT Simulation Language (DSL). Using DSL, the user can create particular models for his/her test system. When analysing transient behavior of power systems, PowerFactory combines the DSL model equations with those describing the dynamic behavior of power system components. An integrated transient simulation is then presented for the analysis.

PowerFactory adheres to a hierarchical modeling philosophy. This means that user proceeds to model the system in a step by step manner. There are four elements of DSL called Composite Frame, Composite Model, Model Definition (also referred to as Block definition) and Common Model. Understanding these four components of DSL are essential to creating dynamic models in PowerFactory. At its very base, PowerFactory uses composite frames and model definitions to define core structural models. These are like blueprints for each network element and connected device. Since each network element can have a similar structure but different parameters, it makes sense to have such an arrangement. For each network element, the way in which different elements are connected to it could eventually be the same, but the values of each parameter set (e.g. associated to particular location and role of the element within the network) may be different. Students with experience in object-oriented programming can understand them as cases of classes (composite frames and model definitions) and objects derived from these classes (composite models and common models). Figure 15.9 shows how the four components are linked to each other.

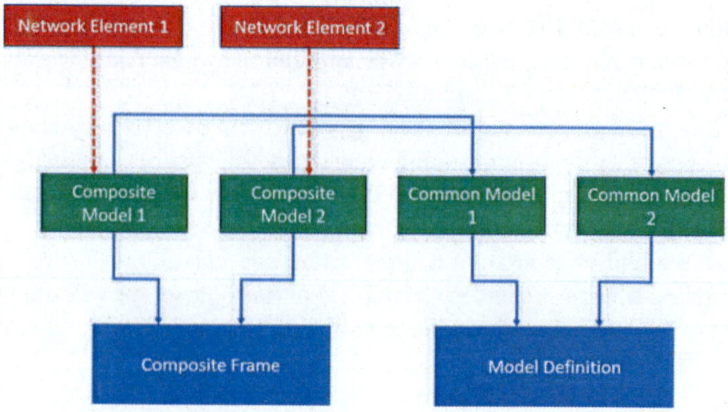

Fig. 15.9 Linking four components of DSL [1]

To get a little more clarity, let us consider a sample system where we have 3 synchronous machines (SM) in a network. Let the synchronous machines in the network also have auxiliary components such as AVRs, PSS, and governor systems. All SMs in the network have a similar structure which defines slots for how the AVR, the PSS and the governor system will be connected with the machine. So, a blueprint for all the SM in the system is defined. This blueprint is called a composite frame. This composite frame is located in the User Defined Models (Fig. 15.8) folder inside project Library. Each of the three SM models are connected to auxiliary models such as AVR, GOV and PSS in a particular arrangement to make up what is called a Composite Model. This arrangement borrows its structure from the composite frame defined earlier. These composite models are located inside Network Data (Fig. 15.8) folder. Thus, we have 3 composite models for 3 machines.

Model definitions are block diagrams that define the skeleton of auxiliary models. AVR, PSS and governor models are defined as model definitions and stored in User Defined Models folder inside project Library similar to composite frames. For each of the three machine composite models, the AVR (and PSS, governors) may have different values of parameters (or coefficients). Common models are objects that fill up the slots in composite model for each machine. The structure of common models is borrowed from model definitions. Thus, each of AVR, PSS, and governor common models picks up its respective model blueprint (or structure) from the model definition. Therefore, for each of the SM composite model, we will have 3 common models each denoting AVR, PSS and governor; 9 common models in total for 3 composite models.

To summarise the method: Composite Models apply the structure from Composite Frames and Common Models borrow structure from Model Definitions. Common Models are filled in slots that are available in the Composite Model. The above example is visualised in Fig. 15.10.

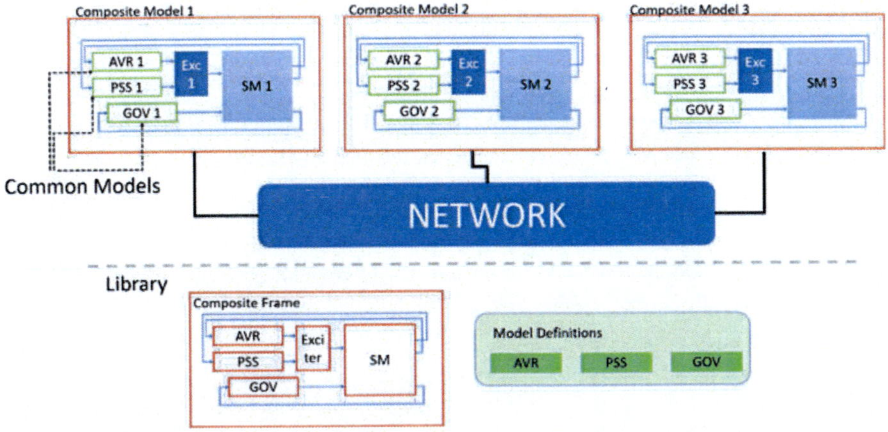

Fig. 15.10 Example of DSL components' implementation

Fig. 15.11 Possible order of creation of DSL components

Composite Model	3	1	1	3	DSL Components
Composite Frame	2	2	3	1	
Common Model	4	3	2	4	
Model Definition	1	4	4	2	

Figure 15.11 shows different orders the user can follow to create dynamic models in PowerFactory. Although the user may choose to create any DSL component first, the order in green area represents the order that is utilised in this course as it follows a more logical way (i.e., top-down design).

To summarise DSL components, Table 15.2 lists some remarks about composite frames, composite models, model definition (or block definition) and common models.

Integration with Python 3.4

PowerFactory also provides support for integration with open source languages such as Python to help in automation of trivial and advanced tasks. For example, one may need to run load flow studies on a system with varying values of system parameters to check at what condition the system load flow calculation fails to converge. Doing this

Table 15.2 Overview and remarks on DSL components

Component	Remarks
Composite model	Located inside Network Data folder (Grid) Common Model object contained inside this Composite Model object
Composite frame	Located inside User Defined Models folder in Library Composite Frame is a structure that contains slots for generator, exciter, governor, PSS etc
Common model	Located inside Network Data Folder (inside respective Composite Model) Common Model object contains the information regarding the parameter values needed (e.g. gain and time constant of transfer function)
Model definition	Located inside User Defined Models folder in Library Contains models in form of transfer function blocks

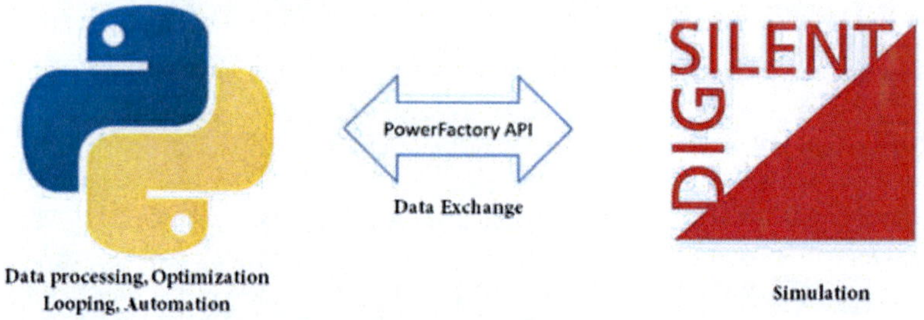

Fig. 15.12 Python—PowerFactory integration. API stands for application programming interface

by hand would require a long amount of time. With Python assisted automation, this task can be performed in as little as 10 lines of code and in a few seconds. During this course, we will learn to use Python to automate such tasks. As for advanced applications, one can use optimisation algorithms developed in Python to optimise various power system tasks on models developed in PowerFactory, such as Optimal Power Flow (OPF), Parameter Identification etc. A simple visualisation of interaction between PowerFactory and Python is shown in Fig. 15.12.

Test System

The SMIB system shown in Fig. 15.13 consists of a synchronous machine radially connected to a strong network (infinite bus) through a step-up transformer and a transmission line. A load is also considered in the system. The data for the SMIB system is summarised in Tables 15.3, 15.4, 15.5, 15.6, 15.7, 15.8, 15.9, 15.10 and 15.11.

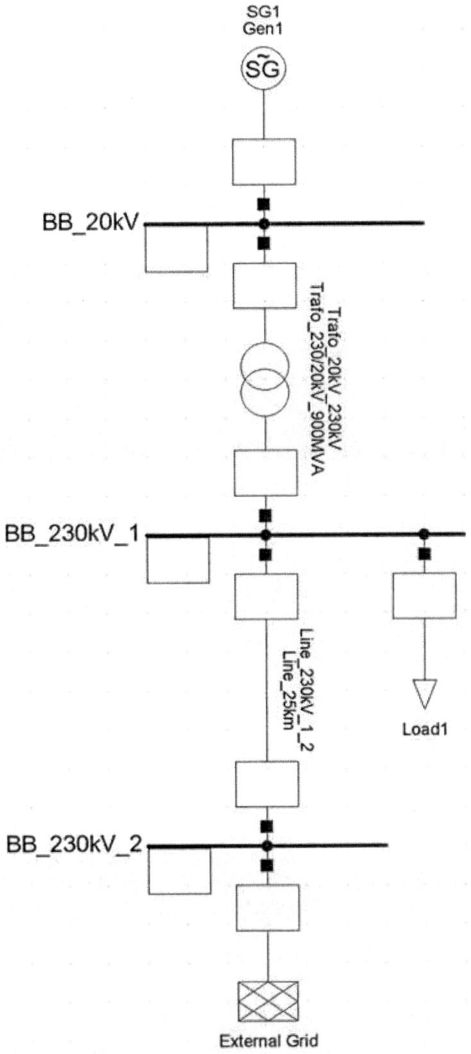

Fig. 15.13 One-line diagram of the single-machine infinite bus system (SMIB)

Table 15.3 Synchronous machine basic data

Nominal apparent power (MVA)	Nominal voltage (kV)	Power factor
900	20	0.8

Table 15.4 Synchronous machine RMS-simulation data (round rotor)

H (s) rated to Sgn			6.5
rstr (p.u.)	0	xq′ (p.u.)	0.55
xl (p.u.)	0.2	xd″ (p.u.)	0.25
xd (p.u.)	1.8	xq″ (p.u.)	0.25
xq (p.u.)	1.7	Td0′ (s)	8
xrld (p.u.)	0	Tq0′ (s)	0.4
xrlq (p.u.)	0	Td0″ (s)	0.03
xd′ (p.u.)	0.3	Tq0″ (s)	0.05

Table 15.5 Synchronous machine operating conditions for load flow

Tab: General			Tab: Operational limits
Active power (MW)	Reactive power (Mvar)	Local controller	max active power (MW)
870	0	Const. V	900

Table 15.6 External grid operating conditions for load flow

Bus type	Setpoint	Angle (deg)	Voltage setpoint (p.u.)
SL	Local	0	1

Table 15.7 Load operating conditions for load flow

Active power (MW)	Reactive power (Mvar)
800	100

Table 15.8 Transformer Basic Data

Rated power (MVA)	Nominal frequency (Hz)	Rated voltage HV-side (kV)	Rated voltage LV-side (kV)	Short circuit voltage (%)
900	60	230	20	10

Table 15.9 Transformer load flow and RMS-simulation data

x.Pos.Seq. HV-side (p.u.)	x.Pos.Seq. LV-side (p.u.)	r.Pos.Seq. HV-side (p.u.)	r.Pos.Seq. LV-side (p.u.)
0.5	0.5	0.5	0.5

Table 15.10 Line basic data

Rated voltage (kV)	Rated current (kA)	Nominal frequency (Hz)	Cable/OHL	AC-resistance R′ (Ohm/km)	Reactance X′ (Ohm/km)	Length (km)
230	2	60	Overhead line	0.1058	1.058	25

Table 15.11 Line load flow and RMS-simulation data

Max. Operational temperature (degC)	Suceptance B′ (uS/km)	Ins. Factor
230	1.654064	0

Creating a Project and Drawing a One-Line Diagram

To create a new project, go to the *File* drop down menu located in the upper left corner of PowerFactory and click on *New* → *Project*. A dialog will pop up. Here you must choose a name for your project (e.g. SMIB_1) and click *OK*. A new dialog will pop up now, where you can enter a name for your grid and select the nominal frequency of the system. **For this session you must enter a frequency of 60 Hz in the *Nominal Frequency*field.** You should now be able to see an empty graphic window where you will be drawing your one-line diagram (see Fig. 15.14).

Before you start drawing your diagram, **make sure that the *Freeze Mode*is not activated.** For this you can click the icon located at the top left corner of the graphic window. The *Freeze Mode* is designed to avoid making undesired modifications to the system by mistake. If *Freeze Mode* is not active, you should be able to see the tool bar displayed in Fig. 15.15. The numbered elements in Fig. 15.15 are the ones you will need in this session.

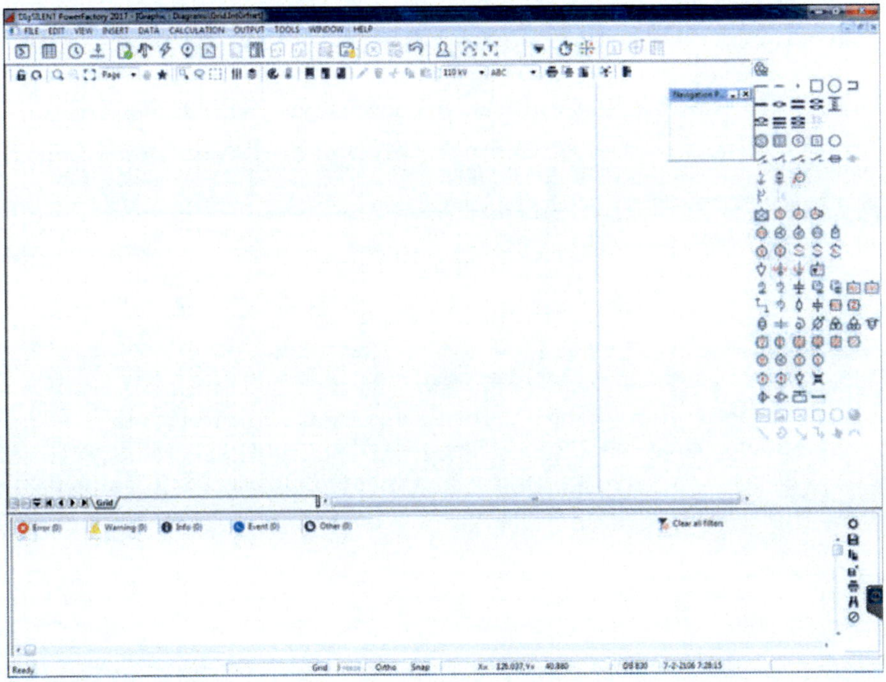

Fig. 15.14 New empty project

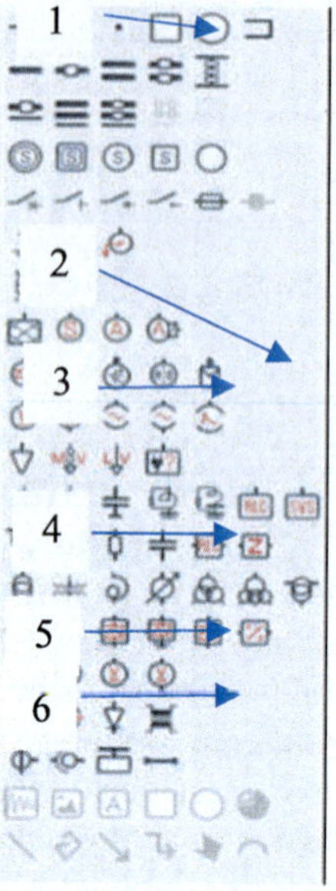

Fig. 15.15 Drawing tool bar. (1) Busbar (2) Synchronous machine (3) External grid (4) General load (5) Line (6) 2-winding transformer

Note: If you want to **modify an existing project** that is already in your database, you must first activate it. For this find the icon at the top left corner of your window and click on it. This will open the data manager. In the data manager find the project you want to activate (it should be under your user name) and right-click on it. In the drop down menu that apears, click on activate. Your project should now apear in the graphic window.

Note: If you want to **import a project** from a source external to PowerFactory, go to *File* → *Import* → *Data* (*.pfd; *.dz; *.dle) and select the project you want to import from its location. If you want to **export a project** to share it or to back it up, go to go to *File* → *Export* → *Data* (*.pfd; *.dz) and choose the destination to where you want to export your project. **More information at the end of this guideline!**

Placing Busbars

To draw your diagram, we recommend that you start by placing the busbars on the graphic window. For this, click on the corresponding element in the tool bar shown in Fig. 15.15, move the mouse to where you want to place the busbar, and click again. You can place more busbars by clicking anywhere on the drawing window. Make sure you are placing ***busbars*** and not ***single busbar systems***. **Once you are done placing your busbars, right-click anywhere on the graphic window or in your 'esc' key before you carry out any other action**. If you need to move a busbar to another location, first select it by clicking on it and then drag it with the mouse. If you want to rotate an element, right-click on it and go to *Rotate*.

Follow this procedure to place the three busbars from Fig. 15.13 on the graphic window.

Connecting Elements Between Busbars

The next step is to connect elements between busbars, such as lines and transformers. In order to do so, click on the corresponding element in the tool bar from Fig. 15.15 and then click on one of target the busbars. You should now see a line that extends from the chosen busbar to the mouse pointer. To complete the connection, click on the second target busbar.

Follow this procedure to connect the transmission line and the transformer from Fig. 15.13.

Connecting Elements to Individual Busbars

The final step is to connect elements to individual busbars, such as generators and loads. In order to do so, click on the corresponding element in the tool bar from Fig. 15.15 and then click on the target busbar.

Replicate the one-line diagram from Fig. 15.13 following the previous instructions. Once you have finished drawing it, double-click on each element in the diagram and enter a more suitable name for it in the *Name* field of the dialog that pops up. You can copy the names from Fig. 15.13 or you can develop your own naming convention.

Defining Models and Entering Model Data

As it is, your one-line diagram cannot be used for any type of simulation because the elements in it are not associated with a model. Follow the next steps to define the models.

Busbars

Two types of busbars are needed in this project: one 20 [kV] busbar and two 230 [kV] busbars. First, double click on each busbar and make sure the Nominal Voltage Line-Line field is set to the corresponding value in the Basic Data tab, then click OK each time. Now we can define the busbar types. Let us start with the 20 [kV] one. Double click on the 20 [kV] busbar and go to Basic Data → Type → New Project Type. In the dialog that pops up choose a name (e.g. Bus_20kV) and set the Nominal Voltage field to 20 [kV]. Do the same for one of the 230 [kV] busbars. Once you have defined a 230 [kV] busbar type for one of your 230 [kV] busbars, all you have to do to select the bus type for the remaining 230 [kV] busbar is to double click component → Basic Data → Type → Select Project Type. This will open the data manager. In the data manager find your project, go to Library → Equipment Type Library and select the 230 [kV] bus type you previously created (see Fig. 15.21). Once you are done click OK twice.

Synchronous Generator

A double click on the synchronous generator opens a dialog. Find the *Basic Data* tab and then click on the ▾ icon next to *Type* (see Fig. 15.16). You should now see three options: *Select Global Type* (for selecting models predefined by PowerFactory) *Select Project Type* (for selecting models that you may have already created in your project) and *New Project Type* (for creating a new model). Select *New Project Type* this time.

> **Hint**: Every time you define a new model type through *double click component → Basic Data → Type → New Project Type*, enter a name for your new type in the *Name* field of the *Basic Data* tab to make the new type easy to identify in the future (e.g. SG_900MVA_20kV for your synchronous generator). Choose a naming convention and remain consistent within your project. In the case of the synchronous generator this must be done in the dialog shown in Fig. 15.12 (*Basic Data* tab).

You should now see a pop-up dialog like the one in Fig. 15.17: Parameter definition dialog for synchronous machines. Make sure to complete the fields in the *Basic Data*, *Load Flow* and *RMS-Simulation* tabs with the data provided in Tables 15.3, 15.4 and 15.5. Special attention is to be given to the values of transient time constants and sub-transient time constants. **ATTENTION**: By default, PowerFactory asks to input values of $\mathbf{Td}'$, $\mathbf{Tq}'$, $\mathbf{Td}''$ and $\mathbf{Tq}''$ as in Fig. 15.17. However, these are Short circuit time constants. The values provided in Table 15.4 are for open loop time constants (**Notice the slightly different variable notation $\mathbf{Td}'$vs. $\mathbf{Td0}'$**). To change to open loop time

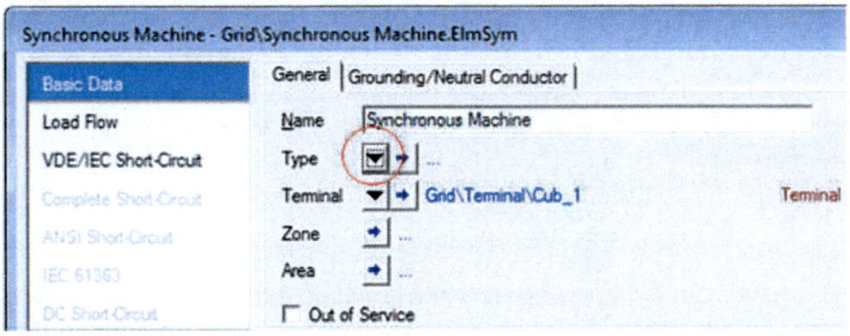

Fig. 15.16 Synchronous generator dialog

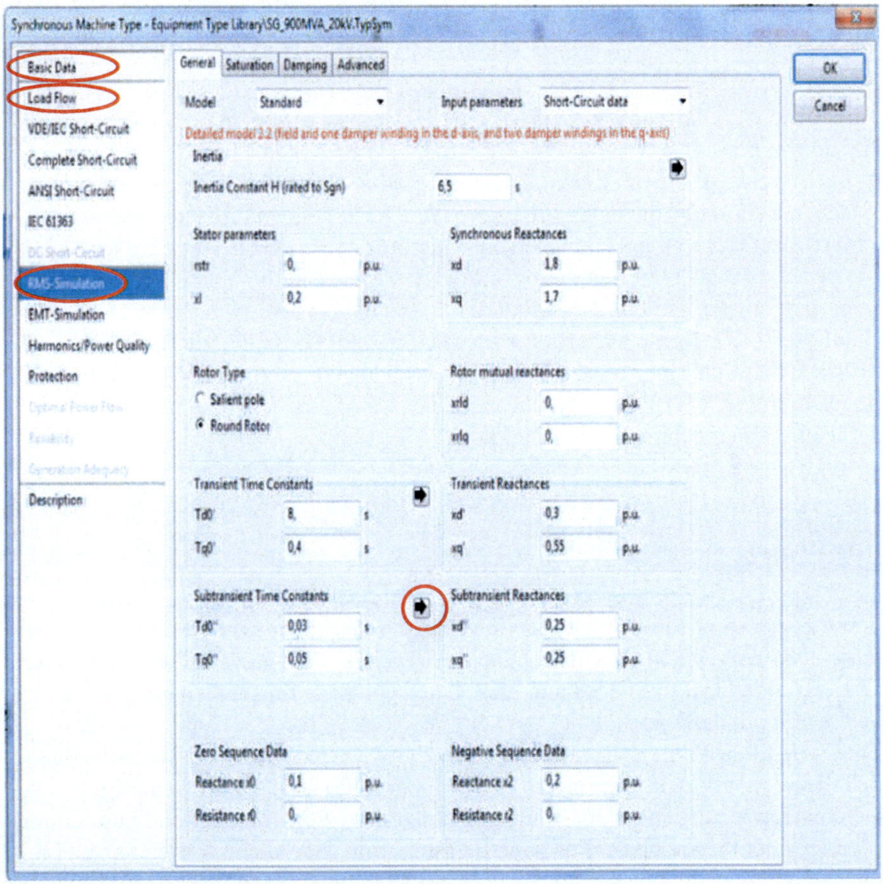

Fig. 15.17 Parameter definition dialog for synchronous machines

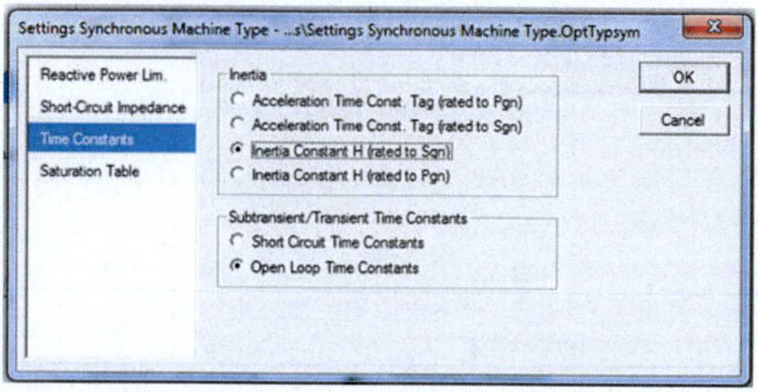

Fig. 15.18 Changing short circuit time constants to open loop time constants

constants input scenario, click on the small black arrow as shown in Fig. 15.17. This opens up the pop-up box shown in Fig. 15.18. Select Open Loop Time Constants from here to change the input parameters as required. Note that the data in the *Load Flow* tab is a subset of the data in the *RMS-Simulation* tab and it is not necessary to enter it twice. Once you have finished entering the data, click *OK*. You are now back to the previous dialog. In this dialog make sure to define the operating conditions of your generator in the *Load* Flow tab using the data provided in Table 15.5. Now click OK again. The same goes for the Inertia Constant H. By default, PowerFactory choses Acceleration time constant. You should change this to Inertia Constant, in the same menu in Fig. 15.18. your synchronous generator model should now be defined and linked to the machine in your diagram.

Transformer

Follow the same procedure that you followed in the case of the synchronous generator (*double click component* → *Basic Data* → *Type* → *New Project Type*) and enter the data provided in Tables 15.8 and 15.9. Once you have finished entering the data, click *OK* in both dialogs.

Once you fill in all the data, make sure that the HV side is connected to the HV bus [230 kV] and the LV side is connected to the LV bus [20 kV], as shown in Fig. 15.19. If the connections are inverted, you can click on the *Flip Connections* button and this will correct the situation. For good measure, you should check the connections again.

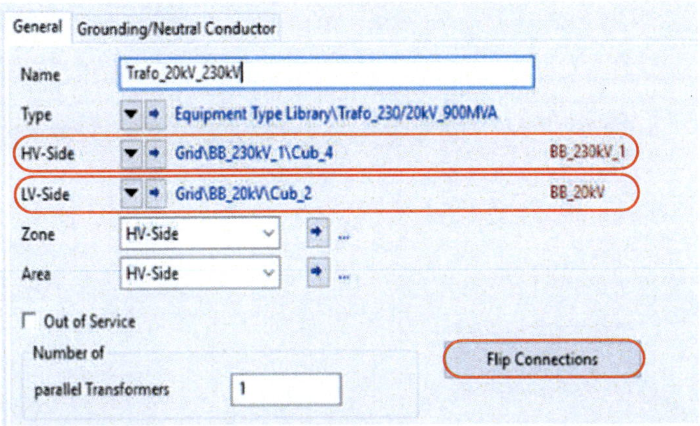

Fig. 15.19 2-Winding transformer's connections

Transmission Line

Follow the same procedure you used previously after double clicking on the transmission line to go to *New Project Type → Line Type (TypLne).* Enter the date provided in Tables 15.10 and 15.11 in the *Basic Data, Load Flow* and *RMS-Simulation* tabs and then click *OK* to go back to the previous dialog. In this dialog enter the line length provided in Table 15.10.

External Grid

In the case of the external grid only the operating conditions need to be entered. For this, double click on the external grid and enter the data provided in Table 15.6 (for self-testing, the angle can be set to 30° to ensure all models initialise correctly). Click *OK* when you are done.

Load

For the load we will follow a similar procedure (*double click component → Basic Data → Type*) but in this case we will choose *Select Global Type.* This action will open the data manager (see Fig. 15.20). In the data manager go to *Library → Types → Loads*, select *Lod-1* and click *OK*. In the remaining dialog enter the data provided in Table 15.7 (Fig. 15.21).

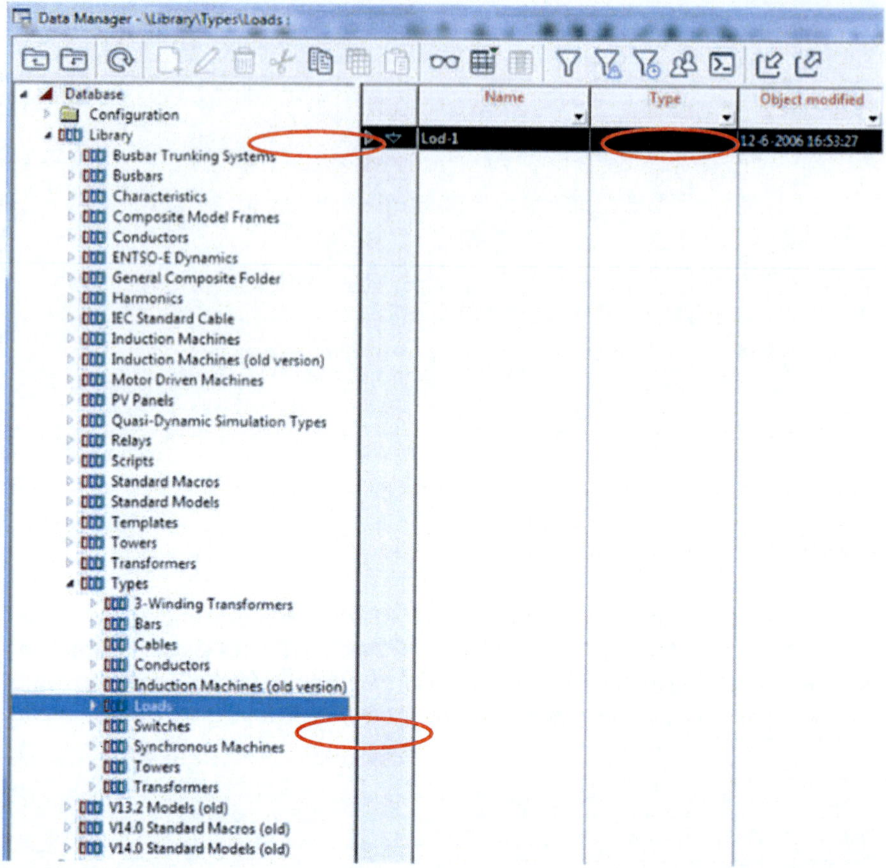

Fig. 15.20 Choosing a load model using the data manager

Running a Simulation and Visualising Results

If you have drawn and set up your models correctly, you should now be able to simulate your system. Follow the next instruction in order to do so.

Load Flow

Load flow calculations provide a snapshot of the power system in steady-state under given load and generation conditions. Run a load flow calculation by clicking on the Calculation menu at the top of the window and selecting Load Flow → Load Flow. In the dialog that pops up click on Execute. Alternatively, you can press Ctrl + F10

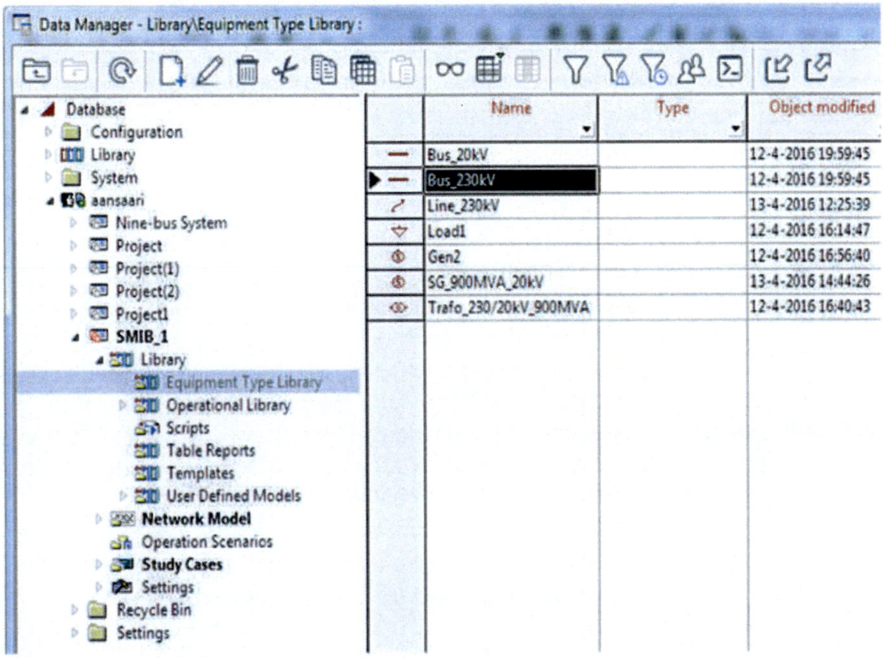

Fig. 15.21 Choosing a bus type from your own project library

or click on the icon. If your load flow is successful, your one-line diagram should look like the one in Fig. 15.22. Note how components that are close to or above their maximum capacity are colored red. The result boxes show power, current, voltage and loading values in each element. Hover over a result box with the mouse to see the results in more detail. Generate complete result reports by selecting Output → Output

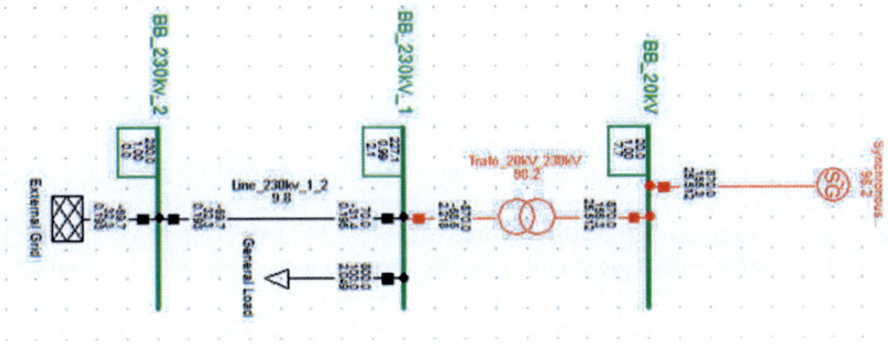

Fig. 15.22 One-line diagram of the SMIB system after a load flow calculation

Calculation Analysis, choosing one of the available options and clicking on Execute. The results will be printed to the output window in tabular form. Try different options and see which results are shown each time. You can reset a calculation by going to Calculation → Reset Calculation or by pressing F12 .

Transient Stability (RMS)

As opposed to load flow calculations, RMS simulations are time domain simulations, that is, they show how the behavior of the power system evolves over time. The focus of these simulations is on electro-mechanical phenomena. In this type of simulation, the power system is modelled through differential algebraic equations which need to be solved numerically over a discretised time interval. In order to do so, it is necessary to select an appropriate integration step size, which in the case of transient stability studies is typically in the order of 0.01 [s] or smaller.

Run an RMS simulation by selecting Calculation → RMS/EMT Simulation → Initial Conditions or clicking on the icon at the top of the window. If this icon is not available, press the icon and select "RMS/EMT Simulation".

In the Initial Conditions dialog that pops up make sure that the Basic Options tab has RMS values (electromechanical transients) selected and the Step Size tab has 0.01 [s] under Integration step size for Electromechanical transients, then click on Execute. This procedure initialises the power system for an RMS simulation through a load flow calculation. The load flow provides the necessary initial conditions to start the numerical integration of the underlying power system equations. Once the power system has been initialised, start the RMS simulation by going to Calculation → RMS/EMT Simulation → Start Simulation or clicking on the icon at the top of the window, entering the Stop time and clicking on Execute. In this case we will run the simulation for 10 [s]. If you want to re-run a simulation, it is recommended that you first reset your calculation (press F12).

Result Visualisation

PowerFactory is capable of generating plots for many internal variables, but for memory efficiency purposes, variables are not monitored by default. This means that it is necessary to select the variables of interest before running a simulation in order to plot them. **Before selecting variables make sure you have reset your calculation (press F12)**. We will illustrate the process for the case of the synchronous generator, but the same process applies to any other element. Right-click on the synchronous generator to open up a menu. In this menu go to *Define... → Results for RMS/EMT simulation*. In the dialog that pops up select the object associated with your generator and double click on the leftmost column (unnamed column left of the *Name* column,

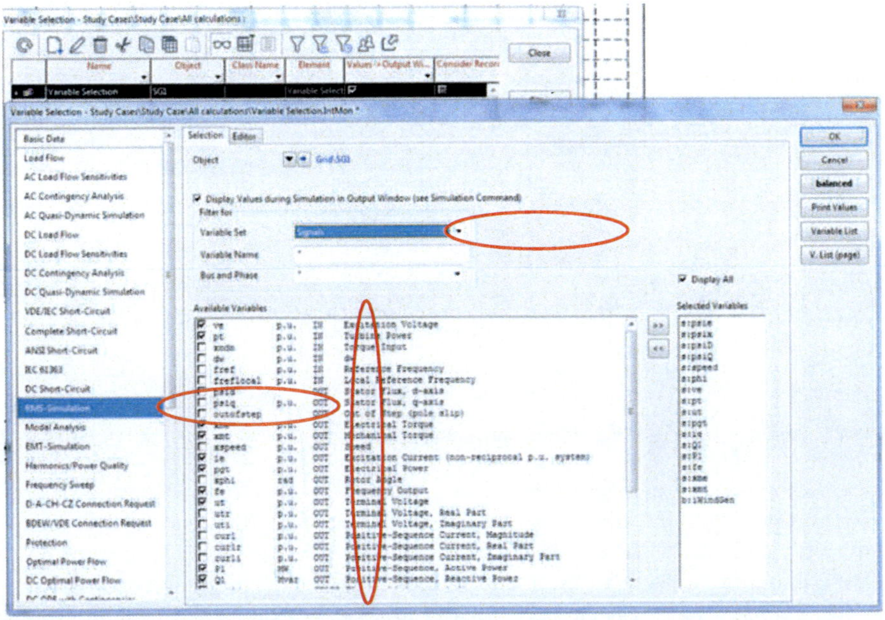

Fig. 15.23 Choosing results for an RMS simulation

background dialog of Fig. 15.23) to open a new dialog. In this new dialog go to the *RMS-Simulation* tab and select the variables you want to monitor by clicking on them. Variables are organised in different variable sets (see Fig. 15.23). Explore what each set contains to get an overview of the available options. **Select the variables: rotor angle, speed, mechanical and electrical power, excitation current, terminal voltage, frequency, active and reactive power and mechanical and electrical torque**. Once you have chosen these variables, run your RMS simulation again.

To create plots, we first need to create a new graphic by clicking on the icon at the top of the graphic window. In the dialog that pops up enter a name for the new graphic (e.g. Plots) and choose the option Virtual Instrument Panel (virtual instrument or VI is the term PowerFactory employs to refer to plots). In the empty graphic that appears you can add as many plots (or VIs) as you like by clicking on the icon and selecting the option Subplot (VisPlot) and enterning the number of VIs that you need in the dialog that pops up. You should now have a set of empty plots.

Hint: Click on the icons and see how your plots are rearranged.

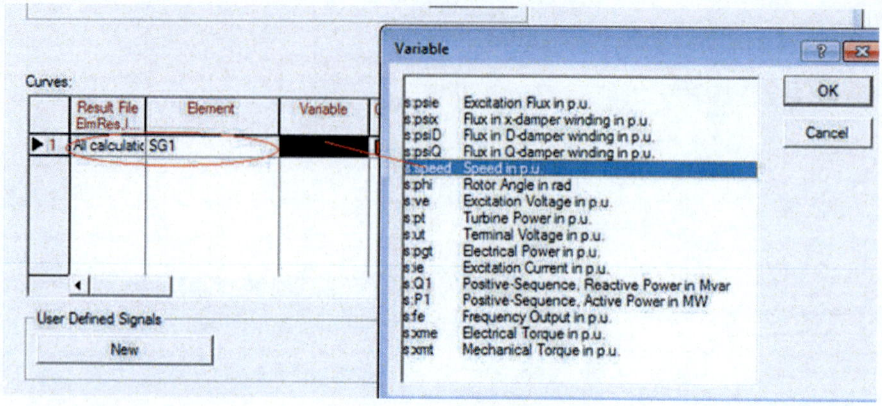

Fig. 15.24 Adding curves to a plot (virtual instrument)

To add results double click on an empty plot. In the dialog that pops you should see an empty table under the title *Curves*, like the one in Fig. 15.24. Double click on the first column and select *All Calculations*, then double click on the second colum and select your synchronous generator. Finally, double click on the third column and select the variable you want to plot. If you want to add more curves to the same plot, right-click on the same table and choose *Insert Row(s)*. Repeat the same process for the new row. After clicking *OK* the curves should appear in the plot you just edited. **Create plots for all the variables you selected in the previous step.**

Hint: Click on the icons and see their effect on your plots.

Exporting Results

It is ussually nesessary to export results either to present them elsewhere or to analyse them with a different software tool. In order to export results, right-click on a plot and select *Export to*. Some of the options you will most likely be interested in are *Textfile* (writes your results as a data series in a plain text file), *Comma Separated Values (*.csv)* (writes your results in a tabular file format that can be edited with spreadsheet software such as Microsoft Excel) and *Windows Bitmap (*.bmp)* (image format you can paste on reports and presentations). Try to export some of your results to learn the process.

Creating Events

PowerFactory allows defining different types of events to disturb the power system. Different system elements allow defining different types of events. For example, lines allow for switching events, transformers allow for switching and tap events, busbars allow for short circuit events, and so on.

Now we will create a short circuit event, however, a similar procedure can be used for creating other types of events. We are interested in studying the behavior of the system when a short circuit occurs at the busbar where the external grid is connected (BB_230kV_2 in Fig. 15.13). Right click on this busbar and go to *Define* → *Short-Circuit Event*. In the dialog that pops up enter 1 [s] as the *Execution Time (Absolute)* and select *3-Phase Short-Circuit* as the *Fault Type*. Then click *OK*. Repeat the same process to create a new short-circuit event at the same busbar. This time enter 1.15 [s] as the *Execution Time (Absolute)* and select *Clear Short Circuit* as *Event Type*. This will produce a short-circuit with a duration of 0.15 [s] starting at 1 [s].

If you want to have on overview of all the events you have created in your simulation or you want to modify events, you can click on the icon to access them directly.

Once you have defined your short-circuit event, run your RMS simulation again and compare your results to those from the previous simulation. Does your synchronous generator behave differently? Can you explain the new behavior?

Defining Synchronous Machine Controls in DSL

PowerFactory allows to define different types of custom models and to use them in your power system simulation though DSL (DIgSILENT Simulation Language). In this section we will learn how to create a simple exciter to control the synchronous machine.

Creating a Composite Frame

A *Composite Frame* in PowerFactory defines a template for a dynamic model. You can create as many dynamic models as you like out of the same *Composite Frame* and link them to elements in your one-line diagram to modify their default behavior. A *Composite Frame* is composed of *Slots*. Each *Slot* encapsulates other (simpler) dynamic models. We will now create a *Composite Frame* that describes the dynamic model of a synchronous generator controlled by an exciter. You can see this *Composite Frame* in Fig. 15.25. Each box in this figure is a *Slot*.

To create a *Composite Frame*, click on the icon at the top of the graphic window, give your frame an appropriate name (e.g. Frame_SG_Exciter) and select

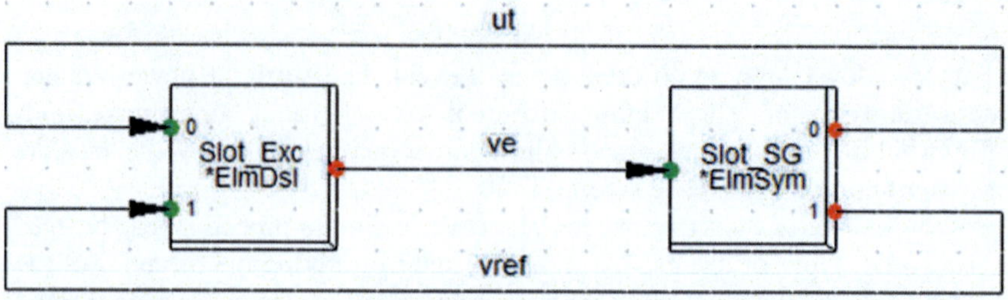

Fig. 15.25 Composite Frame describing a synchronous generator controlled by an exciter

Block/Frame Diagram. Then click *Execute*. From the toolbar that appears to the right of the graphic window click on the icon to place two slots on your graphic. Double click on the *Slot* you will use for your synchronous generator to make a dialog appear. In this dialog give your *Slot* an appropriate name (e.g. Slot_SG), in the *Class Name* field enter **ElmSym*, in the *Output Signals* field enter *ut* (voltage at the terminals of the generator), *vref* (reference excitation voltage) and in the *Input Signals* field enter *ve* (excitation voltage of the generator). Now double click on the *Slot* you will use for your exciter. In the dialog that pops up give your *Slot* an appropriate name (e.g. Slot_Exc), in the field *Class Name* enter **ElmDsl*, in the *Output Signals* field enter *ve* (excitation voltage of the generator) and in the *Input Signals* field enter *ut* (voltage at the terminals of the generator) and *vref* (reference excitation voltage).

Now click on the → icon to interconnect your *Slots* as shown in Fig. 15.25.

Creating a Block Definition

A *Block Definition* in PowerFactory is another kind of dynamic model template that describes a dynamic model through a block diagram, where each block implements a transfer function. We will now create a *Block Definition* to describe the model of our exciter (you will later see that *Slot_Exc* in Fig. 15.25 is only a reference to the *Block Definition* we are about to create). You can see de *Block Definition* we are about to create in Fig. 15.26.

To create a Block Definition, click on the icon at the top of the graphic window, give your frame an appropriate name (e.g. Block_Exciter) and select Block/Frame Diagram. Then click Execute. From the toolbar that appears to the right of the graphic window click on the icon to place three blocks in your graphic (see Fig. 15.26). Place a summation point using the icon as well. The next step is to click on the icon to open the data manager. There, go to PowerFactory's main library and search for Library → Standard Macros → 1st order functions and the transfer function 1/(1 + sT). Right-click on it to copy it and then go to your project library

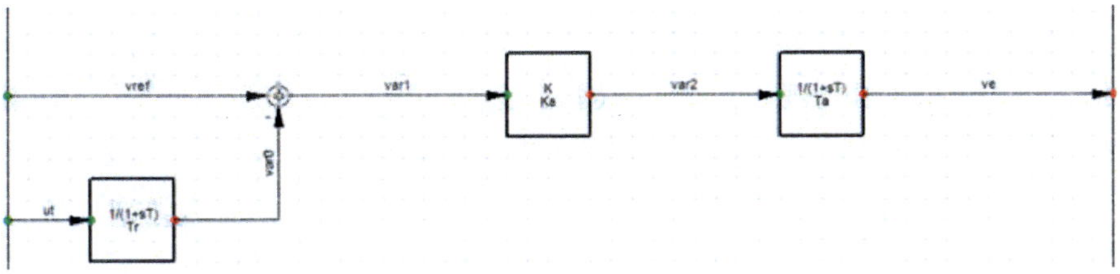

Fig. 15.26 Block definition of the exciter

and find Library → User Defined Models. Paste your transfer function there. Now go back to PowerFactory's main library and find Library → Standard Macros → Gains and copy the gain K to the same location you copied the transfer function. Now go back to your graphic window. Double click on your blocks and then click on the icon next to Type. Choose Select Project Type and select the corresponding transfer function for each block. Edit your Block Definition until it looks exactly like the one in Fig. 15.26. Mind the negative sign in the summation point (Right click on the summation sign → edit data and toggle invert input 1). In the State Variables field of the output block write x_out. In the case of the input block write x_in. Finally, click anywhere on the graphic window. In the dialog that pops up go to Equations and enter de following DSL code:

```
! Initial conditions
inc(x_out)=ve
inc(x_in)=ut
inc0(vref)=x_in+x_out/Ka

! Parameter definition
vardef(Ka)='pu';'[300]'
vardef(Ta)='s';'[0.05]'
vardef(Tr)='s';'[0.01]'
```

The first three lines of code define the initial conditions of your model. The last three define the model parameters (constants). Once you are done, click on the Check button to make sure your block is properly defined. Correct any mistakes you may have made.

Initialising a Block

We will now explain the first three lines of code presented in the previous section.

In power factory Block Definitions are initialised (in most cases) from output to input. All variables that cannot be determined directly from the load flow calculation performed before running an RMS simulation need to be initialised manually in

DSL. In the case of our exciter, all state variables (i.e. x_in, x_out) and all inputs (i.e. ut, vref) need to be initialised. The excitation voltage ve and the generator terminal voltage ut will be determined in the initial load flow calculation (the generator model is initialised first using the load flow results). Based on ve and ut we need to initialise all the remaining variables. To initialise a block, you need to know which equations it implements internally. You can find out by double-clicking on the block, then clicking on the icon next to Type and going to the Equations slot. In the case of the 1/(1 + sT) block you will see something like Fig. 15.27. To understand the DSL code in the Equations tab you are expected to carefully read Sect. 26.11 of [1], which you can access in the Help menu of PowerFactory. The equations in Fig. 15.27 can be written as

$$\frac{dx}{dt} = \frac{yi - x}{T}, \tag{15.1}$$

$$yo = x, \tag{15.2}$$

where *yi* and *yo* are the block inputs and outputs, and *x* is the state variable. Since we assume steady state at the beginning of the RMS simulation, for the initial condition all time derivatives are zero, so

$$0 = \frac{yi - x}{T}, \tag{15.3}$$

and therefore combining (15.2) and (15.3) we get that,

$$yi = x = yo, \tag{15.4}$$

which means that initially, the input, the output and the state variable must have the same value for this block. With this in mind we can analyse the complete Block Definition. Based on (15.4) we know that

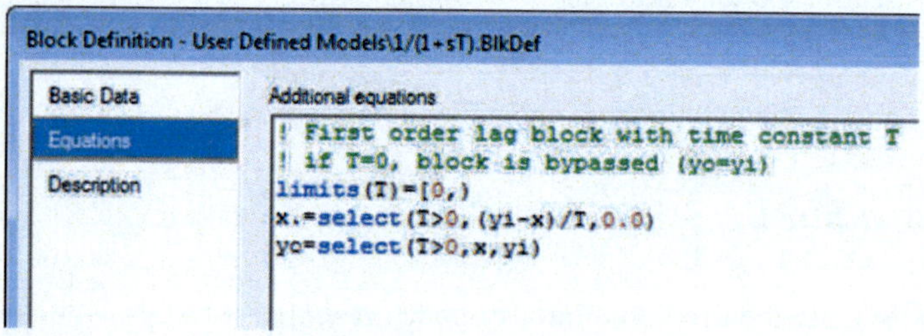

Fig. 15.27 Equations of the 1/(1 + sT) block

$$x\_\mathrm{out} = \mathrm{var2} = \mathrm{ve}, \tag{15.5}$$

$$x\_\mathrm{in} = \mathrm{var0} = \mathrm{ut}, \tag{15.6}$$

and it is also straightforward to derive that

$$\mathrm{var1} = \frac{\mathrm{var2}}{\mathrm{Ka}} \tag{15.7}$$

and that

$$v\mathrm{ref} - \mathrm{var0} = \mathrm{var1}. \tag{15.8}$$

Combining (15.5)–(15.7) in (15.8) yields

$$v\mathrm{ref} = \mathrm{ut} + \frac{\mathrm{ve}}{\mathrm{Ka}}. \tag{15.9}$$

Equations (15.5), (15.6) and (15.9) are thus implemented as

inc(x_out) = ve
inc(x_in) = ut
inc0(vref) = x_in + x_out/Ka.

Creating Composite and Common Models

As it was mentioned before, both *Composite Frames* and *Block Definitions* are model templates, not actual model instances, so you cannot directly use them in your simulation. You can derive as many model instances as you need from the same model template and link these instances to specific elements in your one-line diagram to modify their default behavior. When you create a model instance based on a *Composite Frame*, this is called a *Composite Model*. When you create a model instance based on a *Block Definition*, this is called a *Common Model*. We will now learn how to create both and how to use them to control the synchronous generator.

To create a Composite Model, click on the icon to open the data manager and find your project. Then go to Network Model Library → Network Data → Grid. Make sure Grid is selected and then click on the icon at the top left corner of the data manager. In the dialog that pops up select *Composite Model* and click *OK*. In the new dialog that opens, enter an appropriate name for your *Composite Model* (e.g. Composite_SG_Exciter) and click on the icon next to *Frame*. Now click on *Select Project Type* and then select the *Composite Frame* you previously created. Once you click *OK* you will see how the *Slots BlkSlot* column in the dialog has been

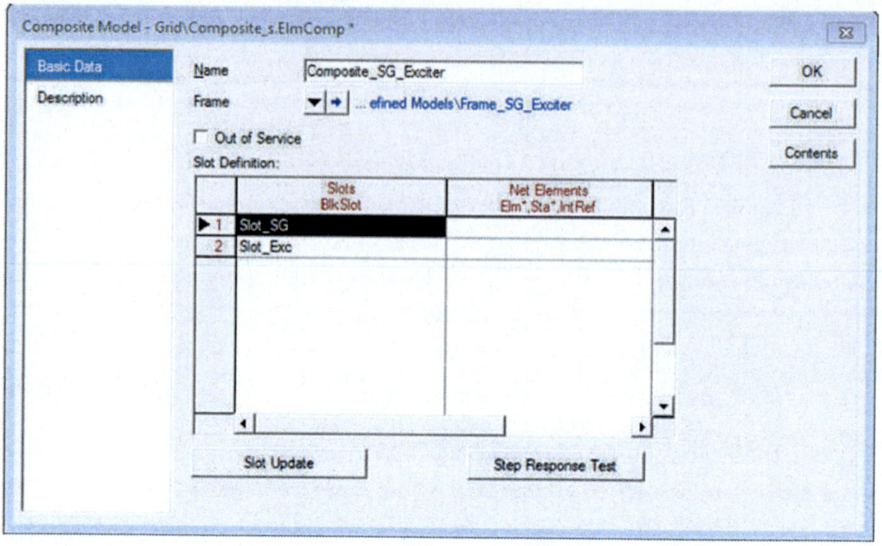

Fig. 15.28 Composite model creation

updated with the *Slots* you created inside your *Composite Frame* (see Fig. 15.28). Click *OK* when you are done.

Now we need to create a *Common Model* for the exciter based on the *Block Definition* we previously created. To do this, select the *Composite Model* you just created in the data manager and click on the icon at the top left corner of the data manager. In the dialog that pops up select *Common Model* and click *OK*. In the new dialog that pops up, select the exciter *Block Definition* you previously created and click *OK*. Another dialog pops up where you can give your *Common Model* an appropriate name (e.g. Common_Exciter) and enter the model parameters. Enter Ta = 0.05, Tr = 0.01 and Ka = 300, then click *OK*. Now find again the *Composite Model* in the data manager, right click on it and choose *Edit*. Update the table in the dialog by clicking on each cell so that it looks like the one in Fig. 15.29. If you did everything correctly, the data manager view of your project should look like the one in Fig. 15.30. PowerFactory automatically determines (based on the defined slots) the slot in which the common model is placed. You also need to manually add the synchronous machine model to the other slot: double-click on the blank cell, in the *Slot_SG* row, and select your synchronous machine.

Run the RMS simulation again and see what happens. Does the behavior of the synchronous generator after the short-circuit improve in comparison to the case without the exciter?

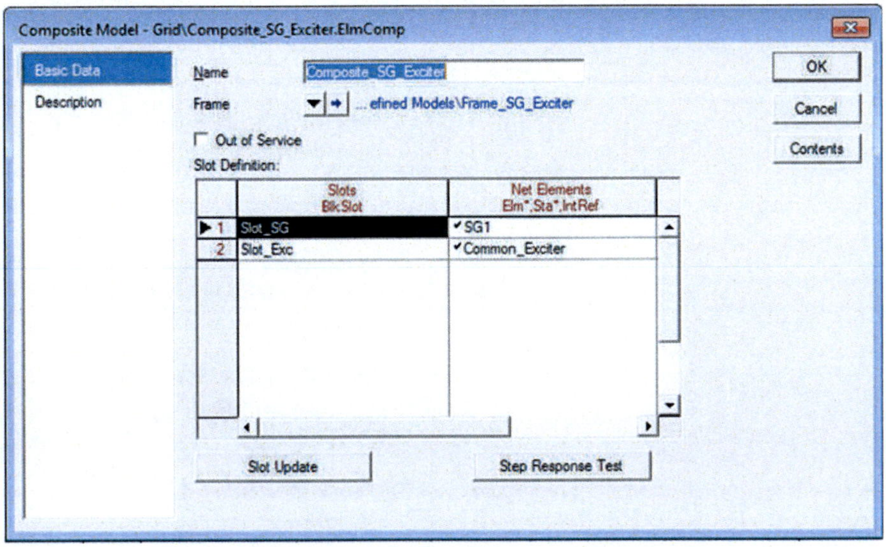

Fig. 15.29 Composite model completion

Modal Analysis

Modal analysis (eigenvalue calculation) is a useful tool for studying the small signal stability of a power system. We will now learn how to execute a modal analysis in Power Factory.

Executing a Modal Analysis

If you click on the icon you will see a menu with all the different kinds of analyses that PowerFactory can run. Choose Modal Analysis and see how the icons to de right change to . Before running the modal analysis click on the icon to calculate initial conditions (make sure they are set the same as for the RMS simulation case) and then click on Execute. To run the modal analysis click on the icon. In the dialog that pops up, under the Basic Options tab select QR/QZ-Method, under the Advanced Options tab select Left Eigenvectors (Controlability), Right Eigenvectors (Observability), Participation Factors, and QR method using system reduction, under the Output tab select all the available options and choose a destination folder for the output files. Then click on Execute. The icon allows you to see a summary of all the calculated modes.

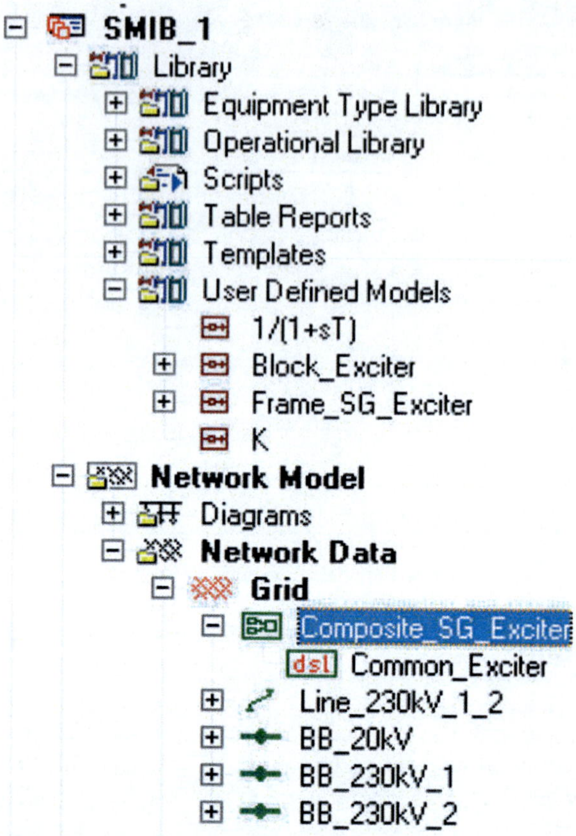

Fig. 15.30 View of the completed project in the data manger

Modal Analysis Visualisation

There are three different visualisation tools you can use under the icon:

is a phasor plot (right eigenvalues).

is a bar plot (participation factors).

is an eigenvalue plot (plot all eigenvalues).

How to Save Your Work—Import/Export

Once you have finished all the instructions, you can save your work. PowerFactory doesn't offer a Save option. In order to save your project you have to export it to a file.

Firstly, you need to deactivate your project by going to FILE → Deactivate or by going to your Data Manager, right-clicking on the active project, the one in bold, and selecting Deactivate.

To export the project go to FILE → Export → Data(*.pfd, *.dz)…. Then, select the project you want to export and click OK.

Afterwards, just select where you want to save your project.pfd file on your PC. Once you have named it and selected the correct folder, just press Save. On the new window that opens up, Fig. 15.31, just press Execute.

If you want to import a project, go to FILE → Import → Data(*.pfd, *.dz)…. Chose the.pfd file you wish to import and click Open. A new window, Fig. 15.32, will open and here you can choose where to import the file to. Click on the downwards arrow next to New Path, and then Select…. It is recommended that you save it under

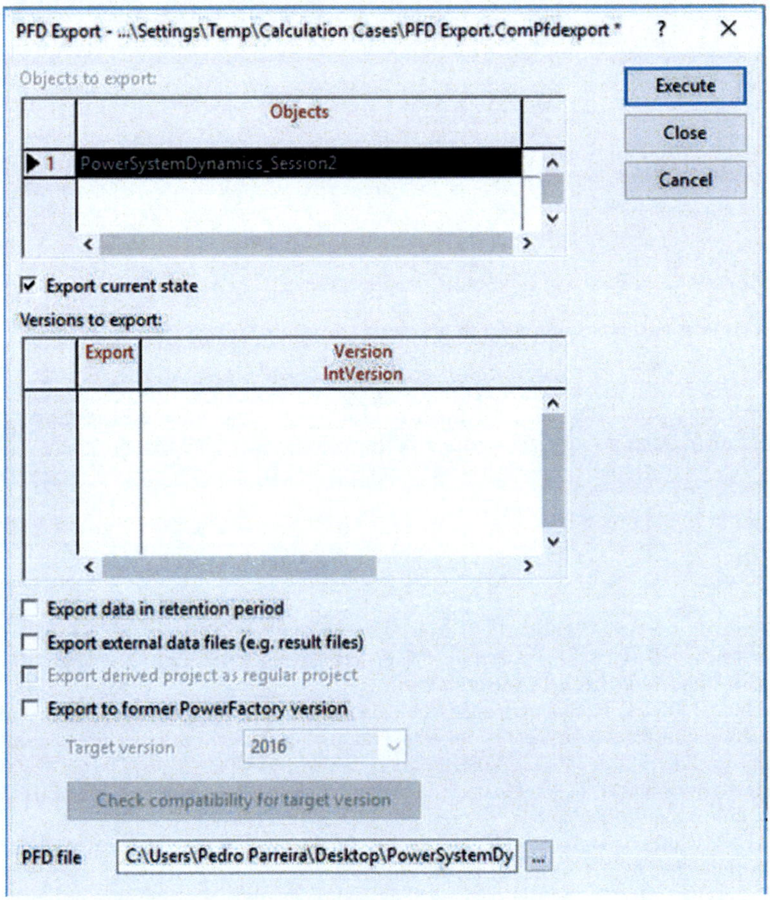

Fig. 15.31 Exporting project to a .pfd file

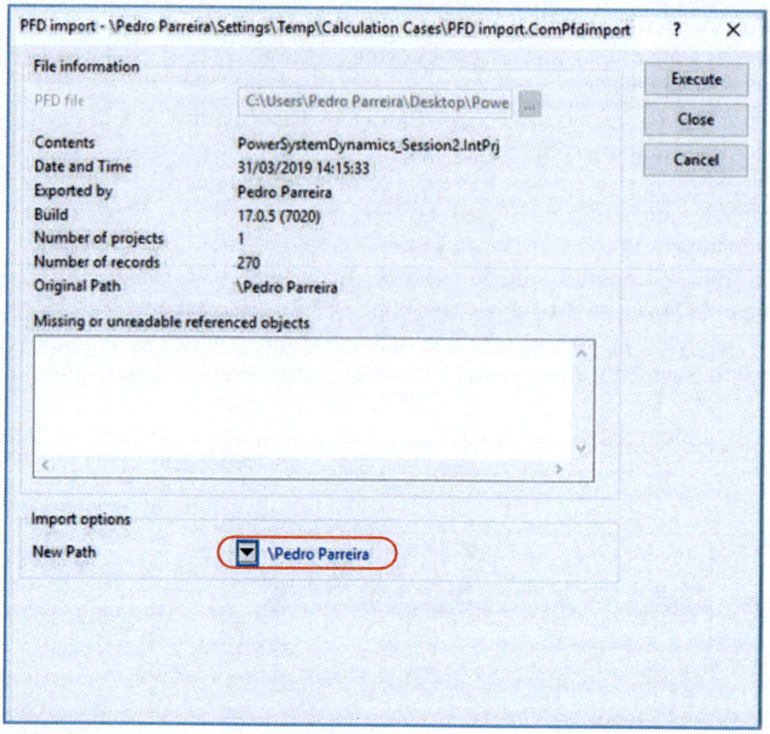

Fig. 15.32 Importing a project from a .pfd file

your profile or a subfolder of your profile. Once you have selected the path, press OK, and then Execute. Once the project is imported, you can work on it.

Reference

1. P.A. Owusu, S. Asumadu-Sarkodie I S. Dubey (Reviewing Editor), A review of renewable energy sources, sustainability issues and climate change mitigation, Cogent Eng. **3**, 1 (2016). https://doi.org/10.1080/23311916.2016.1167990
2. A. Bublitz, D. Keles, F. Zimmermann, C. Fraunholz, W. Fichtner, A survey on electricity market design: Insights from theory and real-world implementations of capacity remuneration mechanisms. Energy Econ **80**, 1059–1078 (2019). ISSN 0140-9883
3. M. Arnaout, R. Rammal, S. Abdulnabi, Electric power system simulator tool in MATLAB, in *Science Education—Research and New Technologies*, InTech (2017)
4. DigSILENT, *PowerFactory—DIgSILENT* (2019). [Online]. Available: https://www.digsilent.de/en/powerfactory.html. [Accessed: 10 Feb 2020].
5. "PF_Renewables_Delft", Tutorial on modelling and simulation of renewable resources and storage in PowerFactory 15.2, Dr. Fransico Gonzalez-Longatt (Delft, 2015)
6. nptel, power-system, Internet: https://nptel.ac.in/courses/Webcoursecontents/IIT-KANPUR/power-system/ui/Coursehome-9.html [14 Sept 2019]

7. S.B. Bhaladhare, A.S. Telang, P.P. Bedekar, P-V, Q-V curve a novel approach for voltage stability analysis, in *National Conference on Innovative Paradigms in Engineering Technology* (NCIPET-2013)
8. P. Kundur, *Power System Stability and Control* (McGraw Hill, 1994)